Functional Fillers and Nanoscale Minerals

Edited by Jon J. Kellar, Marc A. Herpfer, and Brij M. Moudgil

Published by the
Society for Mining, Metallurgy, and Exploration, Inc.

Society for Mining, Metallurgy, and Exploration, Inc. (SME)
8307 Shaffer Parkway
Littleton, Colorado, USA 80127
(303) 973-9550 / (800) 763-3132
www.smenet.org

SME advances the worldwide mining and minerals community through information exchange and professional development. SME is the world's largest association of minerals professionals.

On the Cover: Spherical alumina particles produced at the University of Utah by spray melting at high temperature. Such spherical particles can be obtained from different materials and used as functional fillers, catalysts, and polishing media.

ISBN 0-87335-228-9

Library of Congress Cataloging-in-Publication Data has been applied for.

Contents

Preface

Mineral fillers and nanoscale particles continue to grow in technological importance, and the industrial use of such materials is widespread. Traditional mineral processing plays an important part in the concentration of naturally occurring minerals. These concentrates, in many cases, are the feedstock for fillers and extenders. In addition, traditional mineral processing techniques are valuable in the characterization of fillers and nanoscale particles. Thus, it is fitting that one of the cosponsors of the 2003 Functional Fillers and Nanoscale Minerals Symposium is the Minerals and Metallurgical Processing Division of SME. And given that most mineral fillers are of the industrial type, it is fitting that the Industrial Minerals Division of SME is another cosponsor of this symposium.

The idea for a minerals symposium was first hatched after the recognition and success of Eurofillers '95. This recognition was accompanied by the realization that members of SME, with a long history of mineral development, were an ideal group to organize and host a similar symposium in the United States. Concurrent with the European effort in mineral fillers was the creation of the National Science Foundation Engineering Research Center for Particle Science and Technology at the University of Florida. Because the center is actively engaged in the use and application of particles, it is appropriate that the NSF Center also is a cosponsor of this symposium.

The purpose of the Functional Fillers and Nanoscale Minerals Symposium is to bring together interested researchers and industrial users of mineral and mineral-based fillers (i.e., "performance materials" derived from minerals). Therefore, topics concerning the manufacture of fillers as well as their structure and properties are presented. Specific topics include general purification, synthesis, modification, and characterization of fillers and nanofillers; surface modification; matrix interactions; interfaces; adhesion; filled systems; dispersion and compounding; and effects of fillers on mechanical, electrical, thermal stability, and other properties. Special thanks is given to the session chairs: Dr. Courtney Young (Montana Tech), Dr. Sharad Mathur (Englehard Corp.), Dr. Srinivas Veeramasuneni (USG Inc.), Dr. Jaroslaw Drelich (Michigan Technological University), and Dr. Cesar Basilio (Thiele Kaolin Company).

SECTION 1

Functional Mineral Filler Fundamentals

Planetary Mill Grinding for the Production of Fillers

Raj K. Rajamani*, S. Latchireddi*, and M.K. Abd El-Rehman†

The planetary mill is very much suited for the ultra-fine grinding of mineral fillers. Ultra-fine grinding here refers to grinding powders to a size range 5–10 microns. Such powders enhance the quality of the final product. The operation of a specially built planetary mill is described. It is shown that a target size of 5 microns can be reached in a few hours using the dual drive planetary mill. A centrifugal field of 37G is generated by the gyration arm of the mill. The design allows the desired setting of canister critical speed. The ultra-fine grinding of alumina, titanium dioxide, hematite, magnetite, kaolin and silicon carbide are described. Also, the continuous operation of the mill is shown.

INTRODUCTION

Filler is an inert material that is included in a composition for some useful purpose. Fillers play a major role in improving the quality of the product by means of physical interaction due to their inert nature. The major filler usages industries are: paints, paper, plastics, rubber, bituminus composites, pesticides and fertilizers. The important mineral fillers and their major uses are summarized in Table 1.

World consumption of fillers and pigments by paper and board is over 19.5 million tons in 1995 (Laufmann, 1998). The main (quantity) mineral fillers are kaolin, naturally ground calcium carbonate (GCC), chemically precipitated calcium carbonate (PCC) and talc. Another filler of importance (not in quantity but in effect and cost) is titanium dioxide, which at low dosage provides extra high opacity and brightness to the sheet.

Global consumption of fillers for plastics is estimated at more than 5.5 million tones (Anon, 2002). Calcium carbonate (GCC & PCC), accounts for 40% of this, with glass accounting for 31%, and other mineral fillers and natural fibers 29%. North American and European consumption is estimated to be 60% of the total. Compounding demand

* Dept. of Metallurgical Engineering, University of Utah, Salt Lake City, Utah.

† Central Metallurgical R&D Institute, Cairo, Egypt.

TABLE 1 The important mineral fillers and their major uses

Name	Major Uses	Particle Properties
Barite	Rubber and paint	Generally equi-dimensional
Dolomite	Paints, paper, plastic, pesticides	Unique diatom structure, micro and ultra-micro porosity
Gypsum	Paints, paper, cotton goods, pesticides	Irregular, roughly equi-dimensional
Kaoline	Paints, paper, adhesives, plastic, rubber, ink, pesticides	Thin, flat hexagonal plates, .05–2 microns size and stacks of same
Limestone	Asphalt, fertilizers, insecticides, paints, rubber	Variable size particles, ultimate rhombs
Pyrophyllite	Insecticides, asphalt, roofing, paints, rubber, battery boxes	Minute foliated plates and long particles
Silica	Quartz-paints, bitumens, etc. "Soft" silica-paints, foundry facings, etc.	Variable size, angular and equi-dimensional particles
Talc	Paints, insecticides, roofing, rubber, asphalt, paper, cosmetics, textiles	Lamellar, foliated, or micro- fibrous
Titanium dioxide	Paints, paper	
Iron oxide	Paints, pigments, ferrofluids	

of fillers for plastics will be led by automotive, construction, appliance, business equipment, film and wire and cable applications.

The particle size of the fillers is the most important parameter in their effective utilization. The finer the particles the larger the specific surface as well as the sorptive capacities and other properties largely dependent on surface area. The particle shape and surface characteristics are largely dependent on fineness and method of preparation but are mostly inherent in the natural mineral or rock itself. Fineness and particle size are influenced greatly by the degree of milling and treatment during manufacture.

Ultra-fine grinding process plays an important role in production of fine powder for the manufacture of fillers and modern materials such as ceramic materials, magnetic media, electronic packaging materials and metal ceramic composites (Ding, et al., 1996). Traditionally such fine powders are produced in ball mills, stirred ball mills and vibration mills. However, these mills exhibit a limitation regarding product fineness: a size limit of one micron is often observed after several hours of grinding.

Application of high energy ball milling technique to obtain nanophase materials appears attractive for the potential large scale, economical production of such materials (Matteazzi, et al., 1996). High energy ball milling process has been successfully used to synthesize nanometric ferrite materials (Niviox, et al., 1998). Ferrites with spinal structures, which have nanometric sizes, are very important material in electronic applications such as transformers, choke coil, noise filter and recording heads.

In the field of sub-micron grinding, turbo milling, attrition mill, ball mill and plateray mill are used. A comparative study of the grinding rate using different grinding mills has also been reported by Nilson (1994). It has been shown that the planetary ball mill grinds at a much faster rate than either the ball mill or attrition mill.

The objective of this paper is to show the usage of the innovatively designed dual drive planetary mill (US Patent #6086242) for the production of ultra-fine powder. The ultra-fine grinding of alumina, titanium dioxide, hematite, magnetite, kaolin and silicon carbide are described. The operation of the planetary mill is described.

DESCRIPTION OF PLANETARY MILL

The planetary ball mill described is a specially constructed mill whose schematic is shown in Figure 1a and a photograph in Figure 1b. It consists of two canisters held between a pair of gyration arms and the gyration arm itself attached to the gyration shaft. The gyration motor (5 HP) spins the gyration shaft thus creating very high centrifugal field on the canisters. Likewise the rotation motor produces the self-spinning canisters. The rotation motor being a variable speed motor, the rotation of the canisters can be set to any desired percentage of critical speed.

Since the gyration and rotation shafts are driven by independent motors, any combination of clockwise/counterclockwise rotations can be set. This design feature departs from the fixed gear ratio used by single motor designs, which primarily fixes the critical speed of the mill. Quiet, vibration-free operation is achieved by using transmission belts instead of chains and sprockets or gears and also carefully balancing the weight in the canisters. Table 2 summarizes the technical specifications of the mill.

MATERIALS TESTED AND RESULTS

The following materials types (fillers and pigments) were tested successfully on the planetary ball mill to produce ultrafine ground product.

Alumina

The size distribution of the alumina product subjected for grinding at different times is shown in Figure 2 together with the feed size distribution. It can be noted here that with just 5 minutes of grinding at 37G in planetary mill has reduced the 90 micron (d_{50}) feed size to 2 microns. This illustrates that the planetary mill has grinding rates at least an order of magnitude higher than the traditional ball mill. To illustrate this the planetary mill was operated as a ball mill by setting the gryation speed set to zero while setting the rotation speed to 70% critical speed. The results thus obtained are plotted in Figure 3 which compares the product size distribution of ball mill with that of planetary mill. Figure 3 shows that the ball mill product size does not campare even with the 5 minutes planetary mill product. The higher gravitational field has the effect of confining the particles in the ball mass due to the high force field causing the observed grinding rates (Milin and Rajamani, 1997).

Titanium Dioxide (Rehman et al., 2001)

Figure 4 shows the particle size distribution of the titanium dioxide and the ground products at different time using 0.4–0.8 mm alumina balls at 35% mill filling and 50% by weight solid liquid ratio. It is seen that the fines increase continuously with increasing grinding time. The median size decreased from 0.62 μm in the feed to 0.39 μm but the final product is still far from the target. Usually fines agglomerate and the grinding action simply breaks the agglomerates. A suitable action would be to use a dispersant.

Kaolin

The grinding of kaolin to very fine size using high-energy planetary mill plays an important role in the delamination of kaolin, which make it suitable for paper coating applications. The grinding of kaolin at 40% solid liquid ratio was carried out using glass beads.

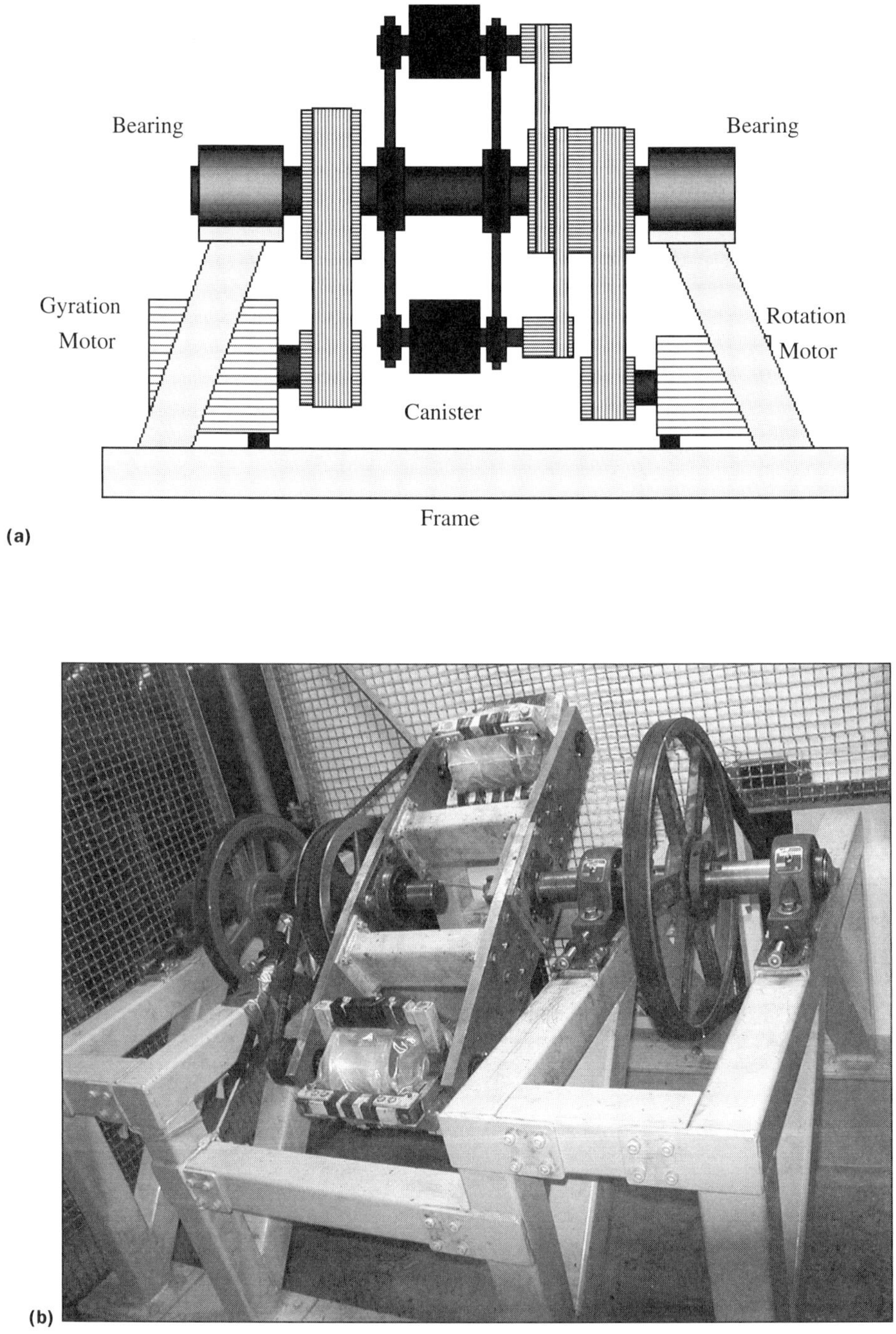

FIGURE 1 (a) Schematic of the dual drive planetary ball mill; (b) photograph of the dual drive planetary ball mill

TABLE 2 Planetary ball mill technical specifications

Gyration motor	Single phase 230VAC 60Hz, 5 HP
Rotation motor	Three phase 0–230VAC, 0–60Hz, 5 HP
Speed of gyration	310 rpm, 345 rpm, 390 rpm
Speed of rotation	Variable—same direction as gyration
Distance between the canisters	69.1 cm
Canisters	12.2 cm diameter and 11.0 cm long with neoprine rubber lining
Grinding media	2–5 mm size beads of alumina, stainless steel, zirconia, glass

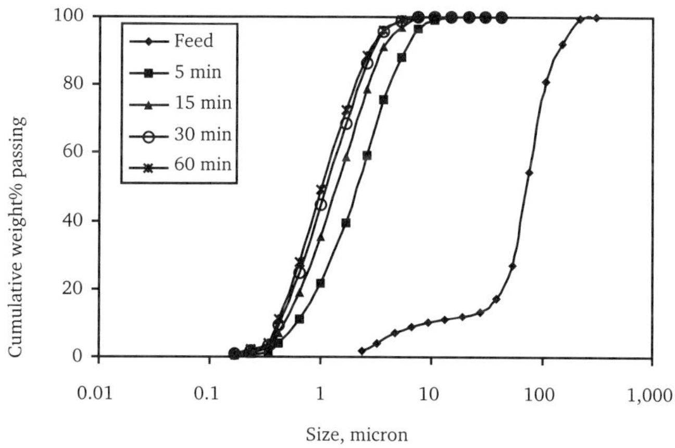

FIGURE 2 Size distribution of alumina at different grinding times

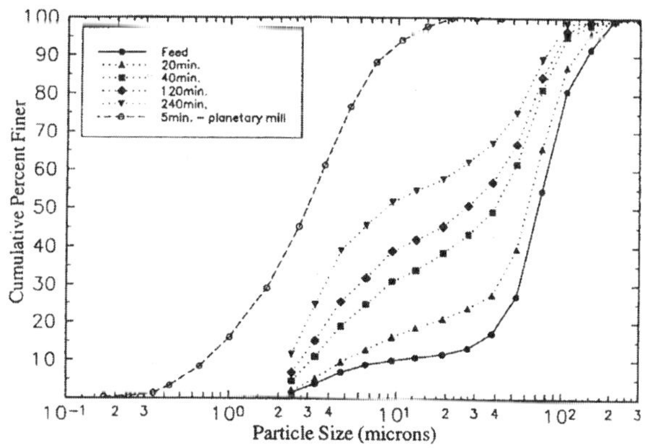

FIGURE 3 Comparison of ball mill and planetary product size distributions

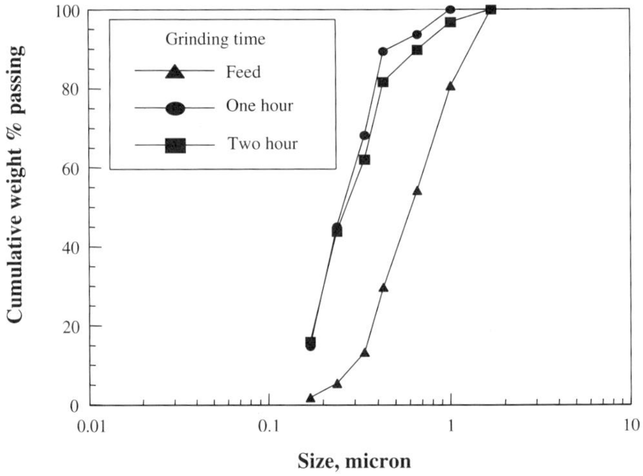

FIGURE 4 Size distribution of titanium dioxide at different grinding times

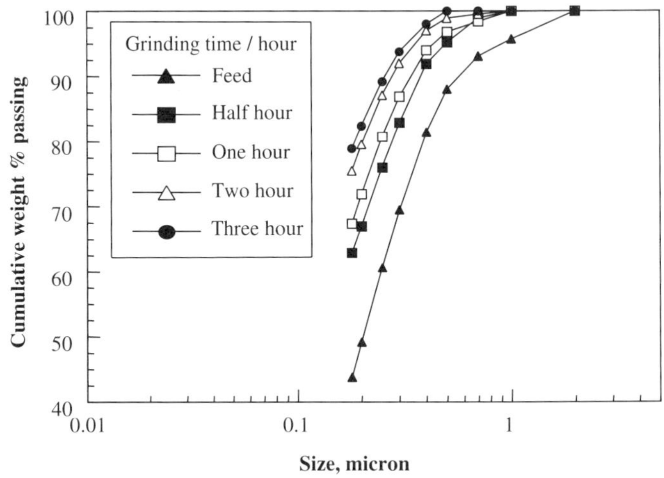

FIGURE 5 Size distribution of kaolin at different grinding times

Figure 5 shows the particle size of ground kaolin. The particle size of kaolin is readily reduced by high energy grinding. In three hours the percent passing 0.2 μm increased from 49% to 82%.

Magnetite (Rehman et al., 2001)

Magnetite oxide is used in different industrial applications such as recording media, ferrofluids and pigment. Figure 6 illustrates the results of dry grinding of magnetite (Fe_3O_4). It is readily seen that the rate of grinding is extremely fast. The fines increase with increase in grinding time and there is a constant reduction size in the first 45 minutes and after that time the size reduction is even faster.

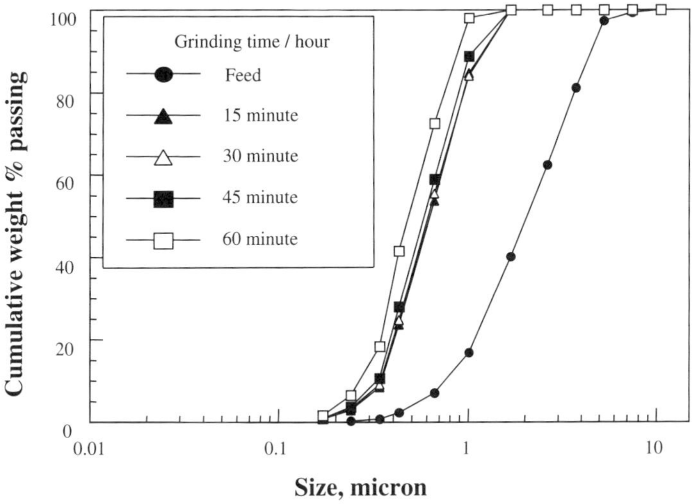

FIGURE 6 Size distribution of magnetite at different grinding times

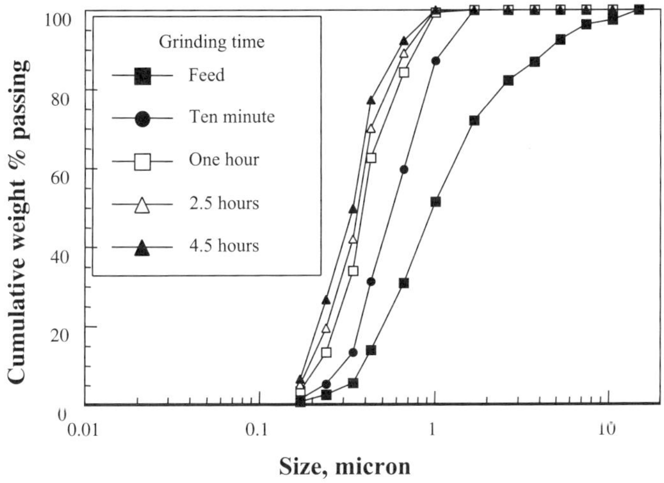

FIGURE 7 Size distribution of hematite at different grinding times

Hematite

Ultra-fine iron oxide powders are suitable for pigments. Figure 7 illustrates the results of wet grinding of hematite. It was noticed that after 10 min grinding time 87% of the hematite sample was less than 1 μm in size and increased to 99% after one hour. The d_{50} reduced from one micron to 0.34 μm during 4.5 hour grinding time. The products obtained at 10 min and 60 min grinding time are suitable to the pigment industry. These results show the ease of reaching sub-micron size using the planetary mill.

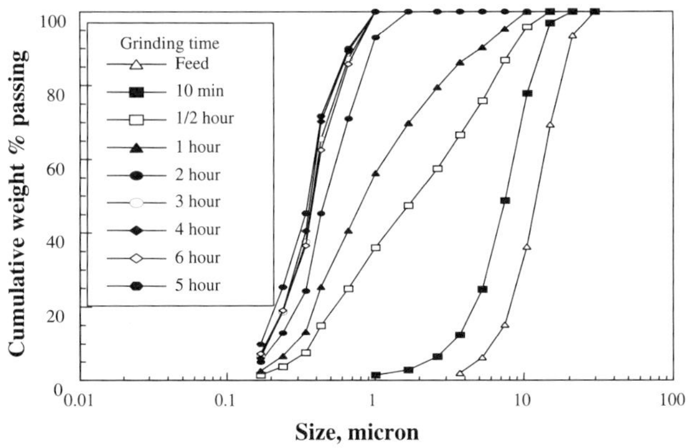

FIGURE 8 Size distribution of silicon carbide at different grinding times

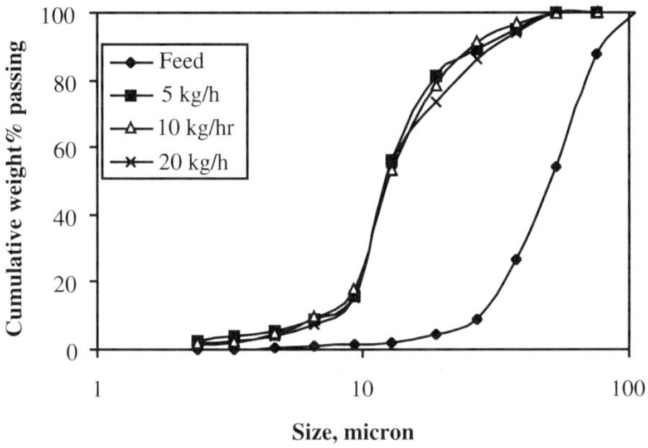

FIGURE 9 Size distribution of magnetite at different feed rates

Silicon Carbide (Ding, 1996)

Silicon carbide is the raw material for many ceramic and refractory products. Figure 8 shows the results of wet grinding of silicon carbide. The particle size distribution reaches to the finest distribution at 5 hour grinding time then it moves negatively to the coarser size range when the grinding time is increased. The rate of grinding in the mill is so fast that the average size of feed silicon carbide particle was reduced from 12.43 µm to 1.08 µm in 60 minutes. The d_{50} was reduced from 12.43 µm to 0.36 µm after 5 hour grinding time.

CONTINUOUS OPERATION OF PLANETARY MILL

The continuous operation of the planetary mill is accomplished by directly pumping the slurry to the canisters. The retention time of the slurry in the canister is controlled by throttling the discharge from the canister. Although the mechanics of feeding and discharging is difficult, it was accomplished by specialized means. Although some problems were encountered initially in feeding and discharge arrangements, they were overcome and tests were successfully conducted. The results obtained at a typical condition are illustrated in Figure 9, which indicate that the planetary mill can be operated continuously to produce the required product size.

SUMMARY AND CONCLUSION

Sub-micron size grinding of oxides and non-oxides were carried out using dual drive planetary ball mill. The oxides used are titanium dioxide, hematite, and magnetite. When reduced to ultra-fine sizes, these oxides exhibit superior properties in respective applications. The other non-oxides used are kaolin and silicon carbide. Kaolin is a very strategic mineral in the filler and paper coating industry. For paper coating, kaolin must be reduced to ultra-fine sizes. The planetary ball mill was selected in this study due to its faster grinding kinetics.

The ball size diameter is the important factor to attain ultra-fine ground products. The size reduction occurs according to the centrifugal field produced by the mill and subsequent stressing of the particles in the interstitial space of the ball mass. The smaller ball size has small interstitial space. In the centrifugal field the ball mass gets pressed against the wall producing a compressive stress. For these reasons the small grinding balls can readily fragment smaller particles. The average size range of the ground products is 0.2-to-0.4 μm, which enhance the properties of these powders in the respective applications.

The grinding results of some of the filler materials illustrate that the planetary mill can be effectively used for production of filler grade sub-micron particles in significantly shorter time. Although the unit capacity of planetary mill is a concern, the test results with continuous operation indicate that it can be achieved.

REFERENCES

Anon., Functional Fillers for Plastics 2002, Toronto, Canada.

Ding, J., Tsuzuki, T., McCormick, P.G., and Street, R., "Ultrafine Co and Ni particles prepared by mechanochemical processing," (1,2) Supplement, J. phys. D: Appl. Phys. 29, 1996, pp. 2365–2369.

Ding, P., Ultrafine Grinding of Silicon Carbide in a Batch Planetary Mill, M.E thesis, University of Utah, 1996.

El-Rehman, M.K. Abd., Ma, N., and Rajamani, R.K., Ultrafine grinding of some oxides and non oxides materials using the planetary mill, Ore Dressing, Issue 5, 2001.

Laufmann, M., Fillers for Paper–A global view, Presented at PTS-Seminar on Wet End Operations, October 1998 in Munchen.

Matteazzi, P., Wolf, F., and Basset, D., "High carbon, hypereutectic, nanostructured white irons by mechanosynthesis: processing and fabrication," (1) Supplement In: Processing and fabrication of advanced material IV, Edited by Srivatsan T.S., and Moore J.J., published by The Mineral, Metals and Material Society, pp. 97 -4-109, 1996.

Milin, L. and Rajamani, R.K., Kinetics of alumina ultrafine grinding in a planetary ball mill, materials and manufacturing processes, Vol. 12, No. 2, pp. 241–260, 1997.

Nilsen, K.J., Milling of ceramic powders in a centrifugal barrel mass finisher, First International Particle Technology Forum, pp. 209–214, 1994.

Nivoix, V., Bernard, F., Gaffet, E., Perriat, P., and Gillot, B., "Mechanical activation conditions of the Fe_2O_3 and V_2O_3 mixture powders in order to obtain a nanometric vanadium spinel ferrite," In: 9th European symposium on comminution, 8–10 September, ALBI, France, Vol. 1, pp. 243–252, 1998.

Chemical Additions for Blended Cement: Performance Enhancers that Maximize Production Economics and Minimize Environmental Impact

J.D. Gargulak* **and V.L. Zajakowski***

The current political climate has forced multinational cement manufacturers to consider technology that minimizes CO_2 emissions. Along with this environmental issue, individual manufacturing facilities are under constant pressure to increase productivity and lower costs. The use of processing or functional additions can help address these difficult demands by maximizing the extender content in blended cements. Recently, cement standards have been modified such that additions can be employed to improve quality. By using these "cement performance enhancers," the cement manufacturer can maximize extender content of cement thus optimizing plant economics, while meeting local quality standards and minimizing greenhouse gas emissions.

INTRODUCTION

Lignosulfonates (LS) have been used in the construction industry for many years. They are known as water reducers, functioning as dispersants in hydraulic cement systems (1). By lowering viscosity and yield point, they reduce the amount of water needed to obtain a given flow. The ability to reduce water gives the opportunity for enhanced strength and control of the rheology characteristics of fresh concrete. Based on these well documented attributes, it seems that the opportunity exists for lignin based products to act as manufacturing additions that can be interground with cement to improve performance and allow cement manufacturers to optimize the non-clinker portion of their product.

This paper gives examples of LS based additions tested with limestone and pozzolan cement in a lab setting where the additive is combined in the test mortar, as well as

* LignoTech USA, Rothschild, Wis.

under trial conditions in the manufacture of a Portland Type 1P cement where the additive is interground with the cement. Both examples demonstrate the potential of these additives to improve cement quality.

Environmental Concerns

Few individuals are unaware of the political pressures associated with the limitation of greenhouse gasses (2). It has been well documented that manufacturing of Portland cement is one of the most CO_2 intensive industrial processes. It is estimated that the cement industry generates 7% of global CO_2 emissions. Recently, leaders of the top ten global cement manufacturers initiated a joint program to minimize CO_2 emissions (3). One possible way to minimize emissions is substitution of fillers (aka extenders) or supplementary cementitious materials (SCM) for clinker in the production of cement.

Provided that the cement produced meets local standards and market expectations, this is an ideal solution. While accounting for CO_2 emissions is complex, it is generally assumed that 1 ton of CO_2 is produced for every ton of clinker consumed to make Portland cement. Thus, every ton of material that replaces clinker in the final cement represents 1 potential ton of CO_2 tax savings (4).

Economic Concerns

Five types of extenders are commonly added to cement: (i) limestone, (ii) cement kiln dust (CKD), (iii) natural pozzolan, (iv) granulated blast-furnace slag (GBS), and (v) pulverized fly ash (PFA). A distinction can be made between fillers (limestone and cement kiln dust), and SCM (pozzolan, granulated slag and PFA). The inclusive term "extender" is used for all materials. The term "filler" is rarely used within the industry.

Apart from the above-described environmental issues, there are productivity gains to be had from optimizing the use of these extenders. They tend to cost much less than clinker. In addition, they tend to be "softer" to mill than clinker. Often these extenders mill faster causing increased throughput, maximizing production capital (5). In general, the more extender/SCM interground the more efficient the milling process becomes. Unfortunately this is not true for GBS, which tends to be harder and more energy intensive to mill. Thus, this paper will concentrate on LS based additions to optimize limestone and pozzolan cements.

Testing Standards

Current cement standards fall into two broad categories based on either the European EN 197, or US ASTM 595 and ASTM 1157. Both standards allow for the manufacture of modified cements containing extenders, however the market acceptance of these cements varies from region to region. The manufacturing standard applied in each particular country or region therefore constrains the ability of a cement company to add or increase their addition of extenders to cement.

ASTM 465 explains the standard specification for processing additions. For this use lignin based additions have been known for some time. They do not find common use as they tend to be less effective as a grinding aid compared to traditional materials like TEA (triethanolamine). The interest in maximizing extender content, particularly the 5% addition of limestone to OPC (Ordinary Portland Cement), may make a process addition of a modified LS attractive.

Mention should be made of ASTM 688, which offers a standard specification for functional additions in blended cements. Here, *functional additions* is defined as a material that beneficially changes the properties of hydraulic cement when the material is interground with clinker during the manufacture of cement. This paper suggests using LS formulations as additions to optimize limestone as a cement extender, and pozzolan as a SCM.

Cement Quality Considerations

Cement quality is judged by a combination of hydraulic properties: (i) compressive strength (CS) development, (ii) workability, (iii) setting characteristics, (iv) color and (v) durability. These properties are inter-related and add up to the concrete making potential of the cement. The use of extenders/SCM in making cement impacts on all the properties and can constrain the amount of extender a cement company can add to their cement. Lignin-based products can favorably impact on these properties and offset most of the detrimental effects of the use of extender (6).

For example, addition of extenders to cement predominantly reduces workability. This in turn means that additional water must be added to the concrete mix in order to place and compact the concrete. This additional water in the concrete mix adversely affects the strength development of the concrete. Lignin-based products, alone or part of a formulation, can offset the detrimental effects of extenders by improving the workability of the cement. Additional water does not need to be added to the concrete mix and strength development is therefore restored to the levels achieved prior to the addition of extenders/SCM. The increased limestone content can also reduce workability retention. It should be possible to minimize this using LS-based additives as well.

EXPERIMENTAL

The mortar testing described in this paper was performed in accordance with EN 196 or ASTM 109. The testing carried out for comparison was flow, air content, and 1 day, 3 day, 7 day and 28 day CS. Other appropriate data, such as dose and deviations from the standard methods, are included in the data section when necessary, except for the following variation. Set time was measured via an LTUS internal method that uses the onset of heat development in a 450 g sample of mortar in an insulated, covered, 0.5 liter thermos. A multiple thermocouple apparatus collects data on 8 mortar samples simultaneously and converts temperature versus time data to an Excel format. The set time was determined by drawing intersecting lines on the hydration curve and using the onset of hydration as the initial set time.

RESULTS AND DISCUSSION

Laboratory Examples with Limestone Extended Cement; Addition Added to Mortar Mix

Suppose that a cement producer manufactured a "General Use" cement blended with limestone. This cement would be manufactured as CEM I (EV 197) or CEM 10 (CSA). Both cement standards allow for an addition of 5% limestone to be interground with the cement. However, it is more common to see a partial addition, as the full addition of 5% tends to cause the cement to have increased water demand. While there is typically no unique specification on water demand, local markets will not accept a product that fails to meet their slump expectations. Adding limestone at the 5% rate could potentially

TABLE 1 Effect of modified LS on CEM 1 with 5% limestone interground (ENV 196)

LS (dose)	Flow (%)	Density (kg/m^3)	Air (%)	Compressive Strength (N/mm^2) 1 Day	7 Days	28 Days	Initial Set	Final Set
Control	220	2,210	5.7	20.9	45.5	54.4	4' 40"	5' 15"
Borrecem DP1 (0.10%)	235	2,160	9.2	18.1	43.7	50	4' 45"	5' 30"
Borrecem DP1 (0.20%)	245	2,010	13.9	13.1	35.2	43.5	7' 00"	7' 45"
Borrecem DP4 (0.10%)	240	2,190	6.9	20.6	45.5	53.6	4' 40"	5' 45"
Borrecem DP4 (0.20%)	250	2,130	9.6	18.8	44.7	51.7	6' 30"	7' 15"

lower CO_2 emissions by 2–3%, as well as increase throughput, reducing the final cost of the cement produced.

A LS-based addition can remedy this problem. To demonstrate, a limestone blended CEM I cement was tested using EN 196 to determine if lignin based additives could have a positive effect on slump (7). The cement selected contained 5% limestone interground. Data collected on two lignin based products at an addition rate of 0.1 and 0.2% owc is displayed in Table 1. Borrecem DP1 and DP4 are developmental product supplied by Borregaard LignoTech. Both materials are modified LS. In this study the additives were dosed to the mix water used in mortar testing. The data clearly show the increased flow, or potential slump enhancement, of these products. The magnitude of flow increased from 220% to 250% with a dose of 0.2% DP4. At a dose of 0.1%, DP4 provides a good flow (slump) increase with little negative impact on air content, CS or set time. Of course, plant trials would need to be undertaken with the specific cement manufacturer to determine the full utility of this concept.

Laboratory Examples with Pozzolan Extended Cement; Addition Combined with Mortar Mix

Natural pozzolan is a cost effective SCM to use in the manufacture of blended cement. This material is abundant geographically wherever volcanic activity has been present. Cement manufacturers use this material by intergrinding after mining and crushing, with some manufacturers using a pre-calcination step prior to introducing the material to the grinding mill. Type 1P cements usually offer other benefits, such as sulfate resistance. Usually manufacturers are limited by local requirements for 1-day strength of Type 1P cement. This strength requirement results in an under-utilization of pozzolan, and a potential lost opportunity for savings.

Table 2 displays the data from an addition of a LS formulation to a low strength pozzolan cement. In this data set, the formulation was added to the mix water at a dose of 0.16% wt/wt. The primary constituent of the formulation is a modified LS. The composition of this Portland Type 1P cement is 27% pozzolan, 63% clinker, 5% limestone and 5% gypsum. ASTM 595 allows for up to 40% pozzolan in Type 1P cement, provided it meets the required performance criteria.

Note how strength is increased at all ages, and water demand is brought more in line with a standard ASTM Type 1 cement. With this formulation there is little negative effect from retardation of the LS, a common issue associated with its use.

Table 3 shows data for similar testing using Type 1P cement with a lower pozzolan content compared to the previous example. Here the blend ratio is 20% pozzolan, 70%

TABLE 2 Strength increase by ASTM C-109 for 595 Type 1P cement; high pozzolan content cement

Sample	w/c	Set, Hrs.	% Water Reduction	% Flow	Compressive Strength, Mpa (% of Control) 1 Day	3 Days	7 Days
Control	0.52	3.4	—	113	4.17	14.38	20.74
Borrecem 200 formulation, 3.6L/MT	0.49	3.5	5.8	113	6.00 (143.9%)	18.74 (130.3%)	25.82 (124.5%)
Borresperse Q formulation, 3.6L/MT	0.49	3.6	5.8	114	5.76 (138.1%)	18.34 (127.5%)	25.68 (123.8%)

TABLE 3 Strength increase by ASTM C-109 for a 595 Type 1P cement; low pozzolan content cement

Sample	w/c	Set Hrs.	% Water Reduction	% Flow	Compressive Strength, Mpa (% of Control) 1 Day	3 Days	7 Days
Control	0.485	2.8	—	115	6.4	14.6	18.7
Borrecem 200 formulation, 3.6L/MT	0.455	3.4	6.2	114	8.1 (127%)	17.8 (122%)	22.8 (122%)
D-748 formulation, 3.6L/MT	0.455	3.4	6.2	115	8.1 (127%)	17.6 (121%)	22.6 (121%)

clinker, 5% limestone and 5% gypsum. This cement has higher early CS than the previous example, but again addition of the lignin formulation has a highly positive effect on strength at all ages.

With both examples, it is assumed that a similar result would be obtained had the addition been interground with the cement. The manufacturers of these cements should consider field trials to determine if the use of LS-based additions could be a tool to increase the pozzolan substitution rate closer to the maximum allowed under ASTM standards.

Plant Trial with Pozzolan Cement; Addition Interground with Clinker

A plant trial was run at a cement manufacturer with a cement of the following composition: 4% gypsum, 61% clinker, 25% pozzolan and 10% limestone. The additive tested was a formulated, modified, sodium LS. In this study, 3.4 liters/MT Borrecem 200 (Borregaard LignoTech) corresponds to a dose of 0.1% owc LS and 0.05% of a proprietary accelerator.

The mill was a horizontal ball mill with a feed rate of about 50 MT/hour. Additive was dosed on the clinker feed belt and samples were taken at 1-hour intervals. A composite sample was prepared and tested. After the addition was stopped, the mill was equilibrated for 4 hours and a final control sample taken. This was the control sample tested. Values for 1 day CS with additive showed an increase of 13.2% (Table 4). Values for 3 and 7 day CS were higher than the control (12.3% and 7.9% increase, respectively).

Product performance and degradation were evaluated by addition of the additive at the same dose as the trial, but to the control cement sample in the mortar mix. The CS difference between the *inter-ground* sample and the *control + additive* sample is small. The CS values are increased by 12.1% (inter-ground) and 13.8% (control cement + additive) over control cement. Data suggest the additive maintained 88% of its effectiveness. The loss of performance is most likely due to degradation of the lignin polymer while milling.

TABLE 4 Field trial data; functional additive interground with cement; ASTM C-109 test method; 595 Type 1P cement

Dose Additive	w/c	Flow	1 Day CS, KPa (% of control)	3 Day CS, KPa (% of control)	7 Day CS, KPa (% of control)	Set Time Initial/Final
Control	0.515	112	5.8	16.2	23.9	2:30/4:00
3.4 liters/MT interground	0.48	110	6.5 (112.1%)	18.2 (112.3%)	25.8 (107.9%)	3:05/4:15
Control w/ 3.4 l/MT added to mortar mix	0.46	108	6.6 (113.8%)	18.5 (114.2%)	26.0 (108.7%)	

At this particular plant, it was claimed that a 10% increase in early strength translated to a potential increase in pozzolan by 5%. Thus, the use of this additive package could facilitate an increase in pozzolan by approximately 5%.

Along with a potential 5% reduction in CO_2 emissions, there are strong economic incentives from the direct substitution of pozzolan for clinker, as well as the potential energy savings and realization of increased throughput.

CONCLUSIONS

When studying additive formulations like these, other considerations need to be carefully weighed. The additives can have a strong effect on milling conditions, particle size and power consumption. These parameters need to be studied more closely in future trials. Also, since each cement plant has different raw materials, manufacturing standards and market expectations, it is expected that each formulation would need to be optimized on a case-by-case basis.

The subsequent use of LS based admixtures to concrete is a concern, and seems to be in conflict with our concept.

However, this has not been our experience. Given the low dose of LS suggested in these formulations, their subsequent impact on the performance of low range water reducing admixtures is negligible.

While much work still needs to be done, LS based additions offer a way to maximize extender/SCM content given current cement manufacturing standards. We acknowledge that some individuals would like to increase the SCM content of cement far beyond the levels permissible by today's standards, with the goal to minimize CO_2 emissions even further and improve cement quality. These organizations recommend the use of superplasticizers to achieve this goal (8). These are admirable objectives, but giving the conservative nature of the cement industry it will take years for change to occur. The technology described in this paper offers an immediate solution that fits the guidelines of current practice and standards today.

REFERENCES

1. Northey, R.A. 1992. Low-Cost Uses of Lignin. In *Emerging Technologies for Materials and Chemicals from Biomass*. Rowell, R.M., Schultz, T.P., Narayan, R., editors, Washington, D.C: ACS symposium Series 476, 146–175.
2. Halhotra, V.M. 1998. Role of Supplementary Cementing Materials in Reducing Greenhouse Gas Emissions. *Proceedings from International Workshop on Concrete technology for Sustainable Development in the 21st Century.*
3. Cement Sustainability Initiative. See http://www.wbcsdcement.org/.
4. Ibid ref 2.

5. Tsivilis, S., Batis, G., Chaniotakis, E., Grigoriadis, Gr. and E. Theodossis. 2000. Properties and Behavior of Limestone Cement Concrete and Mortar. *Cement and Concrete Research.* 30:1679–1683.
6. Whitehopleman, Clark M. 2001. Internal Marketing Report prepared for Borregaard LignoTech.
7. Construction Materials Testing Labs in Derby, UK., 2000. Set time method Vicat EN 196/3.
8. Pagé, M. and N. Spiratos. 2000. The Role of Superplasticizers in the Development of Environmentally-Friendly Concrete. *Proceedings from the CANMET/ACI International Symposium on Concrete Technology for Sustainable Development.*

Colloidal Stability of Bentonite Clay in the Presence of Polymers

Ashoke SenGupta*, Jerry Darlington*, and David McGregor*

The stability of bentonite clay suspensions against flocculation in the presence of certain anionic polymers was investigated. These polymers included a purified sodium lignosulfonate, a sodium oxylignosulfonate, a potassium salt of humic acid, and a sodium polyacrylate. The polymers differed from one another in molecular weight as well as in carboxylic, phenolic, and sulfonate group contents. The scope of this study is in evaluating the extent of thinning achievable in a bentonite clay suspension in the presence of these anionic polymers under varying electrolyte concentration, presumably as a result of deflocculation of the clay platelets by the polymers.

Under certain conditions of polymer dosage and electrolyte concentration, the lignosulfonate seemed to have flocculated the clay suspension, while the other polymers served fairly well as a deflocculating/dispersing agent. The lignosulfonate, however, showed some deflocculating ability when the electrolyte concentration of the suspending medium (water) was increased by adding sodium chloride to the clay suspension. Among the four polymers tested, the polyacrylate was found to be the most effective in deflocculating clay particles, followed by the humate and the oxylignosulfonate.

INTRODUCTION

Nature dictates that any particulate material dispersed in a liquid should readily end up being flocculated, in order to minimize the surface free energy. However, some suspensions can and do remain deflocculated for months or years, meaning that they are reasonably stable against flocculation. This ability of suspended particles to resist flocculation is often called the colloidal stability of suspensions. The origin of colloidal stability lies in the mechanisms that allow the suspended particles to strongly repel each other, overcoming the van der Waals attraction that relentlessly acts between them.

* Amcol International Corp., Arlington Heights, Ill.

Suspended particles can be mutually repulsive if their surfaces have an electrical charge, either anionic (negative) or cationic (positive). Particle surface may become electrically charged if any ionic species dissolved in the suspending medium adsorbs on it. The ionic species may be simple ions and/or they may be ionic surfactants or polymers. The dissociation of any ionic group on the particle surface can also render the surface electrically charged. Regardless of the charging mechanism, particle surface charge can only help in slowing down flocculation, but cannot completely eliminate it. As a result, suspension stability due to particle surface charge is merely kinetic in nature, meaning that it is only such that the particles with a higher surface charge may flocculate slower than the particles with a lower surface charge. Accordingly, only when the suspended particles strongly repel each other due to their surface charge can they remain deflocculated to a large extent over an extended time. The addition of an electrolyte, however, results in "shielding" of the electrical charge on a surface, minimizing the electrical repulsion that another material body with a like electrical charge as the surface would face while closely approaching the surface. The other material entity may be either a surface or an ionic moiety.

Polymers can bring about yet another interparticle repulsion, described as the steric repulsion in literature [1], if (i) they adsorb on the particle surface, covering at least half of the total particle surface area, and (ii) upon adsorption, the polymer segments dangle out off the particle surface, generally, in the form of loops and tails. Suspended particles may remain stable against flocculation so long as the steric repulsion forces continue to act between them, implying that sterically-stabilized suspensions can be thermodynamically stable. Also, these suspensions are likely to remain stable over a much wider range of electrolyte concentration as compared to electrostatically-stabilized suspensions. Only when the electrolyte concentration is so high that the solvency of the suspending medium (water) for the polymer is reduced and/or the polymer chains fail to maintain expanded conformations can the steric repulsion be impaired by an electrolyte. The electrolyte concentration at which steric stabilization may fail could vary from one polymer to another. For the above reasons, there is a general interest in industrial practice to use polymeric stabilizers in producing stable suspensions. As mentioned earlier, the surface charge of suspended particles due to any adsorbed ionic polymer may provide for the electrical repulsion between them. Accordingly, ionic polymers may stabilize suspended particles by virtue of electro-steric repulsion.

As noted above, a necessary condition for the polymers to work as a steric stabilizer is that they are able to adsorb on the particle surface. Various mechanisms involving interactions between the polymer and the particle surface may drive such polymer adsorption. Examples of these interactions include van der Waals interaction, hydrophobic interaction, hydrogen bonding, electrostatic interaction, and acid-base interaction. The mechanism that acts as a supplementary mechanism to all other types of adsorption mechanisms is the van der Waals interaction. The van der Waals interaction-driven adsorption generally increases with increasing molecular weight of the polymer. On the other hand, the adsorption, for example, of an anionic polymer on an anionic surface may be enhanced and/or the packing density of the adsorbed polymer on the surface may be influenced in favor of steric repulsion, with the addition of an electrolyte. The added electrolyte could reduce the polymer-surface electrical repulsion, promoting adsorption, and/or could lower the electrical repulsion between the adsorbed polymer chains, facilitating their tighter packing on the surface.

In the study reported herein, our objective was to evaluate three natural polymers and a synthetic polymer, each differing in molecular architecture from the others, for their ability to deflocculate suspensions of a water-swellable, layered magnesium aluminum silicate material such as smectite clays. These clays are a class of inorganic particulate materials that occur as stacks of individual, planar silicate layers referred to as platelets in clay literature. Examples of smectite clays include bentonite, montmorillonite, bidelite, hectorite, saponite, and stevensite.

The face-surfaces of bentonite clay platelets bear an anionic charge counterbalanced by exchangeable cations that remain electrostatically attracted to the anionic charge of the clay surface. The exchangeable cations are generally either sodium ions or calcium ions. Bentonite clay is referred to as either sodium or calcium clay, depending on the type of counterion associated with the face-surface of the platelets. While the anionic charge on the face-surfaces does not vary with pH, the electrical charge on the edge-surfaces, although anionic under alkaline pH, could be cationic under acidic pH [2].

Water diffuses into the interplatelet region when bentonite clay particles are suspended in water, wetting the face-surfaces of the platelets. The stacked-up platelets can then be forced apart from one another, aided by the electrical repulsion between the anionic face-surfaces and any applied shear. The process of delamination of clay platelets is often referred to as exfoliation in clay literature. The exfoliated bentonite clay platelets can readily engage in forming platelet-to-platelet links, producing in turn a particulate network that may pervade through the entire suspension volume. Unless a deflocculating agent is added, bentonite clay suspensions turn out to be viscous due to the formation of the particulate network structure that entraps the suspending medium, even when the concentration of clay solids is relatively low. A deflocculating agent acts to break down the particle-to-particle links by providing for interparticle repulsion, resulting in thinning of the clay suspensions.

In this study, we evaluated four anionic polymers for their ability to deflocculate bentonite clay particles and thereby produce thinning in the clay suspensions, in the presence and in the absence of an electrolyte. These polymers included a purified sodium lignosulfonate, a sodium oxylignosulfonate, a potassium salt of humic acid, and a sodium polyacrylate. The colloidal stability of concentrated suspensions may be assessed from their viscosity, inasmuch as a lower viscosity indicates a more stable suspension with particles that are deflocculated to a greater extent. Accordingly, a suspension may be characterized to have a lower stability, if its viscosity is higher or if the viscosity increases rapidly with time. As indicated earlier, the addition of an electrolyte may induce flocculation in electrostatically-stabilized suspensions, by reducing the electrical repulsion between the suspended particles. Considering the above, stability-testing of bentonite clay suspensions in the presence of an anionic polymer was carried out by measuring the suspension viscosity using a Brookfield viscometer, with and without an added electrolyte.

EXPERIMENTAL

Materials

The sodium lignosulfonate, sodium oxylignosulfonate, and potassium humate were obtained from Borregaard LignoTech. The lignosulfonate was a purified (ultrafiltered) product, while the oxylignosulfonate was a catalytically oxidized form of the sodium lignosulfonate. The polyacrylate was received from BASF Corporation. The polymers were tested in as-supplied forms. Some key compositional parameters for the lignosulfonate, oxylignosulfonate, and humate polymers are given in Table 1, where the functional group contents are based on the weight of the polymer product. The carboxylated polymer, polyacrylate, had a weight average molecular weight of about 8,000. The sodium bentonite clay was obtained from American Colloid Company.

TABLE 1 Compositional parameters of the polymers

Polymer	Carboxyl Group	Phenolic Group	Sulfonate Group	Weight Averaged Molecular Weight
Sodium lignosulfonate	3.1	0.5	6.2	75,000
Sodium oxylignosulfonate	7.4	3.1	2.6	11,300
Potassium humate	15.1	2.5	0	20,000

Method

A 17-gram sample of a sodium bentonite was added to a polymer solution containing 183 grams of water, a given amount of the polymer, a small amount of a defoamer, and in some cases a specified amount of an electrolyte, sodium chloride (NaCl). The resulting slurry was homogenized in a Waring blender at 22,000 rpm for a total mixing time of 4 minutes. The slurry was then transferred to a plastic cup and degassed under vacuum. The Brookfield-viscosity of the slurry was recorded after 15 minutes and in some cases after various time periods measured from the time of completion of degassing. Degassing was not carried out for the control tests where no polymer was added, and the time periods for the measurement of slurry-viscosity were counted from the time of completion of homogenization.

Results and Discussion

The results of the slurry-viscosity tests are shown in Figures 1 and 2, corresponding to 10% and 30% dosage of the polymers on the weight of clay, respectively. As evident from Figure 1, in the absence of any added polymer, the viscosity of the clay suspension increased much with the addition of NaCl, indicating poor stability of the suspension against electrolyte-induced flocculation. The higher the dosage of the electrolyte, the greater was the suspension-viscosity.

The lignosulfonate showed some interesting trends as for the effects of the added electrolyte on its performance (Figure 1). When NaCl was not added, the lignosulfonate-containing suspension had a higher viscosity than the corresponding control suspension. Yet, at the same polymer dosage of 10%, the lignosulfonate lowered the slurry-viscosity well below that of the control suspension, when NaCl was added to the clay suspension. It may be noted that (the data are not presented here), when the lignosulfonate dosage was increased from 10% to 20%, while not adding any NaCl, the slurry-viscosity decreased marginally, but still remaining higher than that of the control suspension.

Based on these results, it may be speculated that the lignosulfonate served as a flocculant in the dosage range of 10–20%, unless NaCl was added to the clay suspension. Supporting this conjecture would be the phenomenon of bridging flocculation by polymers, where, under the condition of low polymer adsorption on the particle surface, a single polymer chain ends up adsorbing simultaneously on more than one particle, drawing the particles into flocculation. An argument against the speculation would be that a greater extent of exfloliation of clay platelets due to the deflocculating effect of a polymer adsorbed on their face-surface could potentially increase the suspension-viscosity by increasing the number concentration of the platelets. Further studies involving the adsorption of the lignosulfonate on clay are therefore needed for clearly delineating the foregoing results.

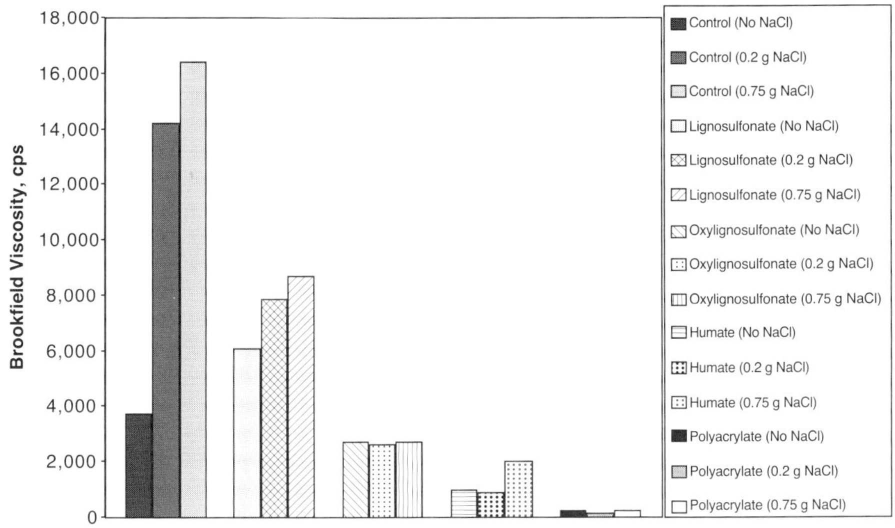

FIGURE 1 Effect of electrolyte addition on slurry-viscosity in the absence and in the presence of the polymers (10% w/w polymer on clay)

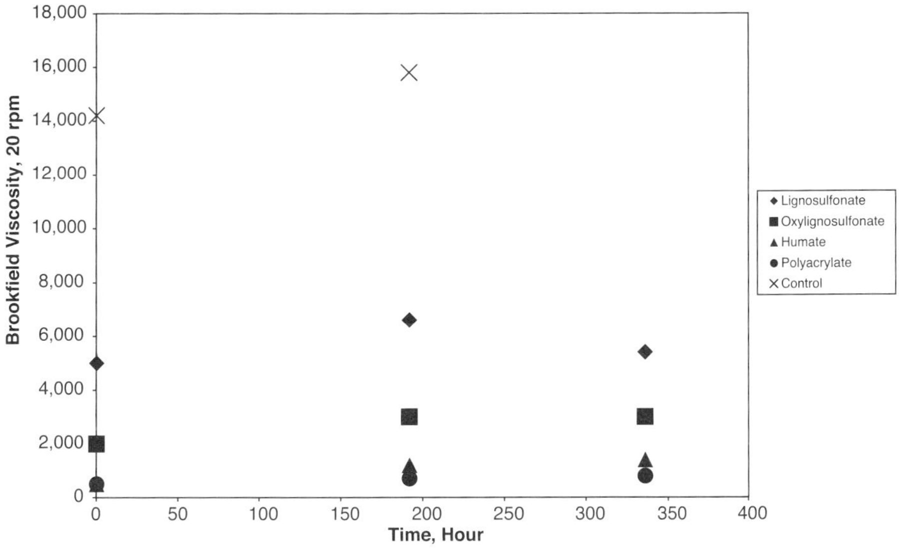

FIGURE 2 Slurry-viscosity vs. time (30% w/w polymer on clay, 0.2 g NaCl)

As per the results presented in Figure 1, the polyacrylate showed the highest deflocculating power (assumed proportional to the level of thinning achieved in the clay suspension), closely followed by the humate. The deflocculating ability of the oxylignosulfonate, although fell short of that of the humate, was considerably higher than that of the lignosulfonate. The extent of thinning achieved in the suspensions that contained either the oxylignosulfonate or the polyacrylate did not vary much with the increasing dosage of the added electrolyte. Even the humate-containing suspension did not show much variation in the level of thinning when NaCl was added at the dosage of 0.2 g NaCl per 183 g of the suspending medium, water. [If it at all must be considered and mentioned, it appears that for all three polymers–polyacrylate, oxylignosulfonate, and humate–there was a slight reduction in slurry-viscosity when NaCl was added at this dosage.] With further addition of NaCl, however, the slurry viscosity approximately doubled, presumably due to a reduction in the deflocculating power of the humate polymer.

Figure 2 shows the level of colloidal stability of the bentonite clay suspension when it contained a given amount of an added electrolyte and any one of the four polymers. In the presence of the polyacrylate, the suspension had a relatively high stability, as indicated by its low viscosity as well as a small increase in viscosity with time. The stability due to either the humate or the oxylignosulfonate, although fell short of that due to the polyacrylate, was appreciable. As evident from Figures 1 and 2, an increase in the polymer dosage from 10 to 30% lowered the viscosity considerably for both the humate and the oxylignosulfonate-containing clay suspensions. The suspension-viscosity increased, however, when the dosage of the polyacrylate was increased from 10 to 30%.

One could draw a general conclusion from the foregoing results that the polymers with higher levels of carboxylic group and/or a combination of carboxylic and phenolic groups performed better as a deflocculating agent for bentonite clay. This in turn raises the question as to whether or not some specific interactions involving these functional groups favored the adsorption of the more effective polymers on the clay surface. These specific interactions may include surface-complexation with multivalent cations, hydrogen bonding, and acid-base interaction. If such a specific interaction facilitated the adsorption of the polyacrylate, humate, and oxylignosulfonate, it is conceivable that the lignosulfonate, the leanest of the four polymers in carboxylic group and with a lower phenolic group-content than the humate and the oxylignosulfonate, had the lowest adscrption among the four polymers. The reason is that had the adsorption of the lignosulfonate not been assisted by any specific interaction, it could have been impaired by the electrical repulsion that an anionic polymer might face in approaching an anionic surface like the clay surface. Such polymer-clay repulsion could be reduced, however, by adding an electrolyte.

Furthermore, the lignosulfonate had an appreciably high molecular weight that is much higher than that of any of the other three polymers. It remains a question, therefore, as to whether or not the relatively large size of the lignosulfonate made it difficult for the polymer to adsorb on the platelet surface, since, for example, a high molecular weight polymer could be prevented due to its bulky size from diffusing into the interplatelet region. The hydrodynamic size of the polymer could decrease, however, due to an electrolyte-induced reduction in the intramolecular repulsion between the polymer segments.

The foregoing discussion would support citing low adsorption of the lignosulfonate as a probable reason for the lignosulfonate-containing suspension to have a higher viscosity than the corresponding control suspension, in the absence of an added electrolyte. As mentioned earlier, a polymer may flocculate a suspension by what is known as bridging flocculation under the conditions of low polymer adsorption on the particle surface. The same polymer may end up deflocculating the suspension when the surface coverage by the polymer is increased, for example, by increasing the polymer dosage.

CONCLUDING REMARKS

The study reported herein shows that certain strongly anionic polymers, including sodium polyacrylate, can serve as a deflocculating agent for bentonite clay suspensions, even though the bentonite clay platelets have an anionic charge. The deflocculating ability could vary, however, from one polymer to another. A common aspect of the three more effective polymers is that they have reasonably high levels of carboxylate and/or phenolic groups. Further studies, especially the ones involving polymer adsorption, and interfacial conformation of the adsorbed polymer, are needed to clearly explain the effects of polymer type (in terms of molecular architecture) and dosage, as well as of electrolyte dosage on the colloidal stability of bentonite clay suspensions in the presence of these polymers.

REFERENCES

1. Napper, D.H., "Polymeric Stabilization of Colloidal Dispersions," Academic Press, 1983.
2. Guven, N., Low, P.F., Mitchell, J.K., Sposito, G., and van Olphen, "Clay-water Interface and its Rheological Implications," CMS (The Clay Mineral Society) Workshop Lectures, Vol. 4, Editors: Guven, N., and Pollastro, R.M., Boulder, Colorado, 1992.

Precipitated Calcium Carbonate: Myth vs. Reality

Rasik H. Raythatha*

Calcium carbonate is perhaps the most widely used mineral in the world. With total consumption of all calcium carbonate exceeding over 25 million tons per year, within this class of filler, a relatively smaller segment of precipitated calcium carbonates (PCC) is often misunderstood as a commodity product and bundled in the same category as ground calcium carbonate. However, each material has a distinctly different role and characteristics. Precipitated calcium carbonates are used to influence specific performance in a specific application whereas ground calcium carbonate is used primarily to reduce cost in many applications. In this paper, attempts are made to demonstrate the role of nano-fine-PCC in improving shear-thinning characteristics of common and specialized sealants and automotive underbody coating and how such "nano-particle" calcium carbonate properties can be exploited to process PVC plastics while reducing overall cost.

We also review the role of precipitated calcium carbonate as "nano" material and why it must be viewed as a specialty additive.

INTRODUCTION

Basically, fillers are divided into functional and non-functional. Historically, non-functional fillers are used to reduce final product cost whereas functional fillers must meet critical product specifications. With the ever-increasing competitive environment, both types of fillers are widely used to lower cost and to improve product performance. For example, non-functional fillers are key components in reducing levels of expensive binders and pigments in paints, thus allowing cost reduction. Similarly, functional fillers are used in plastics to improve mechanical properties of the end parts while reducing processing time and cost by partially replacing more expensive organic modifiers. In sealants and adhesives, functional fillers provide critical rheological control by improving ease of applicability and at the same time providing the added benefit of improved adhesive and cohesive strength to the cured sealant.

* Solvay Performance Chemicals, Houston, Texas.

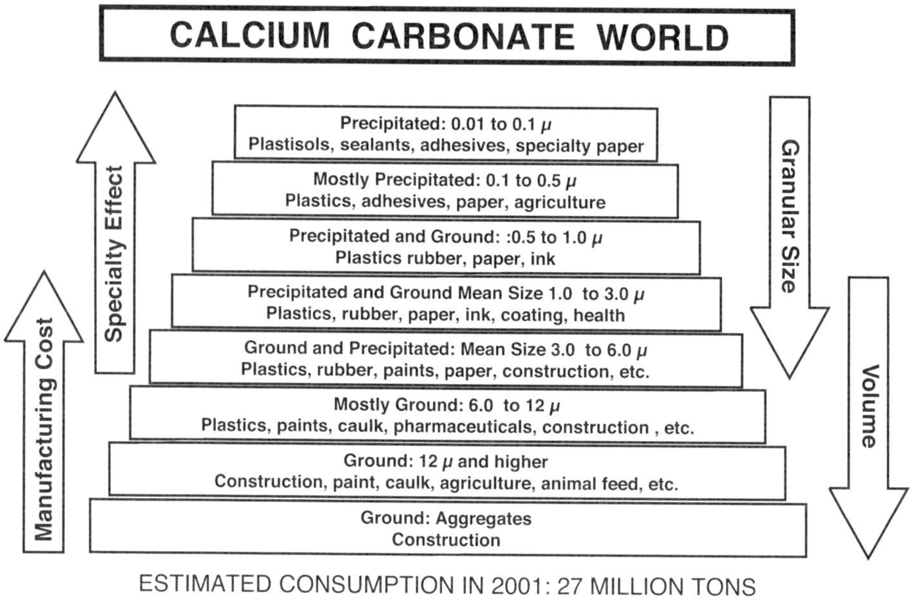

FIGURE 1 Major uses of calcium carbonate

Since non-functional fillers dominate the overall filler market, in many users' minds the delineation of functional vs. non-functional is rather blurred. This problem of differentiating between functional vs. non-functional fillers becomes even more diffuse when it comes to calcium carbonate. The range of calcium carbonate products available varies from large aggregate to nano-particles, purity from a low of 5% to high of 99.999%, sources varying from geological deposit to *seashells* to pure chemical synthesis, all of which define the uses and functionality. Therefore one may wonder about the role of calcium carbonate fillers and classification into functional vs. non-functional. This is especially true for synthetic or often referred to as precipitated calcium carbonate (PCC). One may then question if PCC is just a commodity filler or a specialty product. The answer depends on the choice of the PCC grade and intended end use. We have attempted to show that not all PCC are equal. In fact, seemingly ordinary filler has some extraordinary effects due to its inherent "nano-particle size" and unique crystalline behavior.

Natural calcium carbonate exists in a variety of crystalline forms such as calcite, aragonite and highly meta-stable vaterite. It occurs in nature as marble, soft powdery chalk, compacted limestone or biologically formed corals and *seashells*. Calcium carbonate is used in a range of products such as paper, paint, plastics, adhesives, sealants, construction material, agriculture, and pharmaceuticals just to name a few. The estimated consumption of calcium carbonate, inclusive of all range of products, exceeds 27 million tons per year[1]. A major portion of this volume consists of larger ground gangues and aggregates. Meanwhile, the special identity of the relatively smaller volume precipitated calcium carbonates is often lost and bundled in the same category as ground calcium carbonate. Fortunately, it is becoming apparent that "precipitated calcium carbonates influence specific performance to qualify as a true functional material." The major uses of calcium carbonate, in relation to their relative sizes, are depicted in the Figure 1. As the size decreases, hypothetical transformation of calcium carbonate from

Stage 1. Calcination of limestone	$CaCO_3$	$\longrightarrow$	$CaO + CO_2$
Stage 2. Slaking to quicklime	$CaO + H_2O$	$\longrightarrow$	$Ca(OH)_2$
Stage 3. Precipitation	$Ca(OH)_2 + CO_2$	$\longrightarrow$	$CaCO_3 + H_2O$

FIGURE 2 Production of precipitated calcium carbonate

commodity to specialty occurs arbitrarily around 1.0 μ. To reach a size below one micron either requires energy intensive grinding or chemical precipitation. In practice, physical grinding becomes limiting below 0.5 μ due to prohibitive cost and the inherent limit of the grinding technology. This is not a constraint when synthesizing sub-micron calcium carbonate by controlled precipitation method. In fact, the precipitation not only provides control over the particle size, but also in defining correct morphology, chemical purity, surface characteristics and particle size distribution. Interestingly, in the past indirect precipitation of calcium carbonate, of course unknown at that time, was a key factor in preserving building structures (white wash coating) and old arts (Frescos). In both process, a slow formation of calcium carbonate was responsible for forming opaque and preservative protective coating.

The precipitation of calcium carbonate has been practiced either as a standalone technology or as a by-product for a long time. However, the real application as a functional filler did not begin until the early twentieth century. The explosive growth of PCC consumption is a relatively new phenomenon, which can be attributed to the conversion of acid paper to alkaline paper making and by convenient and cost effective delivery of PCC via on-site satellite production plants. With this advancement, PCC consumption has risen to over 5 million tons in the last twenty years[2].

As indicated above, precipitation allows control of three main characteristics: particle size, shape and crystalline lattice structure. Several processes are known in the literature to produce PCC including gaseous carbonation of hydrated lime, carbonation of calcium chloride and lime-soda-ash processes. Essentially, in each precipitation process, operating parameters can be modulated to control the above properties that can be later exploited to either derive or increase target functionality.

Virtually all of the known commercial production of PCC is carried out by theoretically close loop cycle. In this process, natural limestone is calcined to quick lime, which is later slaked to form calcium hydroxide and the recombination of hydrated lime with recovered carbon dioxide from the calcination stage, Figure 2.

Depending upon the manufacturing process control, one can generate trigonal calcite or acicular needles[3], Figures 3 and 4. The size of these particles tends to be between 0.2 μ to as large as 2.5 μ. A rather involved process allows formation of nano-size particles with distinct rhomohedral morphology, Figure 5. Depending on the intended uses, PCC surfaces are treated with fatty acids or rosin to improve dispersion in organic medium or control surface reactivity, Figure 6. In comparison, ground calcium carbonate tends to be featureless with broad particle size variations, Figure 7.

In this paper attempts are made to provide glimpses of inherently nano-particle size PCC in improving shear-thinning characteristics of sealants and automotive underbody coatings and to demonstrate how "such nano-particle" calcium carbonate property can be utilized to process PVC plastics while reducing cost. Relatively coarser particle size, but well engineered PCC, especially of scalenohedral morphology, has become the dominant filler in the paper industries to enhance opacity, brightness and longevity of the paper, at the same time reducing cost of wood fiber and expensive titanium dioxide and

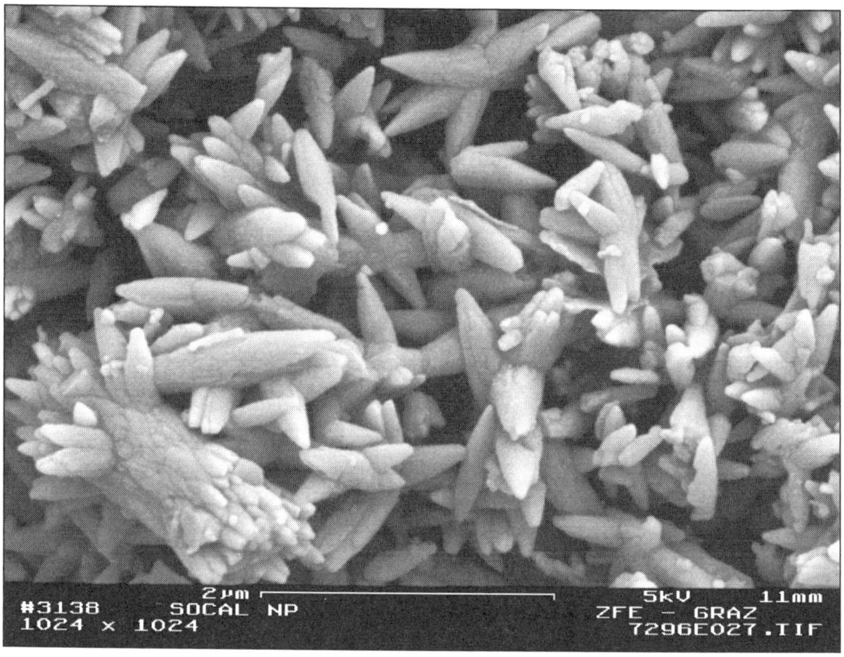

FIGURE 3 Calcite (trigonal scalenohedral PCC)

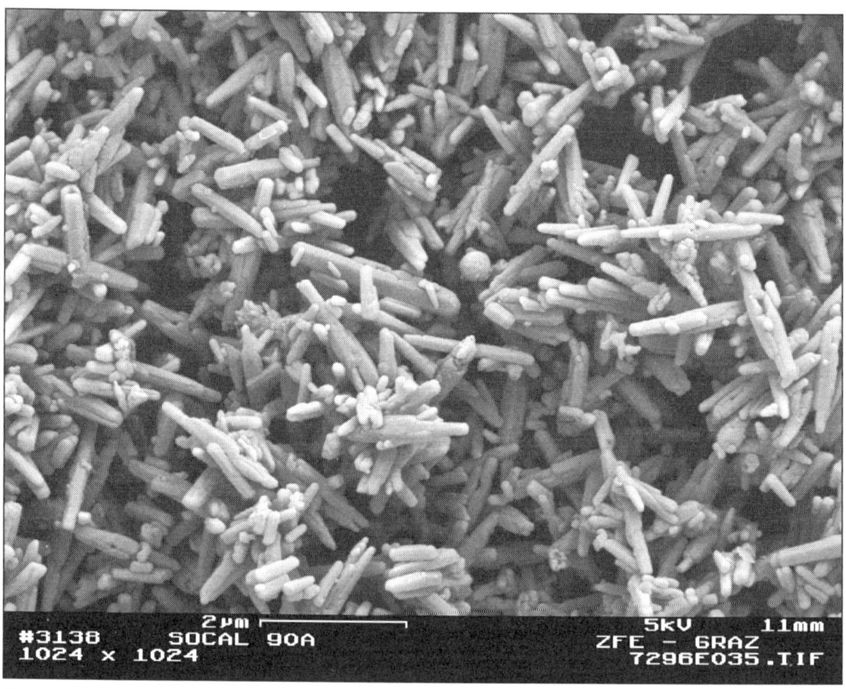

FIGURE 4 Aragonite (ortho-rhombic)

FIGURE 5 **Calcite (trigonal rhombohedral PCC)**

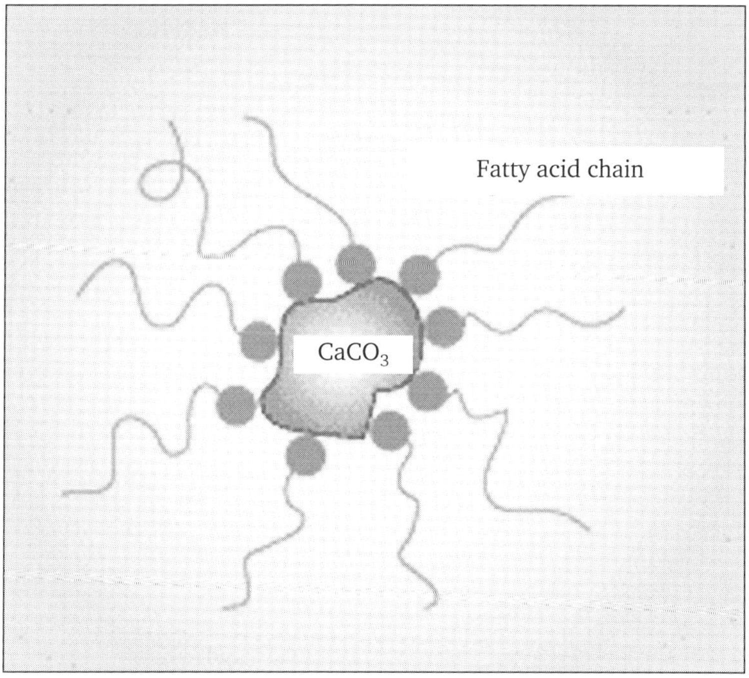

FIGURE 6 **Schematic of PCC surface coating with fatty acid**

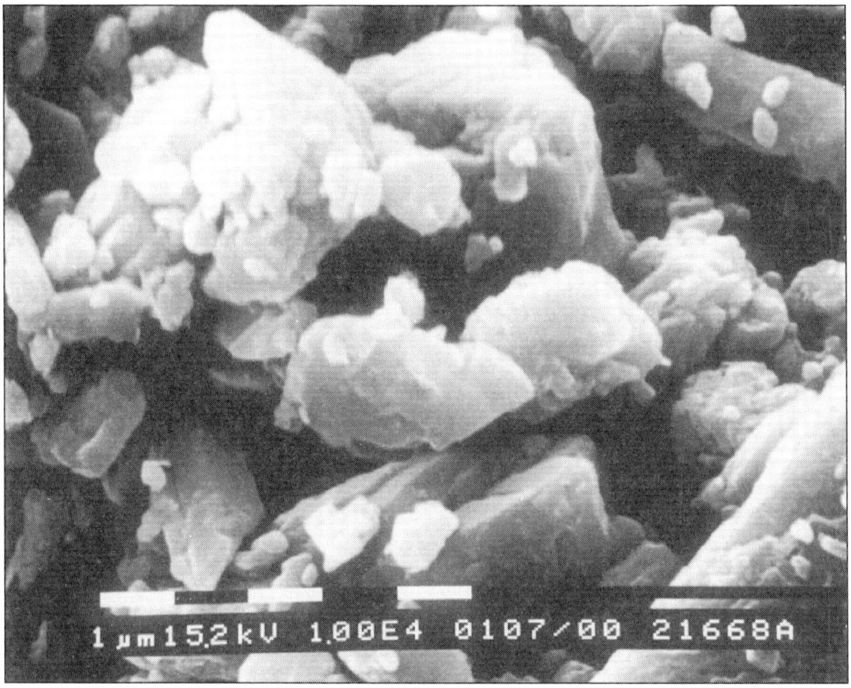

FIGURE 7 Natural fine ground calcium carbonate

reducing mechanical wear because paper making can be conducted under milder conditions. Several authoritative articles and patents have described the role of such PCC in paper application and will not be covered in this article.

APPLICATIONS

Automotive Plastisol

The automotive underbody coatings, commonly referred to as plastisol, are designed to provide corrosion protection along with soundproofing and resistance to coating breakdown under harsh chemical or physical forces. The plastisols are applied to automotive parts under varying shear forces. So it is imperative that the formulation conforms to very demanding rheological properties. On the other hand, during transportation this coating must remain stable, must not show phase separation during either the transportation or storage, and upon application to the designated substrate surface it must restructure very rapidly to a uniform film with minimal slump. Several thixotrope compounds, rheology modifiers, have been used to impart such characteristics, such as fumed silica, hydrogenated oleo compounds and precipitated calcium carbonate. However, it has been found that Solvay nano-PCC is among the most cost effective and superior in providing the desired degree of performance, Table 1 and Figure 8.

As can be seen from Table 1, at equivalent loading of fillers, nano-PCC provides much better performance, specifically, after low-high-low shear cycle. Nano-PCC not only affects the rheological properties, but also structural strength of the actual plastisol joint, Table 2. For a good seal, plastisol properties of low modulus, high adhesion and

TABLE 1 Comparative performance of various thixotrope in PVC plastisol

Thixotrope	Content (%)*	Low Shear Viscosity at 0.5 rpm	Low Shear Viscosity at 1.0 rpm	High Shear Viscosity Visc.	High Shear Viscosity Shear Rate	Low Shear Viscosity at 0.5 rpm	Low Shear Viscosity at 1.0 rpm
Winnofil SPT	24	202	97	7.84	10,144	148	82
Organo clays	24	221	117	6.18	12,949	213	105
UFPCC-A	24	113	61	6.79	11,544	98	52
UFPCC-B	24	155	76	6.97	11,536	114	60
UFPCC-C	24	141	70	6.55	12,388	110	53
Fume silica	3	146	85	5.13	9,513	130	59
UFPCC-A +	24						
Fume silica	2	229	115	7.07	11,338	194	96

* Added to master batch of PVC plastisol formulation.

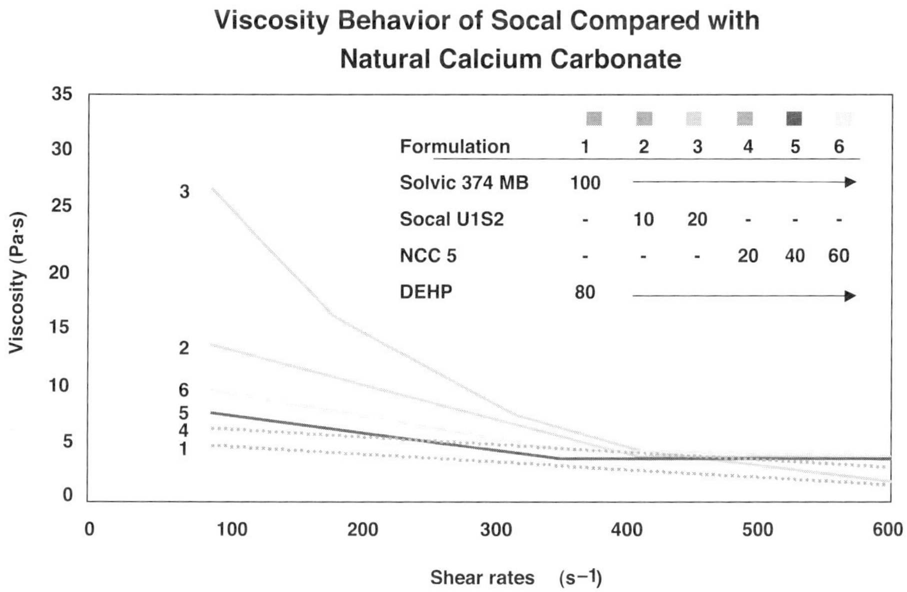

FIGURE 8 Effect of nano-fine PCC on the rheological properties of plastisol

good elongation are generally positive factors. As shown, plastisol containing alternate thixotrope, fumed silica or combination of fumed silica with another PCC shows unacceptable adhesion failure along with poor modulus and elongation.

Construction Sealant

The construction sealant consists of a range of binders, such as silicone, polyurethane, acrylics, polysulfide and butyl rubber. Among these sealants, demand for one critical

TABLE 2 Comparative performance of various thixotrope in PVC plastisol

		Lap Adhesion Strength*		
Thixotrope	**Content (%)**	**Stress, MPa**	**% Elongation**	**Failure Mode**
Winnofil SPT	24	32	25	Cohesive
Organo clays	24	27	14	Cohesive
UFPCC-A	24	27	18	Cohesive
UFPCC-C	24	26	17	Cohesive
Fume silica	3	24	21	Adhesive
UFPCC-A +	24			
Fume silica	2	28	26	Cohesive/Adhesive

* Plastisol was gelled at 165°C for 17 minutes. Lap adhesion strength was measured using an Instron 1122 with cross-head speed of 50 mm/min.

TABLE 3 Effect of Solvay nano-fine PCC on the performance of alkoxy cured silicone sealant

	Formulation		
	A (%)	**B (%)**	**C (%)**
Composition			
Silicone system	55.25	55.25	55.25
Alkoxy cross linker	2.50	2.50	2.50
GCC	33.00	0.00	0.00
Solvay PCC-A	0.00	39.00	0.00
Solvay PCC-B	0.00	0.00	39.00
Fume silica	6.00	0.00	0.00
Silane	1.00	1.00	1.00
Additive	2.25	2.25	2.25
Performance			
Boeing jig, mm	>10	3	2
Extrusion, g/min @ 36psi/6mm	199	237	266
Shore A hardness	17	19	20
Tensile strength, MPa	0.76	0.84	1.00
100% modulus, MPa	0.38	0.29	0.36
Elongation at break, %	320	683	640

property, "rheology," remains a constant factor. The construction sealants are applied by a range of methods from squeeze tube to hand operated guns to mechanical sprayer guns. Each requires significant different flow properties and restructuring of the sealant. Once again, sealant producers have a choice of several rheological medium, but they have found that nano-PCC provides the desired degree of rheological control as well as aesthetics, long term durability, weatherability and adhesion.

A number of different types of construction sealant formulations containing nano-fine PCC, as shown in Tables 3, 4 and 5, exhibits lower slump (lower Boeing Jig numbers) and excellent shear thinning effect as seen by extrusion rate, low modulus, good tensile strength and high elongation. These are some of the most critical properties for a good construction sealant.

TABLE 4 Effect of Solvay nano-fine PCC on the performance of oxime cured silicone sealant

	Formulation		
	A (%)	B (%)	C (%)
Composition			
Silicone system	50.95	50.95	50.95
Oxime cross linker	5.00	5.00	5.00
GCC	20.00	0.00	13.00
Solvay PCC-B	20.00	43.00	30.00
Fume silica	3.00	0.00	0.00
Silane	1.00	1.00	1.00
Additive	0.05	0.05	0.05
Performance			
Boeing jig, mm	2	2	3
Extrusion, g/min @ 36psi/6mm	113	51	133
Skin time, min.	29	36	34
Tensile strength, MPa	2.36	1.92	1.31
100% modulus, MPa	0.88	0.64	0.43
200% modulus, MPa	1.50	1.12	0.71
300% modulus, MPa	2.04	1.55	0.98
Elongation at break, %	400	430	560

TABLE 5 Effect of Solvay nano-fine PCC on the performance of oxime cured silicone sealant

	Formulation		
	A (%)	B (%)	C (%)
Composition			
Silicone system	50.95	50.95	50.95
Oxime cross linker	5.00	5.00	5.00
GCC	20.00	0.00	13.00
Solvay PCC-A	20.00	43.00	30.00
Fume silica	3.00	0.00	0.00
Silane	1.00	1.00	1.00
Additive	0.05	0.05	0.05
Performance			
Boeing jig, mm	1	3	18
Extrusion, g/min @ 36psi/6mm	123	64	170
Skin time, min.	25	37	35
Tensile strength, MPa	2.18	1.77	1.27
100% modulus, MPa	0.88	0.62	0.41
200% modulus, MPa	1.48	1.02	0.68
300% modulus, MPa	2.05	1.39	0.93
Elongation at break, %	419	483	575

TABLE 6 Solvay PCC and nano-fine PCC

Category	Fine	Nano-fine
Trade name	Socal®	Socal® and Winnofil®
Crystallite shape/state	Scalenohedral calcite, acicular aragonite	Rhombohedral calcite
Mean particle diameter, µ	0.2 to 0.40	0.02 to 0.1
Surface area (m^2/g)	6 to 14	18 to 70
Surface treatment (%)	None	1.9 to 3.3
Typical applications	Paper, paint	Sealant, plastisol plastics, inks, etc.

Plastics. From a very simplistic view, virgin polymers are less useful due to their inherent characteristics. For practical applications, molded or extruded plastic products must meet rigorous structural strength requirements[4]. Typically, this is achieved using either organic additives or functional fillers. The organic additives consist mainly of high Tg (glass transition temperature) acrylic or chlorinated poly-olefins. Manufacturers of plastic products often use a variety of fillers, including both precipitated and ground calcium carbonate, to achieve effects similar to organic additive at a lower cost. The coarser GCC, although providing some rigidity, fails to protect against sudden impact forces (low impact strength). This problem becomes extremely severe as the temperature decreases. To offset these limitations, chemical impact modifiers are often the accompanying necessity. It has been found that nano-fine PCC, by itself or more often in combination with the organic modifiers, is a preferred option for cost and performance reasons. For such applications, Solvay offers a range of complementary nano-fine, surface-treated precipitated calcium carbonates, distinctly different from ground calcium carbonate by its primary particle size, shape and surface coatings, Table 6.

Processing Aid Additive

We have found that by controlling the above characteristics, nano-fine coated precipitated calcium carbonates (CPCC) can significantly reduce the need for more expensive chemical processing aid and impact modifiers. The use of coated nano-fine PCC, especially in polyvinyl chloride (PVC) based plastics, allows shorter fusion and rapid gelation, Figure 9. Thus, allowing the optimization in the use of a processing aid.

Impact Modification Additive

One of the inherent deficiencies of polymeric systems, especially PVC, is the lack of resistance to impact fracture. This is especially evident at low temperature. Typically, this impact resistance can be improved with the use of organic impact modifiers. Alternatively, similar performance can be achieved using nano-fine PCC. We have found that use of about 5 parts/100 parts of resin (PHR) nano-fine PCC can replace as much as 1.5 PHR of impact modifiers and the efficiency of the organic modifier is boosted by the presence of nano-PCC[5], Figure 10.

CONCLUSIONS

Synthetic calcium carbonate properties can be controlled with much greater precision to impart more functional properties then can be achieved by ground calcium carbonate. This is especially true with nano-fine calcium carbonate.

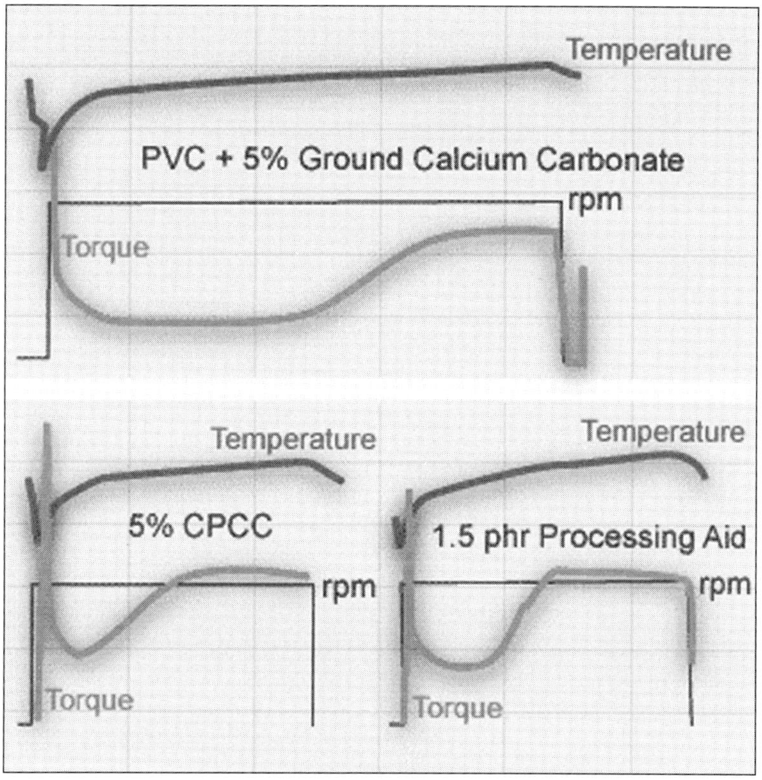

FIGURE 9 Effect of nano-fine PCC on the rate of gelation of PVC

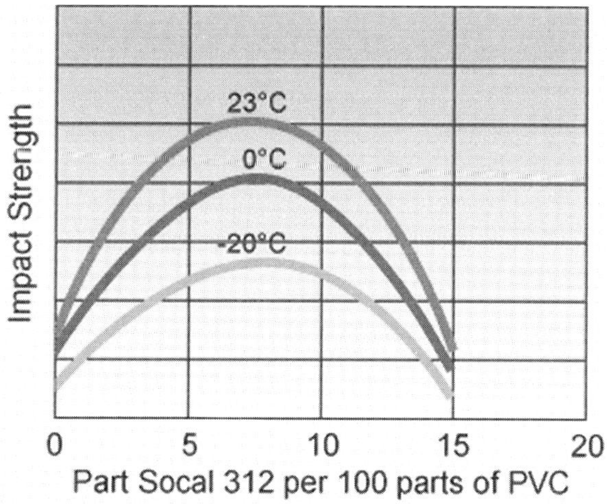

FIGURE 10 Effect of nano-fine PCC on the impact strength

ACKNOWLEDGMENT

The author acknowledges the contribution of Mr. David Cornwell and Dr. David Harrison in diligently reviewing the content of the manuscript and the support provided by the Management of Solvay Minerals, Inc.

REFERENCES

1. Wall Street Journal, October 7, 2002.
2. Chemical Economics Handbook-SRI, July 2000.
3. Solvay Literature.
4. D.W. Cornwell. July 2001. Benefits of PCC as a PVC additive. Industrial Minerals. 35–37.
5. D.W. Cornwell. August 2002. SpecialChem Web Seminar.

Light and Electric Potential Induced Reversible Wetting on Structured Surfaces

Nicola Richards*, Anthony Quinn*, John Ralston*, Geoffrey Reynolds†, and Rossen Sedev*

Derivatised 2,4-diketopyrimidine groups have been anchored to planar gold surfaces as self-assembled monolayers (SAMs) for chemical and photochemical studies. Three derivatives, 5-methyluracil (thymine), 5-trifluoromethyluracil and 5-nitrouracil, were chosen for the specific moiety present at position C-5, which can exert strong or weak inductive effects on the adjacent aromatic ring. The measured acidity constants decrease from 5-methyluracil to 5-nitrouracil as the inductive properties of the group at C-5 position increase. When grafted to the surface the acidity constants increase, albeit still adhering to the expected inductive trend. The functionality also exerts an influence on the surface wetting properties. Surface wettabilities indicate that the orientation of the pyrimidine groups at the surface occurs so that the C-5 position has a direct effect upon the surface chemistry. Upon dimerisation the surface contact angles increase significantly. At pH 11 the contact angle changes reflect that the dimer has a higher pKa than the monomer. The monomer surface is deprotonated at this pH while the dimer surface is partially deprotonated, leading to changes in contact angle of up to 22 degrees. Large changes in contact angle were observed for 5-nitrouracil which were independent of surface charge and were apparently due to conformational changes alone.

Electrowetting of three fluoropolymer surfaces by electrolyte solutions was studied with the sessile drop method. The electrowetting curve (contact angle/potential) is analogous to the electrocapillary curve (surface tension/potential) and may be described by a combination of the Young and Lippmann equations. We demonstrate that for some specific surfaces a systematic deviation occurs at positive potentials. This departure is attributed to double layer effects, namely the adsorption of hydroxide and halide anions. pH, ionic strength and polymer composition can all influence electrowetting behaviour.

* Ian Wark Research Institute, Mawson Lakes, South Australia.

† School of Pharmaceutical, Molecular and Bio-medical Sciences, University of South Australia, Mawson Lakes, South Australia.

INTRODUCTION

Surfaces which respond to external stimuli, such as light, allow various interfacial properties, i.e., wetting, liquid crystal alignment and dispersability, to be altered in a controlled fashion (Abbott et al., 1999; Ichimura et al., 1988; Ueda, Kim and Ichimura, 1994). In our earlier studies, we have shown that wettability can be manipulated reversibly by light when a substrate surface is modified by a photo-responsive monolayer (Abbott et al. 1999).

In 1960 Beukers et al. (Beukers, Ijlistra and Berends, 1960; Beukers and Berends, 1960) found that when frozen aqueous solutions of 5-methyluracil (thymine) were irradiated with UV light, a stable photoproduct was formed which lacked the characteristic UV absorption maximum at 264nm. The photoproduct had a molecular weight twice that of thymine, suggesting that it was in fact two molecules joined by a cyclobutane type ring located between positions C5 and C6. Thymine is thus referred to as a photochromic molecule and adjacent thymines in DNA form these dimers when exposed to UV radiation. The formation of specific photoproducts in DNA is the fundamental adverse effect of UV irradiation and leads to mutagenesis, carcinogenesis and cell necrosis in individual cells and complex organisms (Ruzsicska and Lemaire, 1995). Today it is known that thymine will dimerise reversibly depending upon the wavelength of light to which it is exposed. Molecules possessing this capacity will dimerise in the range of 200–300nm, however below 250nm, the shorter wavelengths of higher energy are able to cleave the photoproduct resulting in an eventual equilibrium (Figure 1). In order to control the position of this equilibrium the most efficient dimerisation is performed at 280nm and the cleavage at 240nm.

The novel aspect of a system where wetting properties are modified by light is that neither bulk phase needs to be altered in order to see a significant change in contact angle. Thin coatings of photochromic molecules can be used as photo-resists, data storage or for molecular recognition (Inaki, 1992). If a surface tension gradient can be established then the flow of liquids can be induced, providing a driving force for the operation of micro- and nano-scale chemical processes (Chaudhury and Whitesides, 1992; Bain, Burnett-Hall and Montgomerie, 1994; Gallardo et al., 1999). Even very small changes in contact angle are important. For example, Ichimura et al. (Aoki et al., 1992; Kawanishi et al., 1992) utilised monolayer systems where a modest, five degree photo-induced change in contact angle led to effective control of large-scale liquid crystal alignment in sandwich displays.

In this investigation, functionalised 2,4-diketopyrimidine groups have been anchored to gold surfaces as self-assembled monolayers (SAMs) for chemical and photochemical studies. We report the results of a light-induced wettability investigation of 5-methyluracil (thymine), 5-trifluoromethyluracil and 5-nitrouracil at various pH. For the specific head groups chosen, the moiety present at position C-5 can exert strong or weak inductive effects on the adjacent aromatic ring, influencing the related acidity constant. We have also investigated the influence of electric potential on surface wettability (electrowetting) using fluoropolymer surfaces. This work is complementary to the light-induced studies.

EXPERIMENTAL

Light Induced Wetting

Reagents. All reagents were of analytical grade and were obtained from Sigma-Aldrich, except 5-trifluoromethyluracil, provided by Pfaltz and Bauer, USA. In this study

TABLE 1 Glass transition temperature, T_g, PDD content, dielectric constant, ε, and critical surface tension, γ_C, of the Teflon AF variants used in this study

Grade	T_g (°C)	PDD (mol %)	ε	γ_C (mN/m)
T1	36	< 1	2.1	15.6
T2	133	61	2.1	13.0
T3	160	65	1.93	12.7

2,4-diketopyrimidine head groups were selected based on the inductive properties of the moiety at position C-5. Water (ultra high quality (UHQ)) used in the experiments was produced by reverse osmosis with two stages of ion exchange and two stages of activated carbon prior to final filtration (surface tension 72.8 $mN.m^{-1}$ at 20°C, k < 10^{-6} μsm^{-1}). Synthetic procedures are described elsewhere (Richards et al., 2002).

Bulk Acidity Constants. Bulk acidity constants were determined spectroscopically. Ultraviolet (UV) spectra measurements were recorded with a Varian Cary 1E UV/Vis. spectrophotometer. Samples were placed in clean 10mm quartz cuvettes with tightly fitting caps (Starna Pty Ltd). Aqueous solutions of 5×10^{-5} ML^{-1} of the pyrimidines were prepared in a background electrolyte of 0.1M NaCl. pH values between 1.5–5.5 were achieved by addition of HCl(aq.) to a freshly prepared 500ml aliquot of the solution. For values between pH 5.5–13 NaOH(aq.) was added to a separate aliquot of solution; both samples were continually stirred with a magnetic stirrer. Small aliquots of solution were removed for immediate sampling and then returned to the stock for further adjustment. The pH was adjusted and recorded at 25°C using a Hanna Instruments pH meter, equipped with a Hanna Instruments HI 1131B-pH electrode calibrated with Merck colour-coded buffers at pH values of 4.00 ± 0.05 and 7.00 ± 0.05 or pH 7.00 ± 0.05 and 10.00 ± 0.05. For each sample spectrum recorded the background was subtracted over the entire wavelength range. As the pH was adjusted, the accuracy of the curves was established by conformity to the trend as well as through the isobestic point.

Preparation of Substrates. Prior to gold deposition, silicon wafers were rinsed in ethanol and water then dried in a stream of nitrogen. Each plate was then plasma cleaned for 2 minutes in a Harrick Scientific PDC-32G plasma cleaner (60W). The wafers were coated with a chromium (99.999%) adhesion layer prepared by resistive evaporation from a tungsten boat. Immediately after, gold was deposited upon this sample surface from a molybdenum boat. The sample was kept at room temperature in an evaporation chamber maintained at 1×10^{-6} Torr. The deposition rate and layer thickness were monitored using a quartz crystal oscillator. Typically the deposition rate of the chromium and gold was 0.3 nms^{-1} and the thickness of the layers was approximately 10 nm and 100 nm, respectively.

Formation of Self-assembled Monolayers. Freshly prepared gold slides were immersed in a thiol solution (1×10^{-3} ML^{-1} in ethanol) and self-assembled monolayers were allowed to form for at least 24 hours, after which point there was no further change in contact angle. Upon removal the slides were thoroughly rinsed in ethanol, rinsed in water and then dried in a stream of nitrogen.

Contact Angle Measurement and Surface Titration. Advancing water contact angles were measured using the sessile drop technique. Images were recorded using a CCD camera with 5× magnification. Small drops of approximately 20 μl in volume were placed directly onto the surface using a micropipette and were illuminated using a diffuse light source. For surface titrations, individual samples were fully immersed in the same pH buffer solution that was used to measure the contact angle. After 45 seconds of spreading, the images of the drops were saved and digitised, then edge detection software was used to determine the sessile advancing angle. Recorded angles are the average of 5 readings and are taken from different areas on the surface. The readings were taken under ambient conditions, 22°C ±1, 50–60% humidity.

Irradiation. The samples were irradiated using a MM3 diffraction grating type illuminator equipped with a 300W Xenon lamp. Photo-dimerisation was initiated by irradiation of the sample at 280 nm (Energy Flux 5.5 $Jcm^{-2}sec^{-1}$). Photo-cleavage was achieved by irradiation of the sample at 240 nm (Energy Flux 0.4 $Jcm^{-2}sec^{-1}$). The bandwidth of the irradiated light was 20 nm.

Electrowetting

The insulating materials studied are amorphous fluoropolymers belonging to the Teflon AF line developed by DuPont (Hung et al., 1996; Bandrup, Immergut and Grulke, 1999). Teflon AF is a random copolymer of 4,5-difluoro-2,2-bis(trifluoromethyl)-1,3-dioxole (PDD) and tetrafluoroethylene (TFE).

These fluoropolymers are soluble in fluorinated solvents and can be deposited as thin films by dip or spin coating techniques. The surface of these films is very hydrophobic (advancing contact angle of water at pH = 5.6 is 120 ± 2°) but also very smooth (RMS roughness ~0.4 nm and peak-to-valley height <2 nm over 10 × 10 μm^2 (Schneemilch et al., 2000)). The good quality of the surface is evidenced by the low (<10° (Schneemilch et al., 2000)) contact angle hysteresis. Experiments have been conducted using three different Teflon AF variants: T1, T2 and T3, whose properties are listed in Table 1.

Teflon AF was dissolved in FC-75 fluorocarbon solvent (perfluoro-2-butyltetrahydrofuran, 3M) and films were deposited onto gold-coated silicon wafers via dip coating. By varying the solution concentrations (1–6 wt %) and the speed of withdrawal (0.1–1 mm/s) final film thicknesses between 0.1 and 10 mm were achieved routinely. Polymer films were dried for 10 minutes in a laminar flow cabinet, followed by an additional 10 minutes drying at 110°C to remove any residual solvent. Freshly prepared polymer films were used in all experiments.

All aqueous solutions were prepared with Milli-Q water (conductivity 5.6 μS/m; surface tension 72.8 mN/m) and analytical reagent grade salts. The normal pH of water was 5.6 and AR grade HCl and KOH were used to adjust the pH of the solutions when desired.

The experimental apparatus was based on the system developed by Verheijen and Prins (1999), where the capacitance of a flat insulator-coated electrode contacted with a grounded droplet was measured as a function of the applied DC potential. The capacitance was then used to determine the solid/liquid interfacial area. For a known drop volume (e.g., 10 μL), the contact angle was obtained from the base area by a calibrated empirical equation. In addition, contact angles were independently determined from images of the sessile drop by numerically fitting a tangent close to the contact line. The data obtained by the two methods agreed to within 3%.

TABLE 2 Acidity constant data

	pKa ± 0.2		
	Solution	N_1-alkyl Derivative in Solution	N_1-alkyl Derivative on Surface
5-methyluracil	9.9	10.5	11.0
5-trifluoromethyluracil	7.3	8.4	9.1
5-nitrouracil	5.0	6.3	8.2

TABLE 3 Contact angle data of monomeric and dimeric surfaces at neutral and alkaline pH

	pH 5.8			pH 11		
	Monomer θ ± 2	Dimer θ ± 2	Δθ	Monomer θ ± 2	Dimer θ ± 2	Δθ
5-methyluracil	67°	76°	9°	59°	69°	10°
5-trifluoromethyluracil	67°	75°	8°	50°	66°	16°
5-nitrouracil	56°	77°	21°	45°	67°	22°

A micro-syringe was used to very slightly advance the droplet at the beginning of each experiment to ensure that the contact angle was advancing prior to the actual voltage application. Small voltage increments (2 or 5 V) were used in order to resolve small contact angle deviations. Each experiment was conducted on a fresh area of surface to avoid effects associated with residual polymer charging (Verheijen and Prins, 1999).

Occasional data involving higher or unstable values of the dissipation factor tanδ were discarded from further consideration. Data with tanδ ≥ 0.05 normally arise as a result of poor polymer film preparation (e.g., the presence of contaminant particles results in leakage currents). Every curve was reproducibly repeated three to five times. Unless otherwise indicated all experiments were conducted at 22°C.

RESULTS

Light Induced Wetting

Acidity Constant (pKa) Constant. The measured acidity constants decrease from 5-methyluracil to 5-nitrouracil as the inductive properties of the group at C-5 position increase. In solution the pKa is 9.9 for 5-methyluracil and decreases to 5.0 upon substitution with a nitro group at C-5. Upon N_1-alkylation, for all species, the pKa increases slightly due to a shift in the electronic distribution in the pyrimidine ring.

When grafted to the surface the acidity constants increase further, still adhering to the expected inductive trend (Table 2).

Contact Angle Data as a Function of pH and Structure. The advancing contact angles at pH 5.8 are similar for 5-methyl and 5-trifluoromethyluracil, decreasing by 10 degrees in the presence of a nitro group at position C-5 (Table 3). Upon dimerisation all the surface contact angles increase to approximately 75 degrees. Surface irradiation results in a ten-degree contact angle change for 5-methyl and 5-trifluoromethyl uracil and a 21-degree change in surface contact angle for 5-nitrouracil.

At pH 11 all contact angles are reduced compared to those measured on the same surface at pH 5.8. Contact angles increase upon dimerisation to approximately 68 degrees. Surface irradiation causes a ten-degree change in contact angle for 5-methyl uracil. In the case of 5-trifluoro and 5-nitro uracils, changes of around 16 and 22 degrees occur.

Electrowetting

A typical electrowetting curve for the T1 Teflon variant (lowest T_g) is plotted in Figure 1.

The points represent the experimental results while the solid line is calculated via (Berge, 1993):

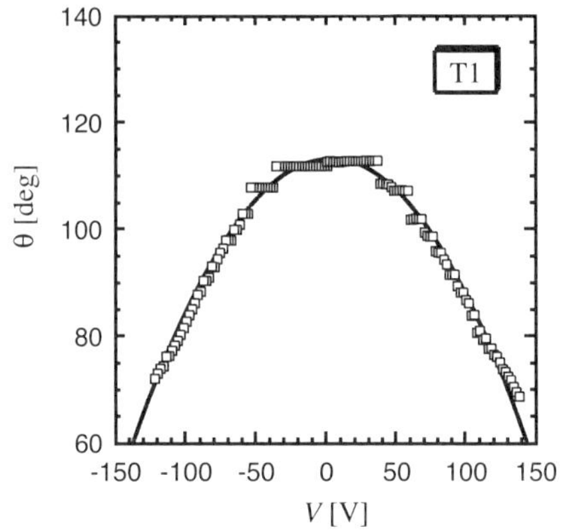

FIGURE 1 Electrowetting data (□) for 0.1 M KCl (pH = 5.6) on 2.0 μm thick T1 (T_g = 36°C). The theoretical prediction (—) is shown.

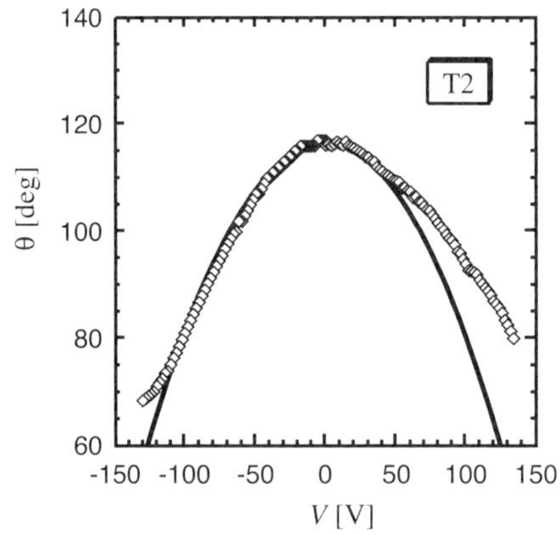

FIGURE 2 Electrowetting response (◇) of 0.1 M KCl (pH = 5.6) on 1.8 μm thick T2 (T_g = 133°C). The theoretical prediction (—) is shown.

$$\cos\theta = \cos\theta_0 + \frac{1}{2}\frac{\varepsilon\varepsilon_0 V^2}{\gamma_d} \quad \textbf{(EQ 1)}$$

where ε is the dielectric constant of the polymer and ε_0 is the permittivity of a vacuum. A typical result for T2 is presented in Figure 2.

In this case the theoretical behaviour is followed only at negative potentials, provided that the saturation limit (~120 V) is not exceeded. At positive potentials however, a marked deviation is seen above 50 V—the contact angle still decreases with external voltage, but more gradually than predicted theoretically. Results for T3 are quite similar (Figure 3).

Since one major difference between the three copolymers appears to be the glass transition temperature (Table 1), electrowetting studies on T1 were conducted at temperatures above and below its T_g, the results of which are shown in Figure 4.

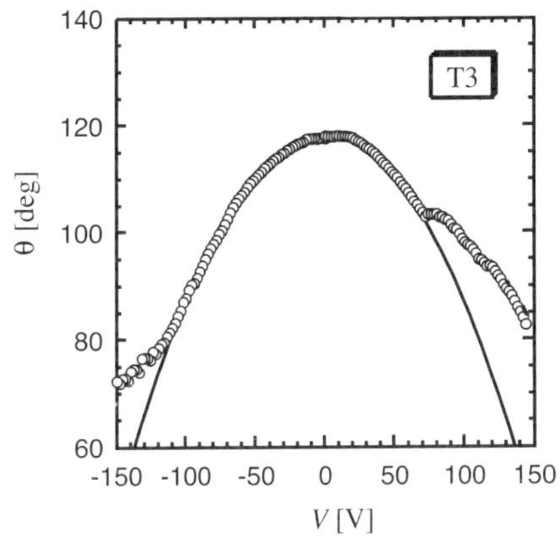

FIGURE 3 Electrowetting response (○) of 0.1 M KCl (pH = 5.6) on 2.4 µm thick T3 (AF1600; T_g = 160°C). The theoretical prediction (—) is shown.

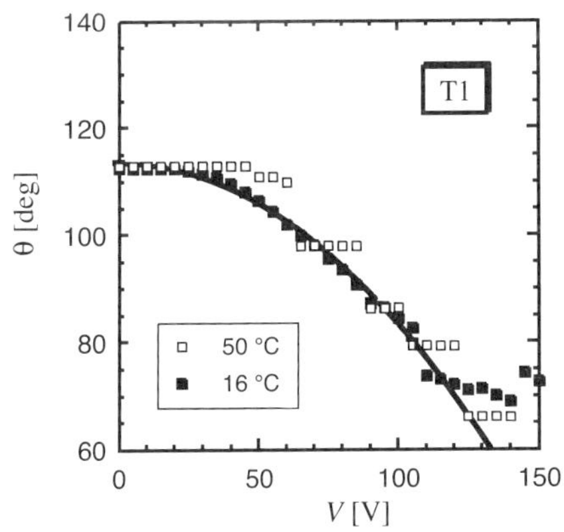

FIGURE 4 Electrowetting response for 0.1 M KCl (pH = 5.6) on 2.1 µm thick T1 (T_g = 36°C) at 16°C (■) and 50°C (□) The theoretical prediction (—) is shown.

Apart from somewhat more sporadic wetting behaviour at the elevated temperature (50°C), the two results are in very good agreement and again follow closely the parabolic decrease described theoretically.

Since the bulk properties of the Teflon variants are quite similar the possible influence of the polymer/solution interface was further explored. The substitution of KBr or KI for KCl had no detectable effect on the electrowetting curve.

However, ionic strength turned out to be an important factor. As can be seen in Figure 5, at increased salt concentration the theoretical prediction is valid over a wider voltage range.

Figure 6 shows the strong dependence of electrowetting on the pH of the liquid at positive voltages.

At the lowest value (pH = 2.1), the correlation with theory is satisfactory up to the saturation limit (~130 V). As the pH is changed through neutral to basic, the deviation from theory becomes progressively more pronounced. At pH = 11.3, deviation from theory is already apparent at approximately 70 V.

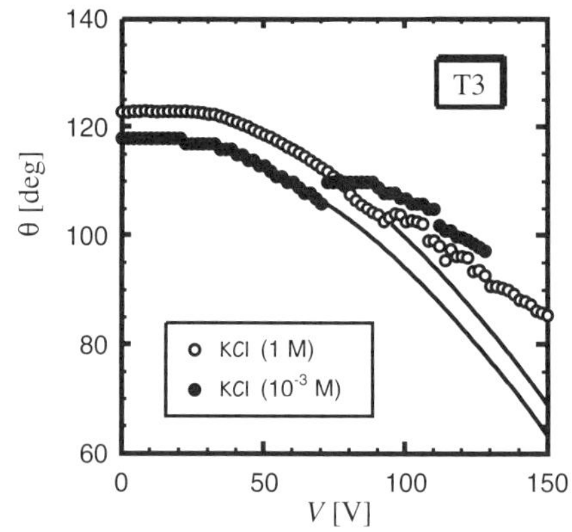

FIGURE 5 Electrowetting response for 1.0 M KCl (○) and 10^{-3} M KCl (●) at pH = 5.6 on 2.2 μm thick T3 (AF1600; T_g = 160°C). The theoretical prediction (—) is shown.

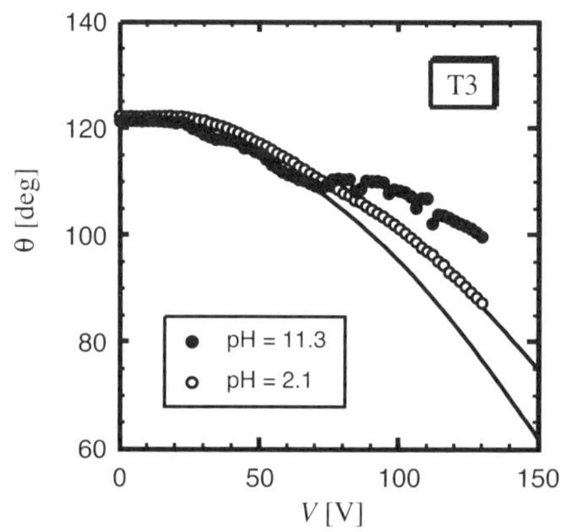

FIGURE 6 Electrowetting response for 0.1 M KCl on 2.2 μm thick T3 (AF1600; T_g = 160°C) at pH = 2.1 (○) and pH = 11.3 (●). The theoretical prediction (—) is shown.

LIGHT INDUCED WETTING

Dimer Formation and Photo-splitting

When irradiated with UV light at 280 nm, dimers form at the surface which then photo-split to yield the monomer at 240 nm. Aqueous solutions of thymine yield four isomeric dimers, illustrated in Figure 7.

In order to maximise the efficiency of the photo-dimerisation it is desirable to orientate the bases so that the cis-syn isomer is the major isomer formed. Since the driving force for cleavage is steric repulsion of the bulky groups at C-5 the cis-syn isomer has the highest reactivity for photo-splitting (Inaki et al., 1999). In DNA, photo-dimerisation yields only this isomer since the conformation of the bases is controlled by connection to the DNA chain and hydrogen bonds through to the base pair adenosine (Inaki et al., 1999). If thymine is grafted to a smooth gold surface as a thin robust film or self-

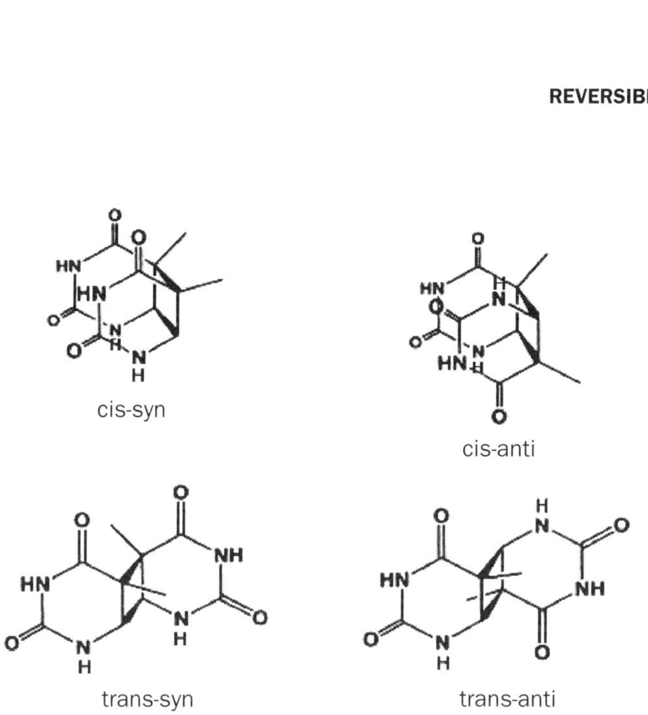

FIGURE 7 Four isomeric dimers—cis-syn, trans-syn, trans-anti, cis-anti—resulting from 280 nm irradiation of thymines in solution (Tohnai et al., 1999)

assembled monolayer (SAM), the fixed orientation should ensure that cis-syn isomers dominate and inert photo-dimer formation is minimal.

Influence of Molecular Structure on pKa

A change in surface wettability is achieved through manipulation of the molecular structure of the self-assembled monolayer. For example, the functionality at position C-5 can be altered from R = CH_3 to include more hydrophobic or hydrophilic species such as R = CF_3 and R = NO_2 respectively. Altering functionality also has an effect on the acidity constant of the compounds which have a dissociable hydrogen at N_3. The lability of the pyrimidine dissociable hydrogen at positions N_1 and N_3 increases as the inductive capacity of the R group at position C-5 is increased. Hence, the pKa decreases from the 9.9 recorded for 5-methyluracil (R = CH_3), to 7.3 when R = CF_3 and 5.0 when R = NO_2 (Table 2). This first pKa leaves the remaining hydrogen free to migrate between the annular nitrogens sites (Richards, Reynolds and Ralston, 2002). If dissociation is forced from position N_3, i.e., when the pyrimidine is functionalised at position N_1, then the first pKa of the compound generally increases, since there is less charge delocalisation available.

When the molecule is grafted to the surface after alkylation at N_1, the pKa alters as the local environment is altered. The reason for the increase in surface pKa is most likely due to a surface hydrogen bond network, combining with attractive electrostatic forces resulting from neighbouring negative charges which increase the energy required to remove the proton. In 5-nitrouracil this is exacerbated by ionisation of the nitro group, raising the pKa even further than expected when compared with 5-methyluracil.

Variation in Contact Angle

Although the photochromic properties of members of the pyrimidine family have been known for some time, Abbott et al. (1999) were the first to realize that the pKa of the two distinct photo-stable species differ. Thus at a given pH the monomer may be charged whilst the dimer remains neutral. Hence irradiation with UV light can influence the surface charge. The effect of surface charge on wetting has been investigated to only a modest extent. In 1989 Fokkink and Ralston (Fokkink and Ralston, 1989) determined that an electrical double layer will form spontaneously at a solid-liquid interface, altering the interfacial tension and thence the contact angle. The system then minimizes the free energy by increasing the area of the solid/liquid interface, i.e., on a charged surface the droplet spreads with the change in contact angle described by equation (2).

$$\cos\theta = \cos\theta(pzc) - \frac{\Delta F_{DL}}{\gamma_{LV}} \quad \textbf{(EQ 2)}$$

At pH 5.8 the thymine monomer is protonated and has a water contact angle of 67 degrees; upon dimerisation this angle increases to 76 (Table 3). At pH 11 the monomer contact angle decreases to 59 degrees for the surface is partially charged. As the pH increases beyond the pKa of the N_1-alkylated monomer, the surface fully deprotonates and the contact angle decreases by a further 5 degrees. The pKa of the dimer is expected to be higher than the monomer due to less available charge delocalisation; therefore at pH 11 there should be a lower proportion of surface deprotonation. Indeed the dimer contact angle at pH 11 is 69. On a 5-methyluracil surface, the largest reversible contact angle change upon irradiation is 10 degrees at pH 11.

A surface SAM with a lower pKa will ionise at a lower pH. Thus the type of substituent present at position C-5 was altered to increase the acidity of the pyrimidine head group. 5-trifluoromethyluracil monomer at pH 5.8 gave a water contact angle similar to that of 5-methyluracil with a similar increase in contact angle after irradiation at 280 nm. With a pKa of 9.1, however, the surface was fully deprotonated at pH 11 and the angle decreased below that of 5-methyluracil, to 50 degrees. The surface contact angle of the dimer at pH 11 was measured as 66 degrees, hence, there is a 16-degree reversible change in contact angle but still at an undesirably high pH. Smaller changes in contact angle were observed at pH values less than 11 but greater than 9.1.

In water at pH 5.8 the contact angle for 5-nitrouracil is 56 degrees, some 10 degrees lower than that observed for 5-methyluracil. Since the structure is similar to 5-methyluracil, this suggests that the nitro group has a direct effect on the contact angle through orientation of the ring at the surface. This reduced contact angle value indicates that the orientation of the pyrimidine groups at the surface occurs so that the C-5 position has a direct effect upon the surface chemistry. After irradiation at 280 nm the contact angle increases to 77 degrees, a value that is very similar to the dimer surfaces of 5-methyl and 5-trifluoromethyluracil. The hydrophilicity of the group at C-5 is obviously lost, suggesting that the molecules reorganise on the surface to shield the C-5 position from the contacting fluid. This radiation-induced conformational change on the surface leads to a large, reversible change in contact angle of 21 degrees in UHQ water. The change in wettability is independent of surface charge; in fact the contact angle changes associated with the deprotonated surface at around pH 9–11 are of a similar magnitude.

Contact angle changes due to photo-induced molecular reorganisation have been reported in the literature for a variety of photosensitive moieties (Negishi, Tsunemitsu and Shinohara, 1981; Hayashida, Sato and Sugawara, 1986; Feng et al., 2001; Siewierski et al., 1996). For example azo-benzene monolayers have been used to form monolayers on planar and colloidal substrates (Feng et al., 2001; Siewierski et al., 1996; Evans et al., 1998; Sekkat et al., 1995). Irradiation of these films effected reversible, photo-induced, changes in water contact angles of around 9–11 degrees (Feng et al., 2001; Siewierski et al., 1996). These films will only photo-switch provided that the azo moiety is not sterically hindered and, in alkane-thiol based systems, where there is little free volume to allow switching between the cis and trans conformer, the monolayer will not respond (Evans et al., 1998). Other groups used to effect much larger changes in contact angle include spiropyrans (Hayashida, Sato and Sugawara, 1986) and photo responsive dyes such as malachite green (Negishi, Tsunemitsu and Shinohara, 1981). Contact angle changes due to apparent conformational changes have been reported to be as large as 40 degrees, however the reversibility is either poor or non-existent and the values reported are suspect since the spin-cast films often degraded in contact with water. In contrast to these reports, 5-methyluracil dimerisation has been shown to be reversible over numerous irradiation cycles, without degradation of the film. Inaki et al. (Inaki, 1992; Sekkat et al., 1995; Inaki et al., 1992) have shown that in thin, spin-cast films the reversibility extends well beyond 10 irradiation cycles with little or no loss of efficiency provided that the thymines are optimally positioned.

Electrowetting has been carried out previously on PTFE (Berge, 1993; Vallet, Berge and Volvelle, 1996; Decamps and de Coninck, 2000) and Teflon AF1600 (Seyrat and Hayes, 2001) surfaces. The agreement between experiment and theory, below the saturation limit, was reportedly quite good. We confirm this for the T1 variant only (Figures 1 and 4). As the PDD content increases, deviations from theory at positive voltages become significant (Figures 2 and 3).

The departure from the ideal electrowetting behaviour described by Equation 1 seems hardly attributable to different bulk properties. The only bulk material parameter related to the polymer which explicitly enters into Equation 1 is the dielectric constant, ε. The values for the three Teflon variants are very similar (Table 1), yet even if they were not they could not account for the type of deviation observed in our experiments (Figures 2 and 3). Since the best agreement between theory and experiment was observed for T1, the Teflon with the lowest T_g, it was hypothesized that chain mobility might be of some importance. The T1 fluoropolymer is indeed softer than T2 and T3, as was qualitatively confirmed by AFM imaging in Tapping Mode. Experiments performed above and below the T_g conclusively show (Figure 4) that compliance with Equation 1 is not affected by temperature changes. Interestingly, electrowetting at elevated temperature is more sporadic. It thus appears that when chain mobility is higher, the stress present at the contact line is strong enough to affect the polymer surface locally and this translates into a pronounced pinning of the contact line.

Since the bulk properties of the three polymers are quite similar it was concluded that the difference between them was interfacial in nature. It is well documented that neutral polymer surfaces in aqueous electrolyte solutions exhibit negative zeta potentials (Garbassi et al., 1994). The negative charge at the polymer/solution interface is most often attributed to specific adsorption of anions (usually Cl^- and OH^-). The latter are more polarisable and less hydrated and this promotes their tendency to partition at the interface (Garbassi et al., 1994). The phenomenon is also commonly observed at the water/air and water/oil interfaces (Kelsall et al., 1996; Exerowa and Kruglyakov, 1998; Karraker and Radke, 2002). Recently Zimmerman et al. (Zimmerman, Dukhin and Werner, 2001) have reported electrokinetic and surface conductivity measurements on Teflon AF1600 surfaces. A negative charge was found and ascribed to the preferential adsorption of hydroxide ions. Thus the existence of a negative charge at the amorphous fluoro-polymer/solution interface in the absence of an externally applied voltage is firmly established.

As already indicated the electrowetting curve is essentially analogous to the electrocapillary curve. This analogy is further stressed by several observations: (i) deviations occur at positive potentials only (Figures 2 and 3); (ii) halide anions have no appreciable effect; (iii) ionic strength is a significant factor (Figure 5); and (iv) pH of the solution is of paramount importance (Figure 6). All these facts imply that the electrical double layer at the polymer/solution interface is the root cause of this non-ideal electrowetting behaviour.

The scale of the deviation can be estimated as follows. The contact angle decrease is shallower than predicted, i.e., in order to reach a contact angle predicted by Equation 1 an excess positive voltage, V^*, is required. Under these conditions Equation 1 becomes:

$$\theta = \theta_0 - \frac{C}{2\gamma}(V - V^*)^2 \quad \textbf{(EQ 3)}$$

where V is the actual applied voltage and $V - V^*$ represents the corrected or ideal voltage at the solid/liquid interface. This voltage shift is (e.g., see Figure 3) $V^* \approx 35$ V at $V = 140$ V. It represents an energy loss during charging. The charge density, σ^*, due to adsorbed charges is:

$$\sigma^* = \frac{\varepsilon\varepsilon_0}{d}V^* \approx 2.5 \times 10^{-4}\ C/m^2 \quad \textbf{(EQ 4)}$$

The magnitude of this charge is rather small, particularly when compared with a metal oxide/water interface (Hunter, 1981). van Wagenen et al. (1981) have performed extensive streaming potential measurements for several neutral polymer surfaces (including polystyrene, polyvinyl chloride and polydimethyl siloxane) and quoted a typical surface charge density of 4.8×10^{-3} C/m^2. Using a very different technique (electrons injected in thin polymer layers), Fowkes and Hielscher (1980) obtained a limiting surface charge density for polystyrene and chlorinated polyethylene of about $1.6–1.8 \times 10^{-3}$ C/m^2. Zimmerman et al. (2001) calculated a charge density of 5×10^{-3} C/m^2 for Teflon AF1600 in 1 mM KCl. Therefore the magnitude of the deviations reported here can be attributed with some confidence to the effect of the double layer existing at the polymer/solution interface.

It should be noted that Equation 3 is formally identical to the one proposed by Verheijen and Prins (1999) to explain contact angle saturation. In their model, beyond a certain threshold voltage (positive or negative) charge trapping occurs "in or on" the polymer layer, beyond the three-phase contact line, i.e., on the "dry" polymer surface. Supposedly this counteracts exactly the electrowetting effect and hence leads to saturation. Our discussion is restricted to deviations, which take place below the saturation limit. Moreover, the deviations we report are asymmetric with respect to polarisation, ion specific, and concentration dependent–in contrast to the picture presented in Verheijen and Prins (1999). Therefore, the above similarity (a diminished electrowetting effect) is only apparent, for the effects that we observe arise from adsorption phenomena at the fluoropolymer/aqueous solution interface.

It is possible to obtain the surface charge density from the electrocapillary curve in much the same way as from electrocapillary curves. The charge density on the electrode, σ, is given by Lippmann equation:

$$\sigma = \frac{\partial\gamma_{SL}}{\partial V} = -\gamma\frac{\partial\theta}{\partial V} \quad \textbf{(EQ 5)}$$

where use of Equation 1 was made. The data from Figure 3 were numerically differentiated and the resulting charge density is presented as function of the applied voltage in Figure 8.

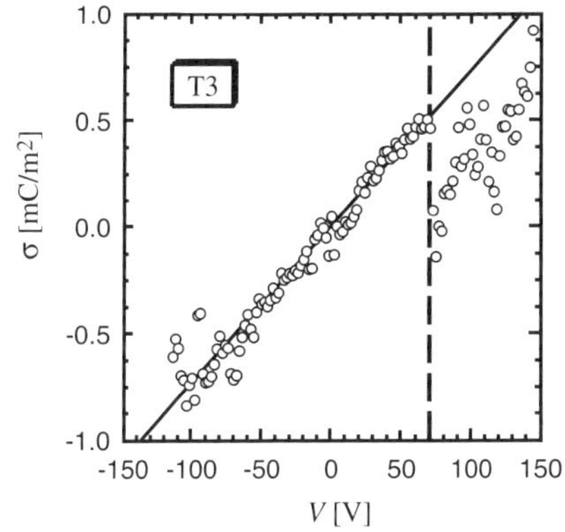

FIGURE 8 Surface charge density on the electrode, σ, as a function of the applied voltage, V (data from Figure 3). The slope of the solid line gives the same capacitance per unit area (7.3 μF/m) as that shown in Figure 3.

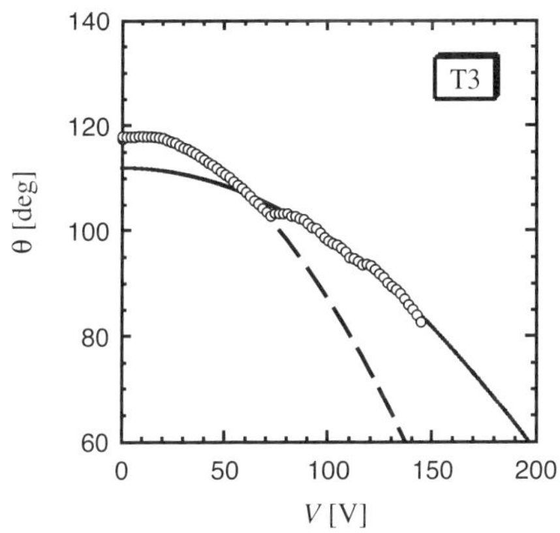

FIGURE 9 Positive potential branch of the electrowetting curve shown in Figure 3. The solid line is given by Equation 1 (the dashed line is taken from Figure 3).

From -100 V to 70 V, σ increases linearly and the slope gives the same value for the capacitance per unit area as the fit shown in Figure 3 (7.1 μF/m). At positive polarization beyond this range charging becomes less effective. In view of the larger scatter the data are approximated with a linear trend, i.e., constant capacitance. This provides a fair description of the deviation as is seen in Figure 9.

Here only the positive potential branch of Figure 3 is presented. The curvature of the solid parabola passing through the deviation is about half that of the dashed one. In this regime the "effective" contact angle at zero external voltage is significantly lower (112° versus 118°).

Specific ion adsorption most certainly occurs at solid/aqueous solution and Hg/aqueous solution interfaces (Bockris and Reddy, 1970; Damaskin and Petrii, 1983; Hunter, 1981; Habib and Bockris, 1980). With this knowledge and the recent picture of specific hydroxide ion adsorption at the water/air interface presented by Karraker and Radke (2002) the following qualitative physical model can be outlined. Some OH^- ions are specifically adsorbed at the polymer surface even in the absence of any external voltage. Most probably they are partially dehydrated (Karraker and Radke, 2002). When positive voltage is applied the insulator is polarized and attracts anions from the solution, i.e., more ions adsorb (Habib and Bockris, 1980). Adsorption of OH^- saturates at a certain voltage (Karraker and Radke, 2002) and further ion adsorption is obstructed. Thus charging efficiency is reduced and deviations in the electrowetting curve are observed at positive potentials (at negative potentials OH^- are electrostatically repelled and no such complications arise).

This picture is consistent with our observations. Anions, which are not specifically adsorbed (Cl^-, Br^-, I^-), do not affect the process. The situation strongly depends on the pH of the solution (Figure 6)–a higher OH^- concentration reduces the contact angle (although marginally) even without external voltage. Charging at high pH, i.e., at high OH^- concentration, is impeded already at 70 V while at low pH the contact angle decrease follows Equation 1 up to 130 V. At normal pH (Figure 9) the deviation occurs at about the same voltage (~70 V) and the value of the contact angle extrapolated to zero external voltage is reduced (reflecting the limiting OH^- adsorption). Ionic strength increases θ_0 (Figure 5) even though the liquid surface tension increases by about 2% (Lide, 1999). This effect may be due to double layer compression (Zimmerman, Dukhin and Werner, 2001), or partial OH^- desorption, or both. Charging at higher ionic strength follows the theoretical trend up to 90 V (as compared to 70 V at lower ionic strength) and then becomes less effective. Additional experiments carried out in the presence of 1 M urea (a well-known water structure breaker (Franks and Reid, 1975)) yielded exactly the same results. Therefore, water structure, per se, does not seem to be a factor.

Finally, we remark on the difference between the different Teflon grades. The T1 variant contains a minimal amount of PDD (which is apparently sufficient to disrupt the crystalline structure of TFE) and its electrowetting response conforms to Equation 1. On the other hand, variants T2 and T3 contain significant amounts of PDD and deviations at positive polarization are always found. Therefore the magnitude of the deviation appears to be related to the amount of dioxole. It may be speculated that hydroxide ions have higher affinity towards the oxygen containing monomers. Such hypothesis is partially supported by the larger zeta potential at basic pH measured for Teflon AF1600 in comparison to PTFE (e.g., −90 mV at pH = 9 in 10^{-3} M KCl (Zimmerman, Dukhin and Werner, 2001) versus −40 mV for PTFE (Werner et al., 1999) under the same conditions). The increasing amount of oxygen in the surface layer is convincingly, even though qualitatively, proven by ToF-SIMS measurements (Figure 10).

However, the dependence of the critical surface tension on the PDD content reveals a somewhat different picture (Figure 11).

The critical surface tension linearly decreases with the amount of dioxole and all values fall within the range of wettability of PTFE (18.5 mN/m (Zisman, 1964)) and closely packed perfluoromethyl groups (6 mN/m (Zisman, 1964)). Thus, PDD is more hydrophobic than the TFE, which implies that its two $-CF_3$ groups point outwards (Zisman, 1964) and quite effectively shield the dioxole ring. Thus the role of the oxygen heteroatom in the surface layer is rather complex.

The dioxole monomer is responsible for the amorphous structure and the limited solubility of the Teflon AF copolymers in fluorinated solvents (Hung et al., 1996; Bandrup, Immergut, Grulke, 1999) which are highly desirable qualities in technical applications. Larger electrowetting response at positive potential and better adherence to theory is found when the dioxole content is lower. Lower PDD content, however,

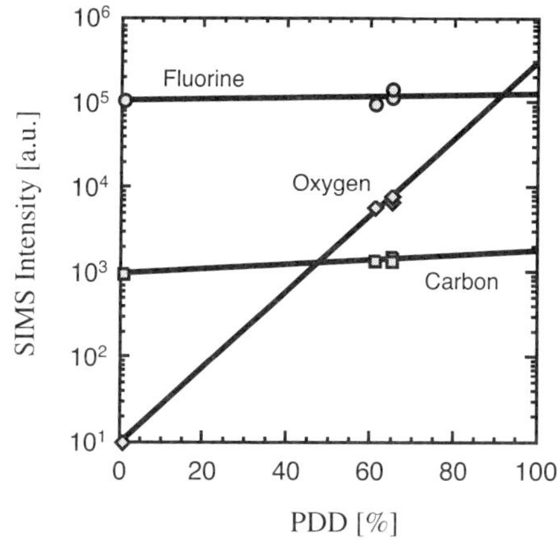

FIGURE 10 ToF-SIMS signal (negative mode) for the three Teflon AF variants

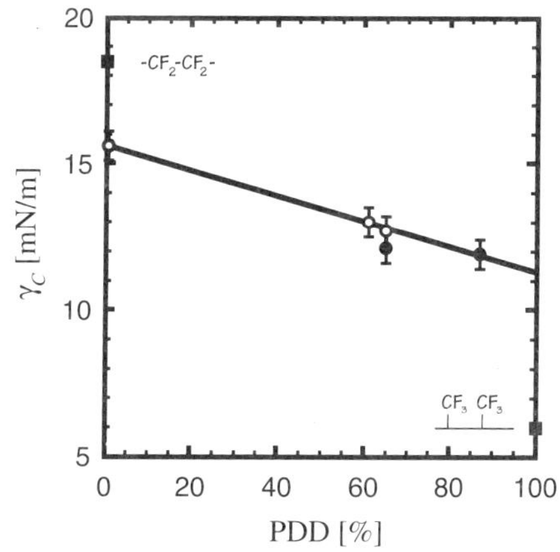

FIGURE 11 Critical surface tensions obtained from Zisman plots of the contact angles of several alkanes (○ this study; ● Drummond and Georgoklis, 1996). Tabulated values (■) for PTFE and closely packed prefluoromethyl groups (Zisman, 1964).

decreases the T_g of the copolymer (Hung et al., 1996) and eventually the electrowetting curve becomes less smooth (cf. Figures 1 to 3). These opposite tendencies must be considered in the specific context of any given application.

CONCLUSIONS

A diminished electrowetting response is observed at positive voltages on amorphous Teflon, which contains larger amounts of dioxole monomer. The effect occurs before saturation of the contact angle and may be attributed to specific hydroxide ion adsorption. The marked influence of pH and ionic strength supports this hypothesis. These findings emphasize the analogy between electrowetting and electrocapillarity.

Self-assembled monolayers incorporating functionalised 2,4-diketopyrimidines have been successfully prepared on planar substrates. 5-methyluracil, 5-trifluoromethyluracil and 5-nitrouracil were anchored as alkylated short chain thiols through the annular N_1 nitrogen position onto cleaned gold wafers. The surface acidity constants increased upon anchoring the molecules, most likely due to electrostatic forces in the fixed orientation environment and the formation of a hydrogen bond network between the closely packed head groups.

Surface wettabilities indicate that the orientation of the pyrimidine groups at the surface occurs so that the C-5 position has a direct effect upon the surface chemistry. Upon dimerisation the surface contact angles increase significantly. At pH 11 the contact angle changes reflect that the dimer has a higher pKa than the monomer. The monomer surface is deprotonated at this pH while the dimer surface is partially deprotonated, leading to changes in contact angle of up to 22 degrees. Large changes in contact angle were observed for 5-nitrouracil which were independent of surface charge and were apparently due to conformational changes alone.

ACKNOWLEDGMENTS

Financial support from the Australian Research Council through the Special Research Centres Scheme is gratefully acknowledged, as well as from Philips Research Laboratories (The Netherlands).

REFERENCES

1. S. Abbott, J. Ralston, G. Reynolds and R. Hayes, *Langmuir*, 15, 8923–8928 (1999).
2. K. Ichimura, Y. Suzuki, T. Seki, A. Hosoki and K. Aoki, *Langmuir*, 4, 1214 (1988).
3. M. Ueda, H.-B. Kim and K. Ichimura, *Journal of Material Chemistry*, 4, 883 (1994).
4. R. Beukers, J. Ijlistra and W. Berends, *Rec. Trav. Chim.*, 79, 101 (1960).
5. R. Beukers and W. Berends, *Biochim and Biophys Acta*, 41, 550 (1960).
6. B.P. Ruzsicska and D. Lemaire, *CRC Press Inc.* (1995).
7. Y. Inaki, *Polymer News*, 17, 367–371 (1992).
8. M.K. Chaudhury and G.M. Whitesides, *Science*, 256, 1539–1541 (1992).
9. C.D. Bain, G.D. Burnett-Hall and R.R. Montgomerie, *Nature*, 372, 414–415 (1994).
10. B.S. Gallardo, V.K. Gupta, F.D. Eagerton, L.I. Jong, V.S. Craig, R.R. Shah and N.L. Abbott, *Science*, 283, 57–60 (1999).
11. K. Aoki, T. Seki, M. Sakuagi and K. Ichimura, Makromol. Chem. 193, 2163 (1992).
12. Y. Kawanishi, T. Tamaki, M. Sakuragi, T. Seki, Y. Suzuki and K. Ichimura, *Langmuir*, 8, 2601 (1992).
13. N. Tohnai, Y. Inaki, M. Miyata, N. Yasui, E. Mochizuki and Y. Kai, *Bull. Chem.. Soc. Jpn.* 72, 1143–1151 (1999).
14. D.L. Wulff and G. Fraenkel, *Biochim. Biophys. Acta*, 51, 332–339 (1961).
15. M.A. Herbert, J.C. LeBlanc, D. Weinblum and H.E. Johns, *Photochemistry and Photobiology*, 9, 33–43 (1969).
16. Y. Inaki, E. Mochizuki, H. Donoue, M. Miyata, N. Yasui and Y. Kai, *Journal of Photopolymer Science and Technology*, 12, 5, 725–734 (1999).
17. N.J. Richards, G. Reynolds and J. Ralston, *Journal of American Society*, in press (2002).
18. L.G. Fokkink and J. Ralston, *Colloids and Surfaces*, 36, 69–76 (1989).
19. N. Negishi, K. Tsunemitsu and I. Shinohara, *Polymer Journal*, 13, 4, 411–412 (1981).
20. S. Hayashida, H. Sato and S. Sugawara, *Polymer Journal*, 18, 3, 227–235 (1986).
21. C.L. Feng, Y.J. Zhang, J. Jin, Y.L. Song, L.Y. Xie, G.R. Qu, L. Jiang and D.B. Zhu, *Langmuir*, 17, 4593–4597 (2001).

22. L.M. Siewerski, W.J. Brittain, S. Petrash and M.D. Foster, *Langmuir,* 12, 5838–5844 (1996).
23. S.D. Evans, S.R. Johnson, H. Ringsdorf, L.M. Williams and H. Wolf, *Langmuir,* 14, 6436–6440 (1998).
24. Z. Sekkat, J. Wood, Y. Geerts and W. Knoll, *Langmuir,* 11, 2856–2859 (1995).
25. Y. Inaki and Y. Wang, *Journal of Photopolymer Science and Technology,* 4, 2, 259–266 (1991).
26. Y. Inaki, Y. Wang, M. Kubo and K. Takemoto, *Second International Symposium on Chemistry of Functional Dyes,* Kobe, Japan (1992).
27. J.O'M. Bockris, A.K.N. Reddy, *Modern Electrochemistry,* Plenum Press, New York, 1970, Volume 2, Chapter 7.
28. I. Morcos, *J. Electroanal. Chem. Interfacial Electrochem.* 62, 313 (1975).
29. B.B. Damaskin, O.A. Petrii, *Introduction to Electrochemical Kinetics,* 2nd Ed., Vyshaya Shkola: Moscow, p. 46 (in Russian) (1983).
30. B.C.R. Berge, *Acad. Sci. Paris,* 317-II, 157 (1993).
31. M. Vallet, B. Berge, L. Vovelle, *Polymer,* 37, 2465 (1996).
32. H.J.J. Verheijen and M.W.J. Prins, *Rev. Sci. Instr.,* 70, 3668 (1999).
33. E. Seyrat, R.A. Hayes, *J. Appl. Phys.* 90, 1383 (2001).
34. M. Schneemilch, W.J.J. Welters, R.A. Hayes and J. Ralston, *Langmuir,* 16, 2924 (2000).
35. C. Decamps and J. de Coninck, *Langmuir,* 16, 10150 (2000).
36. H.J.J. Verheijen and M.W.J. Prins, *Langmuir,* 15, 6616 (1999).
37. F. Garbassi, M. Morra and E. Ochiello, *Polymer Surfaces,* John Wiley & Sons: Chichester, 211 (1994).
38. M.-H. Hung, P.R. Resnick, B.E. Smart and W.H. Buck, *Polymeric Materials Encyclopedia,* CRC, Boca Raton, Vol. 4, 2466 (1996).
39. J. Bandrup, E.H. Immergut and E.A. Grulke (Eds), *Polymer Handbook,* 4th Ed, John Wiley & Sons, New York, V/52 (1999).
40. G.H. Kelsall, S. Tang, S. Yurdakul and A.L. Smith, *J. Chem. Soc. Faraday Trans.,* 92, 3887 (1996).
41. D. Exerowa and P.M. Kruglyakov, *Foams and Foam Films,* Elsevier, Amsterdam (1998).
42. K.A. Karraker and C.J. Radke, *Adv. Coll. Int. Sci.* 96, 231 (2002).
43. R. Zimmerman, S. Dukhin and C. Werner, *J. Phys. Chem. B.* 105, 8544 (2001).
44. R.J. Hunter, *Zeta Potential in Colloid Science,* Academic Press, London (1981).
45. R.A. van Wagenen, D.L. Coleman, R.N. King, P. Triolo, L. Brostrom, L.M. Smith, D.E. Gregonic and J. Andrade, *J. Colloid Int. Sci.,* 84, 155 (1981).
46. F.M. Fowkes and F.H. Hielscher, *Org. Coat. Plast. Chem.,* 42, 169 (1980).
47. M.A. Habib and J.O'M. Bockris, *Comprehensive Treatise of Electrochemistry* (J.O'M. Bockris, B.E. Conway, E. Yeager, Eds), Plenum Press, New York, Vol. 1, Chapter 4 (1980).
48. D.R. Lide (Ed.), *CRC Handbook of Chemistry and Physics,* 72nd Ed., CRC Press, Boca Raton (1991).
49. F. Franks and D.S. Reid, *Water–A Comprehensive Treatise* (F. Franks, Ed.), Plenum Press, New York, Vol. 2, Chapter 5 (1973).
50. C. Werner, U. Konig, A. Augsburg, C. Arnhold, H. Korber, R. Zimmermann and H.-J. Jacobash, *Colloids and Surfaces A,* 159, 519 (1999).
51. W.A. Zisman, *Contact Angle, Wettability and Adhesion* (R.F. Gould, Ed.), ACS: Washington, Chapter 1 (1964).
52. C.J. Drummond, G. Georgaklis and D.Y.C. Chan, *Langmuir,* 12, 2617 (1996).

Characterization of Spherical Alumina Particles Obtained by Melting in a Hydrogen-oxygen Flame

Jakub Nalaskowski*, Marcin Niewiadomski*, Vamsi K. Paruchuri*, and Jan D. Miller*

Spherical oxide particles can be used as mineral fillers, paint pigments, polishing materials and catalysts. They can be prepared using various techniques involving precipitation and spray drying. Melting of particles in a high-temperature flame is a simple technique which produces particles with a high degree of sphericity and high smoothness from non-spherical feed material. In current research the preparation of alumina particles by melting the feed material in a hydrogen-oxygen flame is described. The spherical particles obtained were characterized using scanning electron microscopy, atomic force microscopy, energy dispersive spectroscopy, X-ray diffraction, Fourier transform infrared spectroscopy and electrokinetic measurements. The particles prepared exhibit a high degree of sphericity and a low level of roughness. The chemical composition of the material after preparation is unchanged, however the crystallographic structure is different. It is suggested that alumina changes from pure α-alumina to a mixture of different transitional forms. Changes in the electrokinetic behavior were also observed for alumina which can be attributed to the strong dehydration of the oxide surface and/or change in the crystallographic structure. Also, the possible accommodation of trace amount of impurities in the surface zone during melting cannot be ignored.

INTRODUCTION

Inorganic spherical particles have important technological applications which are related to their well defined and smooth surface, good packing and good rheological properties. Spherical inorganic particles can be obtained using several different methods, including

* Dept. of Metallurgical Engineering, University of Utah, Salt Lake City, Utah.

precipitation from solution (Matijevic and Wilhelmy 1982; Blendell, Bowen, and Coble 1984; Williams, Yocom, and Stofko 1985), ultrasonic spray pyrolysis (Milosevic, Jordovic, and Uskokovic 1994), and melting in a flame (Chen, Gleiman, and Phillips 2001).

Alumina particles are of considerable importance because of their wide use in many applications including use as polishing medium or filler material. It has been found that application of spherical alumina particles instead of irregular ones greatly improves the performance of chemical-mechanical polishing (Basim et al. 2000, 2001), the properties of polymer-filler blends (Chiba, Takahashi, and Otaguro 1988), and the surface properties of magnetic data storage media (Trippel et al. 1988). Spherical alumina particles are also used for the formulation of advanced ceramic materials (Noma et al. 1993; Lopez-Navarrete and Ocana 2001) and even in cosmetics (Tanaka and Miyasaki 1998).

Spherical alumina particles can be obtained using several methods. Hydrolysis of aluminum compounds (Catone and Matijevic 1974; Nishikawa and Matijevic 1994), sol-gel synthesis (Chatterjee et al. 2000), and spray pyrolysis (Murugavel et al. 1998) are the most common. Fused spherical alumina particles can also be obtained by melting non-spherical feed material in a high-temperature plasma flame (Hamano and Asaga 1974; Chen, Gleiman, and Phillips 2001) or by combustion of elemental aluminum particles (Abe, Ogawa, and Kondo 1998). These high temperature processes are especially convenient because they allow a cheap, non-spherical feed material to be transformed into the fused spherical alumina particles of high sphericity and smoothness.

In this paper we describe a relatively simple laboratory setup for the preparation of spherical particles by melting non-spherical material in a hydrogen-oxygen flame. The procedure can be used for preparation of different inorganic spherical particles, especially oxides. This setup was used to transform non-spherical polishing α-alumina into a spherical alumina particles. Feed material and the prepared spherical particles were characterized in detail, with respect to shape, surface morphology, elemental composition, crystallographic structure, and electrokinetic behavior of particles.

EXPERIMENTAL

Materials

Buehler 5 μm polishing alumina (Buehler Ltd.) was used in the experiments. High purity hydrogen and oxygen (Mountain Airgas) were used to create the flame for the preparation of spherical particles. Reagent grade KCl, HCl and KOH (Mallinckrodt) were used for electrokinetic measurements. All solutions were prepared using water deionized and purified using Milli-Q system (Millipore).

Methods

Preparation of Spherical Particles. Spherical alumina particles were manufactured using the laboratory setup shown on Figure 1. Raw alumina particles (feed material) were placed in a gas-washing bottle and oxygen from a cylinder was passed through the alumina bed at a rate to maintain the bed in a fluidized state. The washing bottle was placed in the ultrasound bath in order to prevent clogging of the tube delivering oxygen and to prevent channel formation in the particle bed. The oxygen stream, containing suspended alumina particles, was subsequently directed to the laboratory concentric gas torch where it was mixed with hydrogen. The hydrogen-oxygen flame reaches a temperature of 2700 °C which is sufficient to melt alumina particles (m.p. 2053 °C). The hydrogen-oxygen flame was directed into an 0.6 m quartz tube, the conical end of which

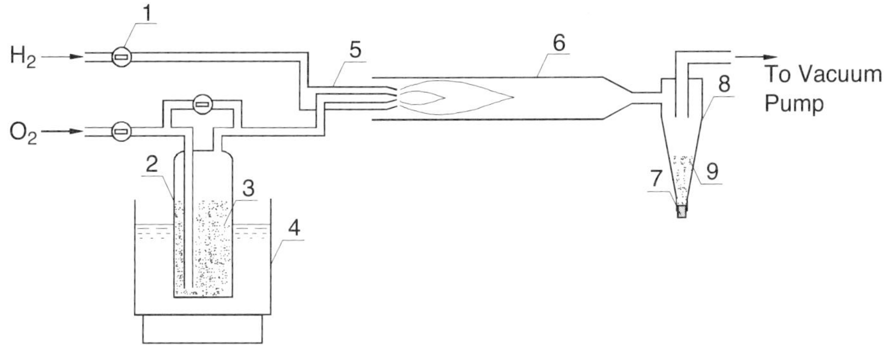

FIGURE 1 Schematic of the laboratory setup for preparation of spherical alumina particles: (1) valve, (2) washer bottle, (3) feed material, (4) ultrasound bath, (5) concentric torch, (6) quartz tube, (7) stopper, (8) cyclone, (9) collected spherical particle

was connected to a small glass cyclone (2 cm diameter). The outlet of the cyclone was connected to a pump which provided sufficient suction for pneumatic transport. Alumina particles, molten in the flame, were carried with the gas stream to the cyclone, undergoing cooling and solidification during transport. The spherical alumina particles were separated from the gas stream in the cyclone due to the centrifugal force and were collected at the bottom of the cyclone. The prepared spherical particles were removed from the cyclone through the port at the bottom.

Scanning Electron Microscopy (SEM). Alumina particles were mounted on a carbon tape and covered with a 20 nm thick gold layer using a Cressington Sputter 108 Auto Coater. Images of the particles were taken by means of Hitachi S-3000N SEM, using a secondary electron detector at a working distance 14.5 mm and 20.0 kV of electron energy.

X-ray Diffraction (XRD). The X-ray diffraction analysis was performed using Rigaku D/MAX x-ray diffractometer. The following scan parameters were used during measurements: step size 0.02°, scan speed 1 deg/min, slit width 1 mm, scatter 1.0°, receiving 0.15 mm. The alumina particles were analyzed as received. They were not grounded for the experiment and their size of approximately 5 μm was found to be sufficient to obtain good quality spectra.

Energy Dispersive Spectroscopy (EDS). The SEM-EDS (Energy Dispersive Spectroscopy) analysis was performed using a Hitachi S-3000N SEM instrument with EDS detector manufactured by EDAX, Inc. and using EDAX SEM Quant ZAF software. The spectra were obtained at a working distance of 14.5 mm working distance and 20.0 kV accelerating voltage. The alumina particles were deposited on the conductive carbon tape prior to the analysis. Spectral information was collected from 150 × 150 μm area covered with particles.

Atomic Force Microscopy Imaging (AFM). The surface of the spherical particles was examined using the Digital Instruments Nanoscope IIIa AFM. Measurements were done in the contact mode (in air) and a triangular silicon nitride cantilever (Digital Instruments, Inc.—spring constant 0.12 N/m) was used.

Electrokinetic Measurements. Electrophoretic mobility of the alumina particles was measured using the Brookhaven ZetaPALS apparatus (McNeil-Watson, Tscharnuter, and Miller 1998). The measurements were taken after 120 h of equilibration in 1×10^{-3} M KCl solution. Prior to measurement the pH was adjusted using solutions of HCl and KOH.

FIGURE 2 SEM image of feed alumina particles

Six consecutive measurements were taken at room temperature and the measurements were averaged and the zeta potential was calculated using the Smoluchowski equation.

RESULTS AND DISCUSSION

Particles of the feed material are shown in the SEM micrograph, see Figure 2. Irregular particles with sharp edges with a size from 5 to 10 μm can be seen. The shape of these particles is typical for fractured material.

Elemental analysis done by EDS shows that except for the main compound, aluminum oxide, only small amounts of impurities are present as can be seen from Figure 3. The quantification of the elements present in the feed material is given in Table 1. Titanium, silicon, and magnesium are the most important impurities which have been found in this sample. The determination of chemical compounds in which these elements are present is impossible using EDS. It can, however, be assumed that they are mostly in the form of oxides, since no other elements were found in sufficient amount. It should be noticed that these impurities are evenly distributed in the feed particles. Scan of a large area was compared with EDS analysis for a single particle and no significant difference was found. Although during analysis several single particles were found to be more contaminated than others, the effect of these impurities on the overall population is negligible.

The X-ray diffraction spectrum, shown in Figure 4, demonstrates that the feed alumina is highly crystalline. The data when fitted to the literature values show a perfect match for the most stable form of aluminum oxide-α-alumina (corundum).

After passing through the hydrogen-oxygen flame in the apparatus schematically shown on Figure 1, feed material particles undergo drastic changes. The temperature of the flame is significantly higher than the melting point of alumina. After melting, the surface tension of molten alumina causes alumina droplets to form the most energetically

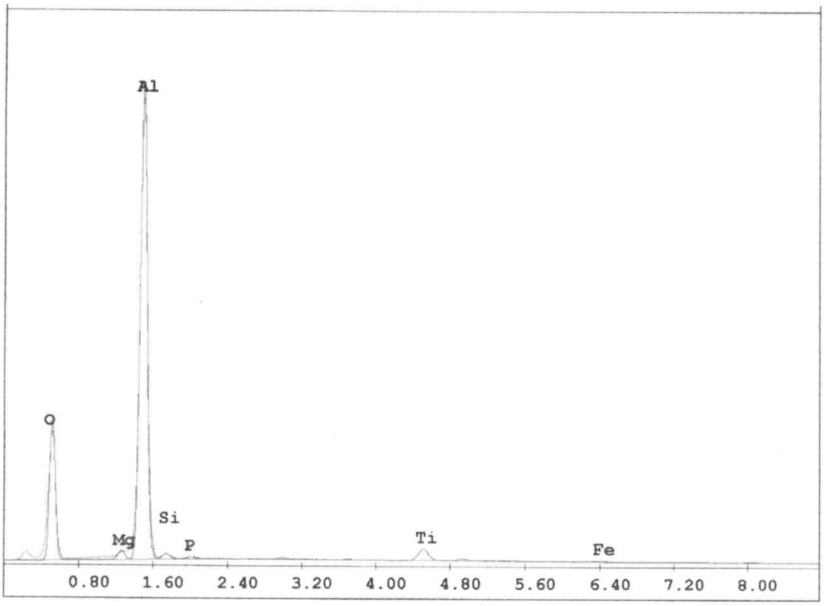

FIGURE 3 EDS spectrum of feed alumina particles

TABLE 1 Composition of feed alumina and spherical alumina particles

Element	Feed Alumina Particles (wt%)	Spherical Alumina Particles (wt%)
Al	54.16	54.10
O	39.25	40.25
Ti	3.17	2.93
Si	1.25	1.25
Mg	1.15	0.82
P	0.53	0.14
Fe	0.60	0.51

favorable spherical shape. An SEM image of these spherical alumina particles is presented in Figure 5. These particles are highly spherical and range in diameter from 2 to 5 μm. It can be deduced from the SEM image that the particles size is directly related to the size of the particles in the feed material. The probability of coalescence of two or more droplets of molten alumina in the flame must be extremely low at the small concentrations of particles in the gas stream. The breakage of the molten drop of alumina into smaller droplets due to the forces in the flame and/or gas stream is also difficult to imagine due to the high values of surface tension and viscosity of molten alumina.

Despite the high sphericity of the alumina particles, AFM analysis reveals that the surface of the particles is not completely smooth. Figure 6 shows a region of the surface for one of the larger spheres. The presence of ripples or terraces can be clearly seen. These terraces are on the order of 500 nm and more in size, while their height is usually around 40 nm. This geometry on the surface results in a mean roughness equal to 21 nm for the 5 × 5 μm scan. The origin of these features at the surface is not completely clear at

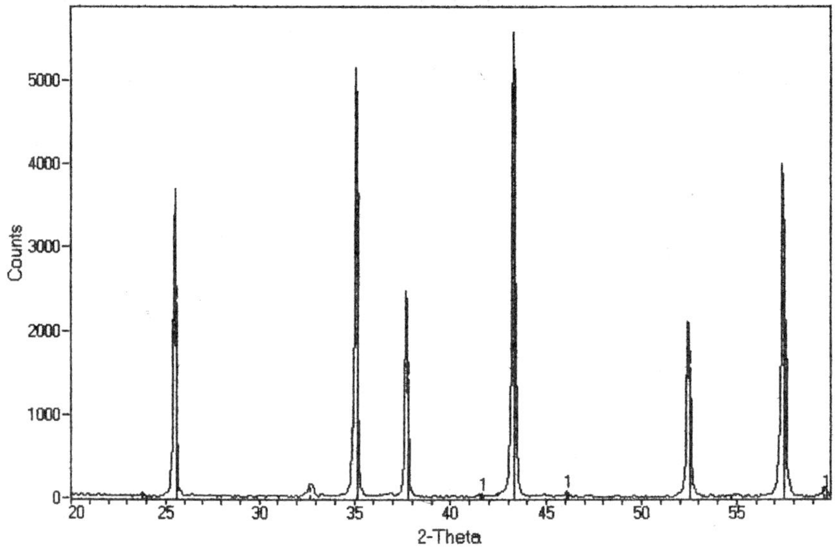

FIGURE 4 XRD spectrum of feed alumina particles

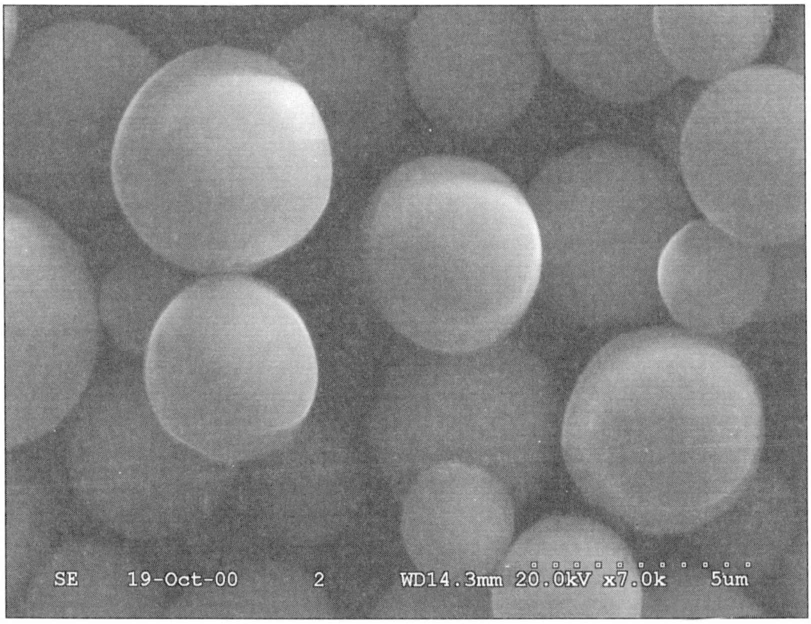

FIGURE 5 SEM image of spherical alumina particles

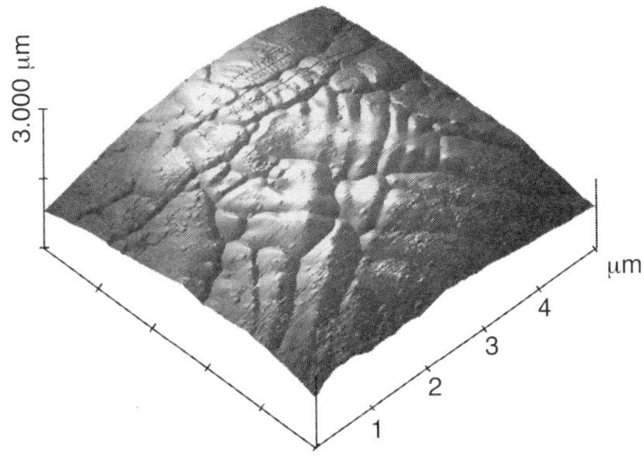

FIGURE 6 AFM image of the surface of a spherical alumina particle. Mean roughness for 5 × 5 µm scan is 21 nm.

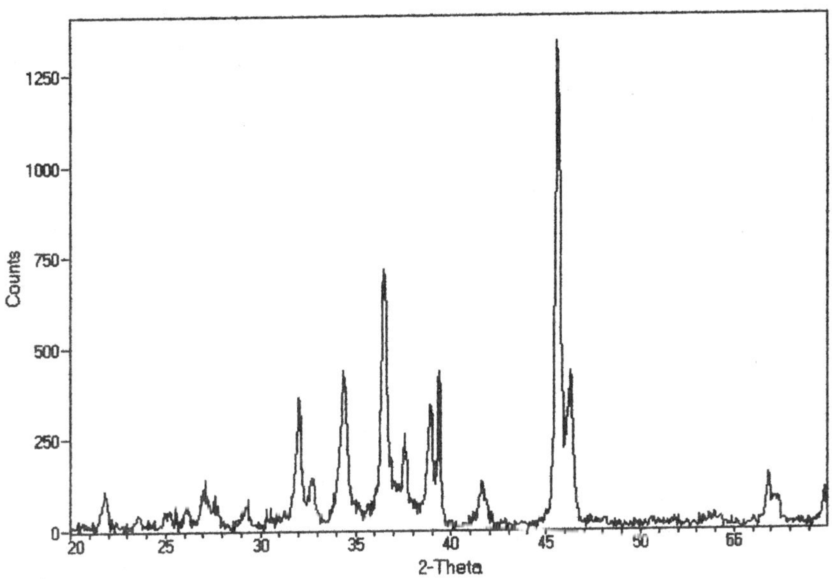

FIGURE 7 XRD spectrum of spherical alumina particles

this point. It could be the effect of quenching in turbulent flow, but also formation of crystallites at the surface cannot be ruled out.

EDS analysis of the spherical particles shows no significant difference when compared with the raw feed material spectrum, see Table 1. The results confirm the previous assumption that impurities are in the high-temperature resistant form, most probably oxides. It can also be concluded that no significant chemical changes are introduced (e.g., reduction) during the preparation of alumina spheres.

Despite almost the same chemical composition of the spherical particles and feed material, XRD spectra shows a huge difference between these two materials, as can be seen from comparing the spectrum in Figure 4 with that in Figure 7. The spectrum in

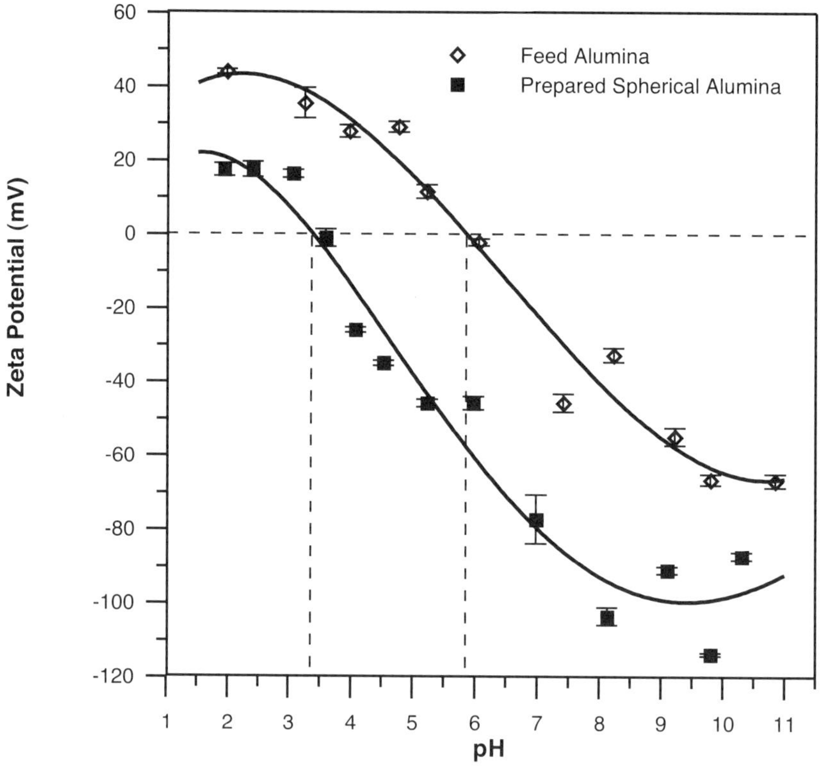

FIGURE 8 Zeta potential as a function of pH for feed material and spherical particles. IEP for feed alumina of irregular shape (feed material) is 5.8, for spherical particles is 3.4

Figure 7 shows that the spherical particles exhibit a high degree of crystallinity, however the diffraction pattern is very different from that of the feed material. It is known that alumina can be present in one of as many as 15 forms, where four of them are the most important: γ, δ, θ, and α. The α-alumina is thermodynamically stable form, while γ, δ, and θ are transitional forms during the transformation from hydrated alumina (boehmite) to the α-alumina (Santos, Santos, and Toledo 2000). These crystallographic forms change with an increase in temperature from the γ-alumina, through the δ, θ, and finally to the stable α-alumina form. However, the XRD spectrum for the spherical alumina particles cannot be fitted exactly to any one of these major phases. Judging from the spectrum shape these spherical particles might be a mixture of different phases, where δ-alumina and θ-alumina may be the predominant forms. Although the feed material is a stable α-alumina form, it is possible that during rapid quenching of molten alumina different, transitional phases can be formed. The presence of δ-alumina in spherical alumina particles has already been reported in the literature (Hiragushi et al. 1982).

The zeta potential of raw feed alumina and spherical alumina as a function of pH is shown on Figure 8. A long equilibration time (120 hours) was used in order to ensure complete hydration of the alumina surface. It was noticed that the surface does not reach the equilibrium and the zeta potential changes slightly even during first two to three days. The isoelectric point (IEP) of raw material was found to be approximately 5.8. Although this is a lower value than usually anticipated, it has to be noted that values reported in the literature are in the range between 6.6 and 9.4. The degree of hydration and the presence of small amounts of impurities may significantly alter the electrokinetic behavior of particles. Surprisingly, after preparation and hydration, the spherical alumina particles became more negatively charged and their IEP shifted to a value of 3.4. Currently we are not sure if this effect is related to the differences between crystallographic structure of the feed and spherical product, differences in the hydration of these two samples or accumulation of impurities at the surface of sphere during the melting process.

CONCLUSIONS

The spherical particles of high-melting point oxides can be successfully prepared in the laboratory using a simple experimental setup. Particles obtained in a hydrogen-oxygen flame from non-spherical feed are highly spherical and relatively smooth. It was found that this process does not change the elemental composition of the particles significantly. A high degree of crystallinity was retained in the particles but the crystallographic forms changed from α-alumina to a mixture of transitional phases (most probably δ-alumina and θ-alumina). The change of IEP to a lower pH value (pH 3.4) after the treatment is not easy to explain without speculation and requires further studies on the influence of small amount of impurities in the alumina matrix and the effect of crystal structure on electrokinetic behavior.

ACKNOWLEDGMENTS

Financial support provided by the Department of Energy, Basic Science Division Grant No. DE-FG-03-93ER14315 is gratefully acknowledged. Although the research described in this article has been funded partially by DoE, it has not been subjected to the agency's required peer and policy review and therefore does not necessarily reflect the views of these agencies and no official endorsement should be inferred.

REFERENCES

Abe, S., Ogawa, M., and Kondo, K. 1998. Synthesis of spherical oxide particles by vaporized metal combustion process. *Silicon for the Chemical Industry IV,* [Conference], Geiranger, Norway, June 3–5, 1998, 329.

Basim, G.B., Adler, J.J., Mahajan, U., Singh, R.K., and Moudgil, B.M. 2000. Effect of particle size of chemical mechanical polishing slurries for enhanced polishing with minimal defects. *J. Electrochem. Soc.* 147:3523.

Basim, G.B., Adler, J.J., Mahajan, U., Singh, R.K., and Moudgil, B.M. 2001. Developing engineered particulate systems for oxide chemical mechanical polishing. *Proceedings–Electrochemical Society.* 2000-26:45.

Blendell, J.E., Bowen, H.K., and Coble, R.L. 1984. High-purity alumina by controlled precipitation from aluminum sulfate solutions. *Am. Ceram. Soc. Bull.* 63:797.

Catone, D.L. and Matijevic, E. 1974. Aluminum hydrous oxide sols. Ii. Preparation of uniform spherical particles by hydrolysis of aluminum sec-butoxide. *J. Colloid Interface Sci.* 48:291.

Chatterjee, M., Naskar, M.K., Siladitya, B., and Ganguli, D. 2000. Role of organic solvents and surface-active agents in the sol-emulsion-gel synthesis of spherical alumina powders. *J. Mater. Res.* 15:176.

Chen, C.-K., Gleiman, S., and Phillips, J. 2001. Low-power plasma torch method for the production of crystalline spherical ceramic particles. *J. Mater. Res.* 16:1256.

Chiba, H., Takahashi, K., and Otaguro, K. 1988. Spherical silica and alumina fillers for polymer compositions. *Jpn. Kokai Tokkyo Koho*:7 pp.

Hamano, K. and Asaga, K. 1974. Preparation of spherical alumina particles by means of a plasma flame and their characteristics. *Yogyo Kyokai Shi.* 82:611.

Hiragushi, K., Fukuoka, H., Matuo, M., Ishibashi, T., Tamehiro, T., and Murakami, K. 1982. Properties of spherical alumina particles obtained by oxy-propane gas flame spraying. *Yogyo Kyokaishi.* 90:105.

Lopez-Navarrete, E. and Ocana, M. 2001. Fine spherical particles of narrow size distribution in the cr_2o_3-al_2o_3 system. *J. Mater. Sci.* 36:2383.

Matijevic, E. and Wilhelmy, D.M. 1982. Preparation and properties of monodispersed spherical colloidal particles of cadmium sulfide. *J. Colloid Interface Sci.* 86:476.

McNeil-Watson, F., Tscharnuter, W., and Miller, J. 1998. A new instrument for the measurement of very small electrophoretic mobilities using phase analysis light scattering (pals). *Colloids Surf., A.* 140:53.

Milosevic, O., Jordovic, B., and Uskokovic, D. 1994. Preparation of fine spherical zno powders by an ultrasonic spray pyrolysis method. *Mater. Lett.* 19:165.

Murugavel, P., Kalaiselvam, M., Renganathan, M.K., and Raju, A.R. 1998. Preparation and characterization of sub-micron spherical particles of al_2o_3, sio_2 and mullite. *Mater. Chem. Phys.* 53:247.

Nishikawa, S. and Matijevic, E. 1994. Preparation of monodispersed spherical silica-alumina particles by hydrolysis of mixed alkoxides. *J. Colloid Interface Sci.* 165:141.

Noma, T., Shoji, N., Wada, S., and Suzuki, T. 1993. Hydroxylapatite ceramics with dispersed spherical alumina particles. *J. Ceram. Soc. Jpn.* 101:923.

Santos, P.S., Santos, H.S., and Toledo, S.P. 2000. Standard transition aluminas. Electron microscopy studies. *Mater. Res.* 3:104.

Tanaka, H. and Miyasaki, I. 1998. Cosmetics incorporated with spherical white pigment particles for good hiding power and natural look. *Jpn. Kokai Tokkyo Koho*:7 pp.

Trippel, G., Steiner, W., Joppien, G., and Hofmann, U. 1988. Process for the preparation of spherical alpha-aluminum oxide particles, the aluminum oxide particles, and their use in magnetic films. *Eur. Pat. Appl.*:13 pp.

Williams, R., Yocom, P.N., and Stofko, F.S. 1985. Preparation and properties of spherical zinc sulfide particles. *J. Colloid Interface Sci.* 106:388.

SECTION 2

Purification, Synthesis, Modification, and Characterization of Fillers and Nanofillers

Synthesis of Organic-inorganic Hybrid Nanoparticles for Use as Polymer Additives

Anna Arkhireeva* **and John N. Hay***

Considerable interest has been shown recently in organically modified silicates (ormosils) because of the wide range of possible applications for these interesting organic-inorganic hybrids. These applications include the use of the hybrids as modifiers of a range of host polymer properties. The ormosils can be synthesised by the conventional hydrolytic sol-gel process, as well as by the non-hydrolytic sol-gel route. To date, however, there are relatively few reports of the synthesis of sub-micron size ormosil particles of well-defined spherical morphology. Here, we report work to date on synthesis of such nanoparticles and also introduce our own work in the area.

INTRODUCTION

Over the past few years, new opportunities in the field of synthesis and application of hybrid organic-inorganic materials have received a significant amount of attention. These compositions have been the subject of extensive investigation because of the remarkable possibilities not only to combine the properties of organic and inorganic phases, but to design materials with truly unique properties.

Ormosils (organically modified silicates) constitute a family of organic-inorganic hybrids in which the organic moieties are bonded to a siloxane or silica matrix. The organic components can be entrapped or impregnated into the silica-based network without impairing their functionalities (Koslova, Viana, and Sanchez 1993; Ueda et al. 1995; MacCraith et al. 1997; Bohmer and Keursten 2000) [class I hybrids according to the classification of Sanchez (Sanchez and Ribot 1994; Judeinstein and Sanchez 1996)] or, alternatively, they can be chemically bonded to the inorganic phase by covalent bonds (Baumann and Schmidt 1994; Ou and Seddon 1997; Prabakar and Assink 1997; Noda et al. 1997; Noda et al. 1999; Noda, Kamoto, and Yamada 2000; Hay, Porter, and

* University of Surrey, Guildford, Surrey, United Kingdom.

Raval 2000; Hay and Raval 2001; Master et al. 2002) [class II hybrids (Sanchez and Ribot 1994; Judeinstein and Sanchez 1996)]. In both cases the siloxane matrix imparts good mechanical strength, and thermal and chemical stability, whereas the organic modification is responsible for introduction of such properties as functionality, flexibility and processability, the desired surface characteristics, as well as the elasticity and the ability to form films, if the organic component is in the form of an interpenetrating polymeric network.

Ormosil materials have begun to attract considerable interest, as the demand to improve and modify a wide range of properties [e.g., mechanical (Iwamoto and Mackenzie 1995), electrical (Nishio et al. 2000), electrochemical (Walcarius 2001), optical (Atkins, Krolikowska, and Samoc 2000) and sensing (MacCraith et al. 1997)] with respect to numerous new applications has increased. However, although many claims for the potential uses of the ormosil materials, such as in the fields of protective and functional coatings (Bohmer and Keursten 2000; Pilz and Romich 1997; Messori et al. 2002), ceramics (Guizard, Julbe, and Ayral 1999), optically active devices (Ueda et al. 1995), etc., have been made, the lack of truly promising commercial applications has been severely criticised (Mackenzie and Bescher 1998). While silica nanoparticles have been investigated for use as fillers in hybrid coating systems (Hou et al. 1997), there has been no systematic study of the use of organic-inorganic hybrid particles as modifiers of the properties of polymer systems and their corresponding composites. In this work, the synthesis of ormosil nanoparticles and control of their properties for use as polymer additives is described.

ORMOSILS

Class II ormosils, in which chemical bonds exist between the organic and inorganic phases, are normally those formed from organically modified precursors, such as organotrimethoxysilanes ($RSi(OCH_3)_3$ with R = $-CH_3$, $-(CH_2)_n-Cl$, $-(CH_2)_n-CN$, $-(CH_2)_n-NR'_2$, etc.), where the organic group does not form a bridge with the siloxane matrix that features only Si–O–Si linkages. The products where each silicon atom is covalently attached to one organic group are the silsesquioxanes with the empirical formula $R_2Si_2O_3$ (Baney et al. 1995). Bridged polysilsesquioxanes are another family of ormosil materials, which are prepared using organosilane precursors with two or more trialkoxysilyl groups bridged by an organic group (e.g., $(R'O)_3Si-R-Si(OR')_3$ with R = $-CH_2-$, $-CH_2-CH_2-$, $-CH=CH-$, $-C_6H_4-$, etc.) (Cerveau, Corriu, and Framery 2001; Shea and Loy 2001). The silica-based networks in such materials exhibit Si–R–Si linkages in addition to Si–O–Si linkages. This class of hybrids, which include the periodic mesoporous organosilicas (PMO) (Asefa et al. 1999; Inagaki et al. 1999; Kruk and Jaroneic 2001), has been extensively studied in recent years because of the unique opportunities to tailor the resulting properties, such as the pore size, the specific surface area and the level of chemical functionality, through the choice of the bridging groups.

Ormosil Synthesis

While some exotic methods for the preparation of organic-inorganic silica-based materials, such as biocatalytic routes (Brott et al. 2001; Clarson 2002), have been reported recently, the well-established sol-gel process and, sometimes, emulsion polymerisation techniques still play a dominant role by providing convenient methods for the production of ormosils. The sol-gel method enables the preparation of numerous types of ormosils directly from molecular precursors under mild conditions. One of the attractive features of this technique is the relative ease with which the hybrid materials can be processed to form fibres, coatings, nanoparticles or bulk products (Schottner 2001), as well as the compatibility of the process with mass production methods.

Hydrolytic Sol-gel Synthesis. The most common approach is the hydrolytic sol-gel route, which involves base- or acid-catalysed hydrolysis and condensation reactions of alkoxide precursors in an aqueous solvent system. The chemistry of this process is presented by the general equations shown below:

$$\text{Hydrolysis: } \equiv\text{Si–OR} + H_2O \rightarrow \equiv\text{Si–OH} + \text{ROH} \quad \textbf{(EQ 1)}$$

$$\text{Water condensation: } \equiv\text{Si–OH} + \text{HO–Si}\equiv \rightarrow \equiv\text{Si–O–Si}\equiv + H_2O \quad \textbf{(EQ 2)}$$

$$\text{Alcohol condensation: } \equiv\text{Si–OR} + \text{HO–Si}\equiv \rightarrow \equiv\text{Si–O–Si}\equiv + \text{ROH} \quad \textbf{(EQ 3)}$$

Several groups (Ueda et al. 1995; Ou and Seddon 1997; Prabakar and Assink 1997; Iwamoto and Mackenzie 1995; Atkins, Krolikowska, and Samoc 2000) have reported the synthesis of ormosil materials via the hydrolytic sol-gel route. A wide range of hybrid sol-gels has been obtained by reacting an inorganic precursor, such as tetraethoxysilane, and an organically modified alkoxysilane ($RSi(OR')_3$ or $R_2Si(OR')_2$) with water in a solvent in the presence of an acid catalyst at relatively low temperatures.

Non-hydrolytic Sol-gel Route. While the hydrolytic sol-gel process has been used extensively as a versatile route to organic-inorganic hybrids, the development of alternative methods, such as the non-hydrolytic sol-gel approach, has been limited until very recently. The non-hydrolytic sol-gel process involves the reaction of a 'metal' halide with an oxygen donor, such as an alkoxysilane or an ether, under nonaqueous conditions. Normally, the ormosils are synthesised by reaction of silicon halides with alkoxides or dialkyl ethers in the absence of a solvent and in the presence of a Lewis acid catalyst (Hay, Porter, and Raval 2000; Hay and Raval 2001; Bourget, Leclercq, and Vioux 1999). This route has been used to prepare a range of ormosils, such as those represented by the general equations shown below:

$$RSiCl_3 + 0.75\ Si(OR')_4 \rightarrow [RSi_{1.75}O_3]_n + 3\ R'Cl \quad \textbf{(EQ 4)}$$

$$0.75\ SiCl_4 + RSi(OR')_3 \rightarrow [RSi_{1.75}O_3]_n + 3\ R'Cl \quad \textbf{(EQ 5)}$$

$$RSiCl_3 + RSi(OR')_3 \rightarrow [R_2Si_2O_3]_n + 3\ R'Cl \quad \textbf{(EQ 6)}$$

$$2\ RSiCl_3 + 3\ R'_2O \rightarrow [R_2Si_2O_3]_n + 6\ R'Cl \quad \textbf{(EQ 7)}$$

Both silsesquioxane structures and ormosils containing only one organic group for every two silicon atoms can be synthesised. Potential advantages of the non-hydrolytic route include the possibility of avoiding various drawbacks associated with the use of solvents, as well as reduction of the number of residual silanol groups in the product, arising from the different mechanistic course of the reaction compared with the hydrolytic sol-gel route. This approach also offers the possibility to avoid phase separation, which occurs during the hydrolytic synthesis of hydrophobic hybrids, even in the case of the organically-substituted silicas containing bulky organic groups, such as decylphenylsilsesquioxane (Hay, Porter, and Raval 2000).

ORMOSIL NANOPARTICLES

While most of the work in this field has been concerned with the synthesis of bulk or monolithic products, the sol-gel route has often been overlooked for the preparation of ormosil particles, particularly nanoparticles. Noda et al. (Noda et al. 1997; Noda et al. 1999; Noda, Kamoto, and Yamada 2000) have described the emulsion polymerisation

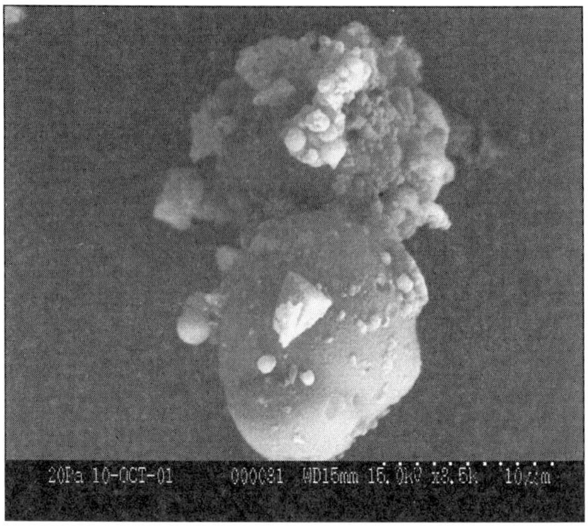

FIGURE 1 Scanning electron micrograph of methyl-modified silica particles synthesised via the non-hydrolytic sol-gel route (Masters et al. 2002)

synthesis of methylsilsesquioxane particles with diameters in the range of 0.2–5.0 μm, and having various topological morphologies, using a range of emulsifiers and sodium hydroxide as a catalyst. Baumann and Schmidt (1994) have shown that polycondensation in microemulsions is useful for the preparation of strictly spherical ormosil particles. Methyl-modified silica particles of controlled size in the range of 3–15 nm were synthesised by polycondensation of methyltrimethoxysilane using benzethonium chloride as a surfactant and sodium hydroxide as a catalyst. Katagiri et al. (1998) have reported the synthesis of phenylsilsesquioxane particles via the hydrolytic sol-gel route. Aggregated spherical particles with diameters in the range of 0.4–1.0 μm were prepared by the hydrolysis and condensation reactions of phenyltrimethoxysilane in the presence of an acid catalyst and a base catalyst, added successively. The synthesis was undertaken at 25 °C over a period of twenty hours. More recently, Masters et al. (2002) have extended the non-hydrolytic sol-gel method to the formation of ormosil particles under 'pseudo-dispersion polymerisation' conditions. Methyl-modified silica particles were obtained by reacting methyltrichlorosilane with tetraethoxysilane in the presence of surfactants using a high shear stirrer. The products were large aggregated, irregular particles with diameters in the range of 1.0–9.0 μm (Figure 1).

In this work, we have investigated the applicability of the hydrolytic sol-gel method to the synthesis of ormosil nanoparticles with potential applications as polymer additives. A range of hybrid nanoparticles (e.g., Figure 2) has been prepared from corresponding organoakoxysilane precursors using this facile, low-temperature route. The morphology of the particles depends on the type of organic group present in the ormosil and the specific reaction conditions. The synthesis of these particles and control of their properties for use as modifiers of selected polymer systems is currently being evaluated by our research group.

ACKNOWLEDGMENT

The authors thank the UK Engineering and Physical Sciences Research Council for 50% funding and the MoD for 50% funding under the Joint Grants Scheme.

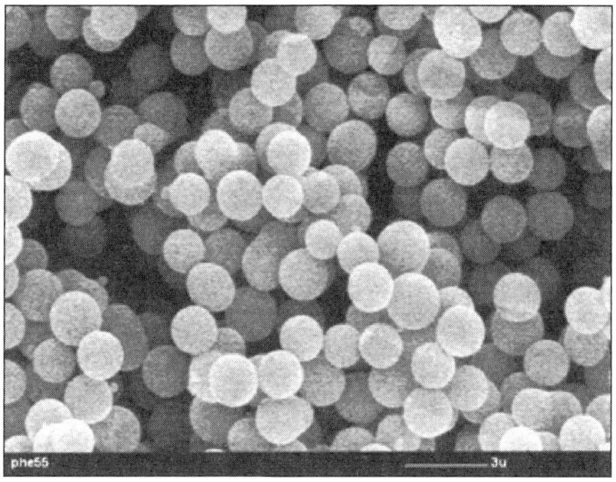

FIGURE 2 Scanning electron micrograph of phenyl-modified silica particles synthesised via the hydrolytic sol-gel route

REFERENCES

Asefa, T., M.J. MacLachlan, N. Coombs, and G.A. Ozin. 1999. Periodic mesoporous organosilicas with organic groups inside the channel walls. *Nature*. 402:867–871.

Atkins, G.R., R.M. Krolikowska, and A. Samoc. 2000. Optical properties of an ormosil system comprising methyl- and phenyl-substituted silica. *J. Non-Cryst. Solids*. 265:210–220.

Baney, R.H., M. Itoh, A. Sakakibara, and T. Suzuki. 1995. Silsesquioxanes. *Chem. Rev.* 95:1409–1430.

Baumann, F., and M. Schmidt. 1994. On the preparation of organosilicon m-spheres: a polycondensation in m-emulsion? *Macromolecules* 27:6102–6105.

Bohmer, M.R., and T.A. Keursten. 2000. Incorporation of pigments in TEOS derived matrices. *J. Sol-Gel Sci. Tech*. 19:361–364.

Bourget, L., D. Leclercq, and A. Vioux. 1999. Catalyzed nonhydrolytic sol-gel route to organosilsesquioxane gels. *J. Sol-Gel Sci. Tech*. 14:137–147.

Brott, L.L., R.R. Naik, D.J. Pikas, S.M. Kirkpatrick, D.W. Tomlin, P.W. Whitlock, S.J. Clarson, and M.O. Stone. 2001. Ultrafast holographic nanopatterning of biocatalytically formed silica. *Nature*. 413:291–293.

Cerveau, G., R.J.P. Corriu, and E. Framery. 2001. Nanostructured organic-inorganic hybrid materials: kinetic control of the texture. *Chem. Mater*. 13:3373–3388.

Clarson, S.J. 2002. Biological routes to novel silica structures. *Proceedings of Organic-Inorganic Hybrids II*. Paper 2.

Guizard, C.G., A.C. Julbe, and A. Ayral. 1999. Design of nanosized structures in sol-gel derived porous solids. Applications in catalyst and inorganic membrane preparation. *J. Mater. Chem*. 9:55–65.

Hay, J.N., D. Porter, and H.M. Raval. 2000. A versatile route to organically-modified silicas and porous silicas via the non-hydrolytic sol-gel process. *J. Mater. Chem*. 10:1811–1818.

Hay, J.N., and H.M. Raval. 2001. Synthesis of organic-inorganic hybrids via the non-hydrolytic sol-gel process. *Chem. Mater*. 13:3396–3403.

Hou, L., B. Hoffmann, H. Schmidt, and M. Mennig. 1997. Effect of heat treatment and additives on the photochromic and mechanical properties of sol-gel derived photochromic coatings containing spirooxazine. *J. Sol-Gel Sci. Tech*. 8:923–926.

Inagaki S., S. Guan, Y. Fukushima, T. Ohsuna, and O. Terasaki. 1999. Novel mesoporous materials with a uniform distribution of organic groups and inorganic oxide in their frameworks. *J. Am. Chem. Soc*. 121:9611–9614.

Iwamoto, T., and J.D. Mackenzie. 1995. Hard ormosils prepared with ultrasonic irradiation. *J. Sol-Gel Sci. Tech.* 4:141–150.

Judeinstein, J., and C. Sanchez. 1996. Hybrid organic-inorganic materials: a land of multidisciplinarity. *J. Mater. Chem.* 6:511–525.

Katagiri, K., K. Hasegawa, A. Matsuda, M. Tatsumisago, and T. Minami. 1998. Preparation of transparent thick films by electrophoretic sol-gel deposition using phenyltriethoxysilane-derived particles. *J. Am. Ceram. Soc.* 81:2501–2503.

Koslova, N.I., B. Viana, and C. Sanchez. 1993. Rare-earth-doped hybrid siloxane-oxide coatings with luminescent properties. *J. Mater. Chem.* 3:111–112.

Kruk, M., and M. Jaroneic. 2001. Gas adsorption characterization of ordered organic-inorganic nanocomposite materials. *Chem. Mater.* 13:3169–3183.

MacCraith, B.D., C. McDonagh, A.K. McEvoy, T. Butler, G. O'Keeffe, and V. Murphy. 1997. Optical chemical sensors based on sol-gel materials: recent advances and critical issues. *J. Sol-Gel Sci. Tech.* 8:1053–1061.

Mackenzie, J.D., and E.P. Bescher. 1998. Structures, properties and potential applications of ormosils. *J. Sol-Gel Sci. Tech.* 13:371–377.

Masters, H., J.N. Hay, J.M. Lane, M. Manzano, and S.J. Shaw. 2002. Hybrid (nano)particles via sol-gel chemistry. *Proceedings of Organic-Inorganic Hybrids II.* Paper 5.

Messori, M., M. Toselli, F. Pilati, E. Fabbri, P. Fabbri, and S. Busoli. 2002. Poly(caprolactone)/silica organic-inorganic hybrids as protective coatings for poly(methyl methacrylate) substrates. *Proceedings of Organic-Inorganic Hybrids II.* Paper 20.

Nishio, K., K. Okubo, Y. Watanabe, and T. Tsuchiya. 2000. Structural analysis and properties of organic-inorganic hybrid ionic conductor prepared by sol-gel process. *J. Sol-Gel Sci. Tech.* 19:187–191.

Noda, I., M. Isikawa, M. Yamawaki, and Y. Sasaki. 1997. A facile preparation of spherical methylsilsesquioxane particles by emulsion polymerisation. *Inorganica Chimica Acta* 263:149–152.

Noda, I., T. Kamoto, Y. Sasaki, and M. Yamada. 1999. Topological morphology of methylsilsesquioxane derivatives. *Chem. Mater.* 11:3693–3701.

Noda, I., T. Kamoto, and M. Yamada. 2000. Size-controlling synthesis of narrowly distributed particles of methylsilsesquioxane derivatives. *Chem. Mater.* 12:1708–1714.

Ou, D.L., and A.B. Seddon. 1997. Near- and mid-infrared spectroscopy of sol-gel derived ormosils: vinyl and phenyl silicates. *J. Non-Cryst. Solids.* 210:18-7-203.

Pilz, M., and H. Romich. 1997. Sol-gel derived coatings for outdoor bronze conservation. *J. Sol-Gel Sci. Tech.* 8:1071–1075.

Prabakar, S., and R.A. Assink. 1997. Hydrolysis and condensation kinetics of two component organically modified silica sols. *J. Non-Cryst. Solids* 211:39–48.

Sanchez, C., and F. Ribot. 1994. Design of hybrid organic-inorganic materials synthesised via sol-gel chemistry. *New J. Chem.* 18:1007–1047.

Schottner, G. 2001. Hybrid sol-gel-derived polymers: Applications of multifunctional materials. *Chem. Mater.* 13:3422–3435.

Shea, K.J., and D.A. Loy. 2001. Bridged polysilsesquioxanes. Molecular-engineered hybrid organic-inorganic materials. *Chem. Mater.* 13:3306–3319.

Ueda, M., H.B. Kim, T. Ikeda, and K. Ichimura. 1995. Photochemical behaviour of tethered dianthryl compounds in sol-gel glass. *J. Mater. Chem.* 5:889–894.

Walcarius, A. 2001. Electrochemical applications of silica-based organic-inorganic hybrid materials. *Chem. Mater.* 13:3351–3372.

Core-shell Nanocomposites—New Materials for Functional Fillers

P. Somasundaran*, T. Chen*, and Diptabhas Sarkar*

We have developed a layer-by-layer deposition technique for preparation of nanocomposite particles. Using this process, it was possible to make nanocomposites of various oxides with desired local composition and microstructure. The attachment of the shell particles to core particles was achieved using controlled polymer adsorption. Removal of excess polymer was accomplished without detaching the pre-adsorbed polymers by a specific stepwise washing scheme. The inherent chemistry of the process makes it easy to incorporate functional fillers over core particles. The nanosized filler materials will replace the shell particles in the present process and the polymer bridge would be chosen by taking into consideration the respective surface chemistries of various shell and core particles.

INTRODUCTION

Nanostructured substances are materials consisting of atomic clusters as building blocks with at least one dimension in the size of 1–100 nm. These ultra fine colloidal particles, which might be called "nanoids" [1], have properties corresponding neither to those of free atoms or molecules that make up the clusters nor to those of bulk solids with identical composition. Their distinct properties arise from the large surface area to volume ratios. Ceramics, personal care and pharmaceutical products developed from these composite particles hold the potential of having tailor-made structural and compositional properties. Fabrication techniques currently used include gas phase condensation, laser pyrolysis, plasma technique and chemical methods such as sol-gel processing, emulsification and liquid phase precipitation. However these techniques cannot provide an efficient way of functionalizing the nanostructures.

It is the aim of this chapter to describe new methods that were developed to prepare novel nanocomposite structures. The composite particles were synthesized using a two-stage process in which nanoparticles were coated onto submicron particles through

* Henry Krumb School of Mines, Columbia University, New York, N.Y.

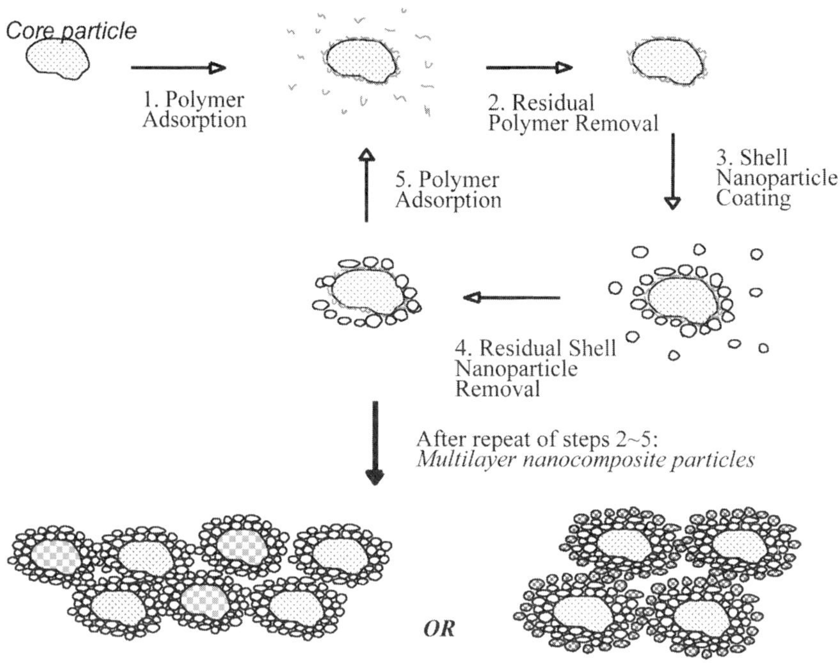

FIGURE 1 Schematic representation of the synthesis of nanocomposite particles

controlled polymer bridging. The present work is aimed towards the synthesis of multilayered composite particles by repeating the polymer adsorption–particle deposition sequence. The adsorption of the polymers on the first generation nanocomposite particles and the subsequent coating of the nanoparticles were monitored using electrokinetic measurements. Results showed the feasibility of synthesis of multilayered nanocomposite particles. The new scheme has been extended to concentrated dispersions, and the materials, thus prepared, have been further processed to prepare bulk nanocomposite green bodies and sintered disks. The effect of compression pressure on the density of these green bodies was studied to determine the flexibility.

MATERIALS AND METHODS

Two types of alumina powders (AKP-15 and AKP-50, Sumitomo Chemical, Tokyo, Japan) were used as the core particles. Both AKP powders were α-alumina, with a mean particle size of 0.7 and 0.2 mm and a specific surface area (BET) of 3.6 and 10.9 m^2/g respectively. Nanosized α-alumina powder (NanoTek, Nanophase Technologies) with a mean particle size of 36 nm and specific area of 47 m^2/g was used as shell particles. Polyacrylic acid (PAA) of two different molecular weights, 2,000 (Aldrich Chemical Co.) and 150,000 (Polyscience), were used as bridging aids. The polymers were used as received and the purity was tested using Fourier transform infrared (FTIR) spectroscopy. Sodium chloride was used for ionic strength control, and hydrochloric acid and sodium hydroxide for pH control.

A schematic drawing of the processing scheme used for preparing the multilayer nanocomposite particles is given in Figure 1 [2]. In the first step of the process, the core shell alumina particles were coated with the polyacrylic acid. Figure 2 shows the effect

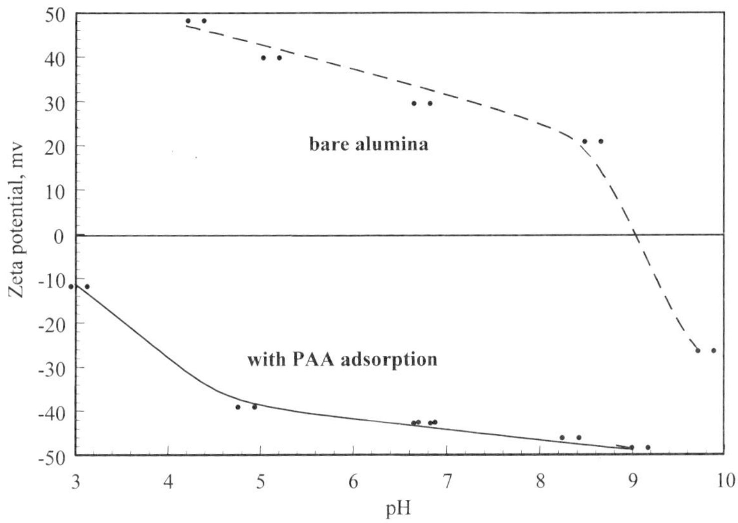

FIGURE 2 Effect of polyacrylic acid adsorption of zeta potential of AKP-50 alumina

of polyacrylic acid adsorption on the zeta potential of alumina. The inversion of the sign of potential on polyacrylic acid coated particles is an indication of the presence of polymer. Since polyacrylic acid is an anionic polymer it is expected to render the positive alumina surfaces negative upon adsorption. After removal of excess free polymers in the supernatant, PAA adsorbed core AKP-50 alumina particles remain highly negatively charged. This causes a strong electrostatic attraction to bare positive Nanotek shell particles that were introduced next into the system. Nanocomposite particles are thus obtained, on exposing the polymer coated particles to the suspension of the nanosized particles. The removal of the excess polymer is one of the two critical steps for the success of the entire procedure. Otherwise, self-flocculation of the nanoparticles, that is detrimental for controlled layer deposition, can take place as the polymer remaining in solution can adsorb on the added nanoparticles.

Given the very nature of the process, it is possible to make a range of nanocomposite particles using a wide variety of the core and shell particle combinations. Thus nanocomposites were also formed using, silicon nitride, titanium dioxide and hematite.

RESULTS AND DISCUSSIONS

The coating of nanoparticles on the core particles can be quantified by electro-kinetic techniques or depletion methods. The adsorption isotherm of nanosized alumina particles on core particles in this case is shown in Figure 3. It is similar to those obtained for high affinity type adsorptions generally seen for the polyelectrolyte adsorption, which is not surprising considering the fact that the adsorption of the nanosized particles is actually taking place on the polyelectrolyte species on the core-particle surface. The isotherm was found to fit the Langmuirian adsorption model (Figure 4) from which the maximum coating density was found to be 62 mg/m^2, which is 93% of a theoretically calculated coating density of 62 mg/m^2, using the assumption that the nanoparticles coat the surface in a planar square packing geometry.

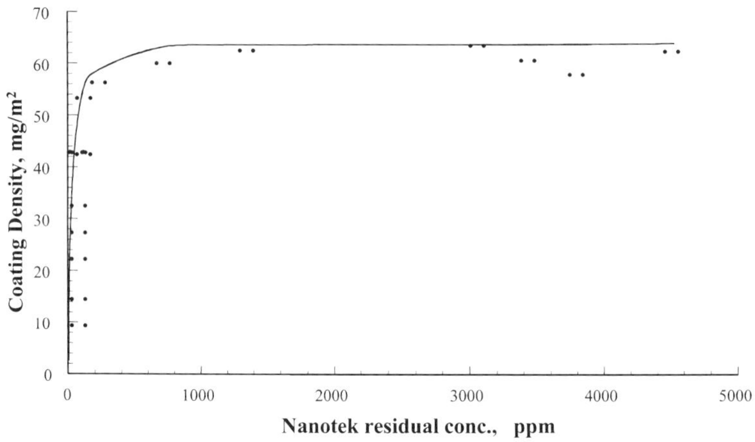

FIGURE 3 Coating isotherm of Nanotek alumina on PAA-covered AKP-15 alumina

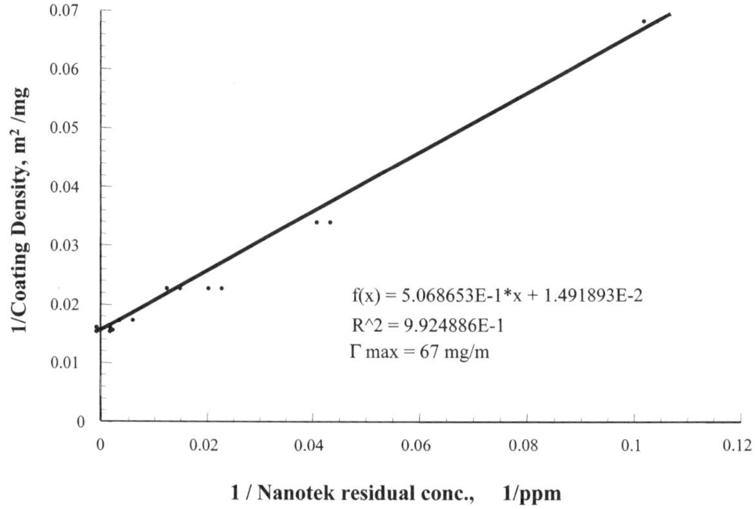

FIGURE 4 Langmuirian fit to the coating isotherm of Nanotek alumina on PAA-covered AKP-15 alumina

In addition to the above, scanning and transmission electron microscopy (SEM/TEM) was performed on these nanocomposite particles to examine the structure of the product. The SEM pictures (Figure 5a) showed uniform coating of the nanoparticles on the micron-sized core particles. Comparison with SEM pictures of a mixture of core and shell particles (Figure 5b) showed the presence of homogeneous uniformly sized particles in case of the nanocomposites, thus confirming the role of the polymer as the bridging aid. TEM micrographs suggested that the nanoparticles essentially formed a single layer coating over the core particles. The presence of multilayered sections was attributed to the formation of nanoparticle aggregates during the coating process.

Nanotek alumina coated AKP-15 composite particles—with polymers

Mixture of Nanotek alumina and AKP-15 particles—no polymer

FIGURE 5 Scanning electron microscopy images of bare and coated AKP-15 alumina particles

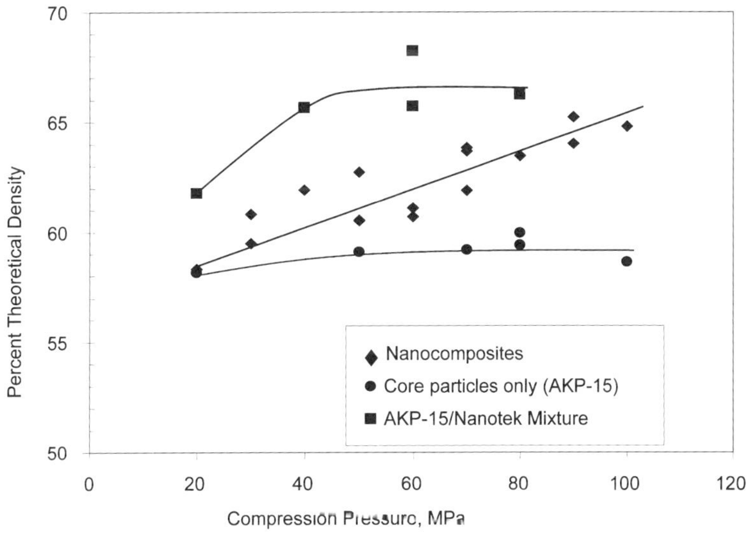

FIGURE 6 Effect of compression pressure on the green density of composites and compacts

The controlled polymer adsorption and coating process was performed successfully at higher solids loading (15 volume percent)[3]. Using the composite particles produced under high solids loading conditions, disc shaped ceramic green bodies were prepared using pressure filtration. The green density was found to be a linear function of the compression pressure (Figure 6). This observation can be accounted for by taking into account the flexible nature of the nanocomposite particles. When the compression pressure is increased, the nanoparticles begin sliding and filling in pores among core particles causing the green body to become more compact with a resultant increase in its green density. Linear continuous increase in the force-density plot is suggestive of the flexible nature of the new composites.

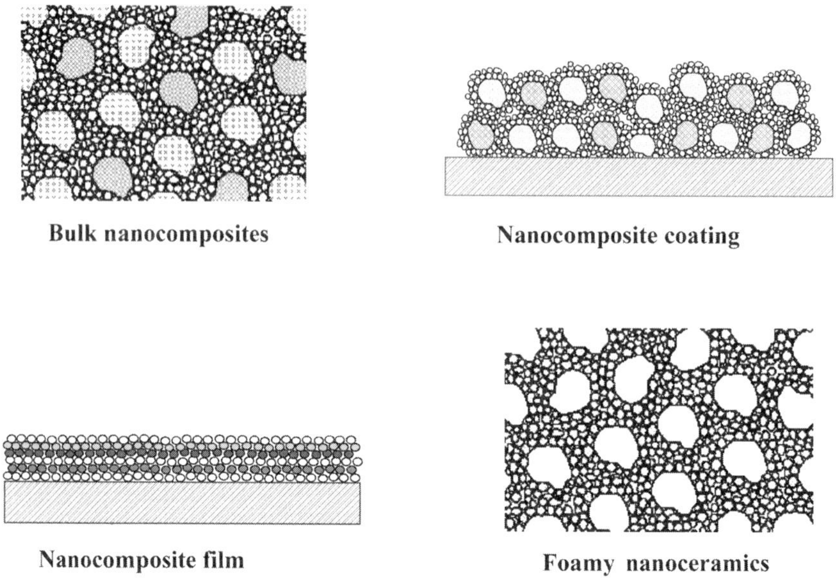

FIGURE 7 Future applications of nanocomposites

FUTURE APPLICATIONS

The current study has demonstrated the feasibility of new processing routes for the preparation of nanomaterials using a multi-stage coating/aggregation process. This layer-by layer deposition scheme in the case of nanocomposites allows control of the microstructure of the nanosystem. By varying the core and shell particles one can develop novel materials composed of any number of different solids with many different properties. With this controlled polymer adsorption–nanoparticle coating process, it would be possible to prepare many type of products with unique structure for special applications, as illustrated in Figure 7. Some of the categories of materials that can be prepared using the proposed processing method include the following:

Nanocomposite Coating. The process gives us an opportunity to use these engineered particles for surface coating. This could be performed using methods such as spray coating, slip casting or tape casting with the composite suspensions.

Nanocomposite Film. The controlled polymer adsorption–nanoparticle coating process could be used to prepare multilayer dense nanocomposite films on flat substrates. Possible applications could be in preparing radiation-sensitive films, for example, in mammography. Simply viewing the substrate as a giant core particle with an infinite radius of curvature, the two-stage coating process can be conveniently utilized to construct layer-by-layer coatings of nanoparticles on the substrate.

Foamy Ceramics. It is possible to prepare foamy ceramics by using the current coating process. This could be accomplished using the polymer beads or decomposing minerals as the core for nanoparticle coating. The polymer or mineral cores can then be removed by calcinations, leaving voids inside the bulk. This could be also done using soluble minerals as the cores and then dissolving by washing with relevant solutions. For both the routes the main challenge would be the sustainment of frame structure during the removal of the "core" components. If successful, these foamy ceramics (sponges) could be applied in various applications such as catalyst support, reagent reservoir and controlled release products.

CONCLUSIONS

Controlled surfactant, polymer adsorption and cross-linking procedures have been used to design and generate special nanocomposites. In this regard, in addition to the amount adsorbed, conformation and/or association of the adsorbed species play a governing role in determining the final structure. Composites prepared by the controlled polymer treatment were found to possess better strength characteristics in terms of flexibility and uniform distribution of particles. Nanoscale science has captured the imagination of scientists and technologists in a major way. For example, recently close to a hundred papers were presented at an international symposium in China [4], which dealt with the science and applications of nanotechnology in various areas of structural engineering, environment, energy and health care. Nanocomposites with immense future potential were developed in this study. It is clear from the current work that using the knowledge of how surface-active materials behave and taking into consideration their structure property relationships, one can prepare nanostructures of desired composition and component distribution.

ACKNOWLEDGMENTS

The authors gratefully acknowledge the support for this work from the Industry/University Cooperative Research Center (IUCRC) for Advanced Studies in Novel Surfactants (NSF/EEC-98-04618) and the Engineering Research Center (ERC) for Particle Science and Technology at the University of Florida (NSF/EEC-94-04168).

REFERENCES

1. Somasundaran, P., "Simple Colloids in Simple Environments Explored in the Past; Complex Nanoids in Real Dynamic Systems to be Conquered Next: Some Enigmas, Challenges and Strategies," *Submitted to JCIS,* 2002.
2. Tsung-yuan, Chen, P. Somasundaran, *J. American Ceramic Society*, 81, 140, 1998.
3. P. Somasundaran, Tsung-yuan Chen, D. Sarkar, *Materials Research Innovations*, 2, 325–327, 1999.
4. Proceedings of the *International Symposium on Nano-Materials and Technology*, Beijing, China, July 2–5, 2001.

Patterning of Gold and Polystyrene Nanoparticles into Mesostructures Using a Laser-based Particle Deposition

J. Drelich*, E.M. Nadgorny†, E.T. Zellers‡, J. Xu§, Ch. Zhou†, C.L. White§, and W.M. Cross**

A laser-based particle deposition (LBPD) technique has recently been developed at Michigan Technological University for patterning mesostructures made of nanoparticles and their clusters. The LBPD technique makes use of laser induced optical forces to transport and pattern nanoparticles on a variety of conductive and non-conductive substrates from nanoparticle-in-liquid suspensions. A weakly focused laser beam is used to confine and propel particles along the laser beam. Confinement is accomplished by forces due to optical field gradients that act on polarizable particles and tend to move them to axis of the laser beam. The force propelling the particles along the direction of the laser beam arise from momentum conservation of photons scattered from the particles. In this paper, this novel LBPD technology is briefly reviewed, and several examples of patterning are presented. The size and morphology of mesostrips produced from gold and polystyrene nanoparticles, under a range of deposition conditions, have been characterized by optical microscopy, atomic force microscopy, and scanning electron microscopy. The results presented here demonstrate that gold and polystyrene nanoparticles deposited by LBPD can form clusters of particles. When deposited at an appropriate rate, such clusters can form multilayer mesostrips of high particle density. One goal of research on such systems is to fabricate arrays of parallel lines made up of colloidal particles having a range of surface functionalities. Such arrays have potential for application as novel microsensors for environmental sensors.

* Dept. of Materials Science and Engineering, Engineering Research Center for Wireless Integrated MicroSystems, Michigan Technological University, Houghton, Mich.

† Dept. of Physics, Michigan Technological University, Houghton, Mich.

‡ Engineering Research Center for Wireless Integrated MicroSystems, Dept. of Environmental Health Sciences and Chemistry, University of Michigan, Ann Arbor, Mich.

§ Dept. of Materials Science and Engineering, Michigan Technological University, Houghton, Mich.

** Dept. of Metallurgical Engineering, South Dakota School of Mines and Technology, Rapid City, S.D.

NEED FOR NOVEL PATTERNING TECHNOLOGIES

A wide range of new techniques generally referred to as "direct-write technologies" have recently emerged in response to needs for rapid prototyping of porous micropatterns made of fine particles.[1] The development of fabrication techniques has evolved from technologies for rapid prototyping based on selective consolidation of powder materials, especially on polymers, with low melting or decomposition temperatures. Technologies such as inkjet printing, micro-pen writing, plasma spray, laser particle guidance, and matrix-assisted pulsed-laser evaporation can directly fabricate microstructures on a variety of substrates using a wide range of materials. None of these approaches require masks or other complications associated with lithographic technologies for microelectronic device fabrication.[1]

Manipulation of individual nanoparticles on surfaces, as well as formation of highly ordered arrays of nanoparticles, are rapidly evolving technologies. Recent reports indicate that several laboratories are close to developing technologies for synthesizing structural materials from highly ordered and homogeneous arrays of micro- and nanoparticles.[2–10] Such new materials, if defect-free, will have mechanical, optical, and electrical properties that are superior to conventional materials of comparable composition. The requirement for perfection of the particle arrangements, however, is likely to impose significant constraints on the flexibility and speed of novel material fabrication.

Inkjet writing and laser-based particle deposition (LBPD) are the only techniques so far that show promise for fabrication of patterned structures by guiding and depositing particles on a substrate. The manufacturing of environmental and biomedical devices is one of many potential applications of such techniques. The LBPD technique deposits particles that are localized near the axis of the laser beam, and line-widths achievable by this process can be less than one micron.[11,12] Because of its significantly greater patterning density and precision, LBPD has significant advantages over the inkjet technique for fabrication of sensors.

In addition to improved patterning density, the LBPD technique also allows deposition of almost any material that can be formed from a colloidal suspension. The technique, therefore, has far fewer constraints on solvent viscosity than inkjet techniques. Finally, LBPD is capable of co-depositing different materials. All such abilities offer a potential to fabricate advanced structures that combine distinct electronic, organic, and biological materials using the same basic fabrication technology.

LASER-BASED PARTICLE DEPOSITION

The LBPD technique is referred in this paper to a proprietary Laser Guided Direct Writing (LGDW) technology developed in the late 1990s[11–13] and Laser Guided Microsensor Patterning (LGMP) technology invented in 2002.[14] Instead of hollow optical fibers used in LGDW, LGMP employs gold-coated laser-proof apertures to channel the laser beam. The apertures reduce clogging and improve accuracy during deposition and permit deposition at considerably lower laser beam intensities. These characteristics offer the prospect of not only improved control of deposition porosity and structure but also allow graded interfaces having specially tailored surface chemistry to be produced.

LBPD enables fabrication of micrometer-scale structures from a wide range of materials, including metals, semiconductors, oxides, ferroelectrics, ionic crystals, and biological cells. These structures can be deposited on a variety of substrates, such as glass, plastic, sapphire, and silicon.[11–14,16]

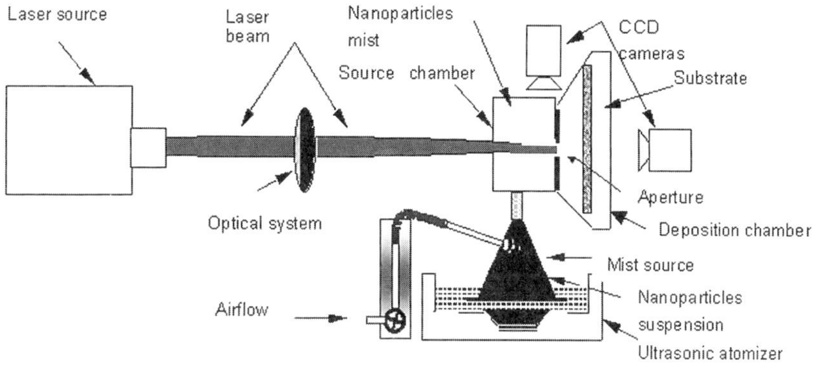

FIGURE 1 Schematic of the laser-based particle deposition (LBPD) apparatus

The basic LBPD apparatus (Figure 1) consists of a source chamber connected to a deposition chamber by a micron-sized aperture (or hollow fiber). Nebulizers or other aerosol generators are used to produce a mist of sub-micron sized liquid droplets in the source chamber. The droplets containing solids are prepared from colloidal suspensions, while liquid solvent or precursor solutions are atomized directly. A moderately intense laser beam directed along the axis of the aperture (hollow fiber) provides guidance for the droplets transporting them from the source to the deposition chamber. The aperture (hollow fiber) orifice diameter is significantly larger than the particles being transported, and the strong focusing effect of the laser light allows micron-size lines to be formed during patterning.

There are two laser-induced optical forces acting on the droplets during the deposition process.[11–13] One force, known as radiation pressure, arises from back-scattering of light from the droplet. Radiation pressure provides the force that propels droplets forward along the path of the laser beam. The second transverse force on the droplets arises from electrical forces on a polarizable particle in a gradient optical field of the laser beam. This gradient force confines droplets to the high intensity region along the axis of the laser beam. During deposition, the droplets or particles can achieve velocities of several hundred mm/s, and can be guided along precise trajectories for several hundred microns after leaving the aperture (hollow fiber). These features enable conformal particle deposition on the substrate, and through translation of the substrate relative to the beam allow high-resolution patterning.

In principle, virtually any material that is suspended in a liquid or formed by decomposition of a precursor liquid can be deposited by the LBPD technique. The principle limitations of the technique are that some deposited particles may undergo unwanted reactions (e.g., premature decomposition) during transport by the laser beam, and that the substrate should be transparent to laser light to avoid excessive interaction with the beam.

ARRANGEMENT OF NANOPARTICLES IN THE LBPD DEPOSITED STRUCTURES: PRELIMINARY OBSERVATIONS

The size and composition of the droplets in the source chamber strongly influences the character of the deposited material and the linewidths achievable by the LBPD technique. As discussed in the previous section, nebulizers are used to produce a mist of

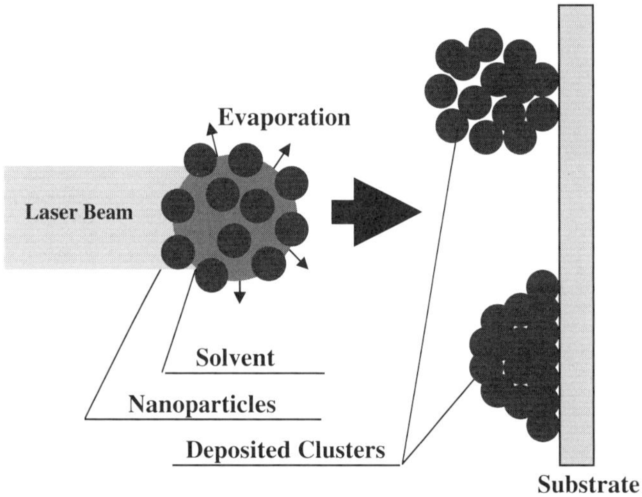

FIGURE 2 A schematic illustration of the shape of particle clusters where the liquid is either partially or completely evaporated during LBPD deposition

micron and sub-micron sized liquid droplets in the source chamber. Models for ultrasonic atomization of single phase liquids predict that droplet diameters depend on the solution surface tension σ, mass density ρ, and transducer frequency ω according to an equation of the following form[15]:

$$d = C\left(\frac{\sigma}{\rho\omega^2}\right)^{1/3} \quad \textbf{(EQ 1)}$$

where C is a numerical factor between 1 and 5, depending on the model. Equation 1 indicates that the transducer frequency is the most controllable parameter affecting the droplet size.

During transport from the source chamber to the substrate, the liquid in the droplets will evaporate partially or completely (Figure 2). Droplets containing suspended particles tend to form porous clusters with dimensions comparable to or smaller than the size of droplets generated by the nebulizer. Since the cluster size depends on the concentration of particles in droplets, the size can be substantially reduced below 1 μm by diluting the original solutions.

The droplets guided by the laser beam experience two major thermal effects including thermal conduction and liquid evaporation.[13] The rate of evaporation depends on such parameters as combined absorption of laser light by particles and liquid, the intensity of laser beam, thermal conductivity of solid particles, liquid and surrounding gas, and liquid vapor pressure in the gas. Whether the liquid evaporates completely or only partially from the particle-in-liquid aggregates has a strong effect on the structure of deposited clusters.

The size and size distribution of droplets and nanoparticles in the droplets, the rate of liquid evaporation during transport from the source chamber to the substrate, and the particle-liquid interactions during transport and deposition all influence the character of the individual deposited clusters and therefore the deposited material as a whole. Figure 3 shows clusters of 20-nm Au-thiolate (a) and 100-nm polystyrene particles (b), as deposited by LBPD[16]. As seen from Figure 3a, the 20-nm gold nanoparticles formed

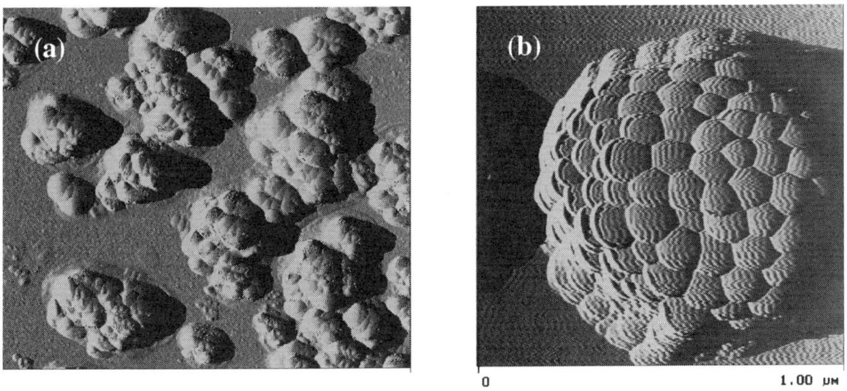

FIGURE 3 Atomic force microscopy images of deposited clusters made of nanoparticles. (a) Gold clusters made of #20-nm gold nanoparticles coated with self-assembled monolayer of $C_6H_5C_2H_4SH$. These particles were deposited from a toluene suspension. (b) A 0.8-μm cluster of 100-nm polystyrene particles deposited from water suspension.[16] Both of the depositions were done with a laser producing a 532 nm wavelength light. The laser power was set at 100 mW for deposition of Au-thiolate and 300 mW for deposition of polystyrene particles.

irregular clusters on the glass slide. This irregularity is caused primarily by the toluene used as a liquid medium for dispersion of the gold nanoparticles. Toluene has a strong affinity for the thiolate moieties (2-ethylphenylsulfide) self-assembled on the gold surface, causing the particles to remain dispersed in the toluene droplet instead of accumulating at the liquid-air interface.

In contrast to the gold nanoparticles, spherical particles of polystyrene deposited from water suspension formed spherical clusters with well-ordered particles (Figure 3b). Water wets polystyrene poorly, and capillary forces drive polystyrene particles to sit at the interface forming a regular pattern with dense and tight packing. Capillary forces can also be exploited to create even more complex 3-dimensional structures such as the one in Figure 4. This deposit is a mixture of 100-nm and 400-nm polystyrene particles, where the core of larger particles tends to be surrounded by the smaller particles.

As indicated in Figure 3a, the size and shape of clusters and arrangement of particles with diameters #20 nm within clusters require better control of deposition conditions. We have already been able to build some patterned structures along this direction, as shown in Figure 5 for 7-nm irregular Au-thiolate particles. Two quite different morphologies of patterned arrays can be seen. Evaporation of toluene before droplet's deposition leads to formation of relatively uniform porous spheres such as shown in Figure 5a. Although the formation of spherical structures in patterning of uniform arrays of particles is often an undesirable effect, this phenomenon can be applied in manufacturing of porous microspheres for medical or other applications.

The other structure of a dense array composed of 7-nm gold particles deposited from a toluene suspension at lower laser intensity is shown in Figure 5b. The particles form a more-or-less uniform dense layer with small local transitions in thickness. A few spheres of clustered particles observed on the top of the dense layer (Figure 5b) suggests that spherical clusters are deposited on a substrate in the first stage of deposition process but they disintegrate to a dense layer in the second stage. A transition from spherical clusters to 2-dimensional structures for nanoparticles on the substrate might result from the spreading effects initiated by a residual solvent present in the deposited clusters. For

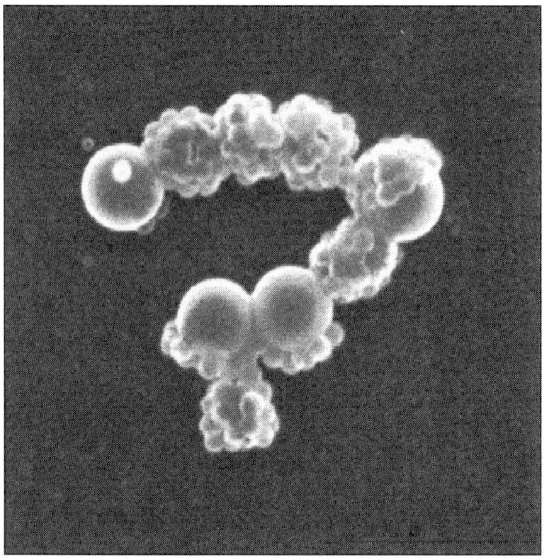

FIGURE 4 Scanning electron micrograph of an array of 400-nm polystyrene particles partially coated with 100-nm polystyrene particles. The mixture of the polystyrene particles in water was deposited using a 532 nm laser source using a laser power of 300 mW.

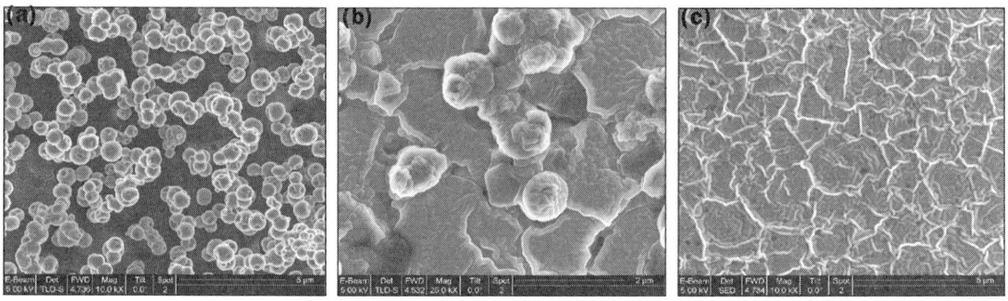

FIGURE 5 Scanning electron micrograph of structures formed by 7-nm Au-S$(CH_2)_7CH_3$ particles. (a) Au-thiolate particles deposited by the LBPD using a 532 nm laser with a power of ~200 mW. (b) Au-thiolate particles deposited by the LBPD using a 532 nm laser with a power <100 mW. (c) A layer of Au-thiolate particles formed by spreading the particle-in-toluene suspension followed by toluene evaporation.

comparison, Figure 5c shows a structure of similar dense array of 7-nm gold particles but prepared just by mechanical spreading of a gold nanoparticle-toluene suspension and evaporation of the solvent. Similarities in morphology of deposited gold particles on Figures 5b and 5c suggest that it is possible, with appropriate control of laser beam parameters, to preserve a certain volume of liquid in deposited droplets, which can promote spreading of the gold particle-in-toluene suspension over the substrate surface.

PATTERNING OF GOLD NANOPARTICLES INTO MICROSENSOR ARRAYS

The electrical resistivity of conducting and semiconducting thin films patterned on electrodes have been shown to depend upon the type and concentration of various environmental organic analytes. Similar "chemiresistors" (CRs) can also be constructed from a variety of other materials including organometallic clusters,[22,23] electrically conductive polymers,[24] and carbon-filled insulating polymers.[25–27]

Nanostructured materials have also played a prominent role in designing gas and vapor sensing devices.[17–21] Metal nanoparticles covered with organic modifiers constitute a new class of materials for the CR sensor fabrication. Such particles have been investigated extensively in recent years because of their unique electronic and electrochemical properties.[28–30] The electrical conduction in such arrays is controlled either by interparticle quantum tunneling or activated hopping of conduction electrons. The electronic transport characteristics of Au-thiolate nanoparticle arrays depend on the structure of the thiolate molecules bonded to the gold surfaces, the structural organization of particles in the arrays, and the particle-substrate interaction.[28–30] Chemical modification of the nanoparticle surfaces may change the mechanism of electron transport in the array from tunneling to activated hopping.[31–33] For example, local tunneling resistance depends exponentially on interparticle spacing when electron-rich π-bonded modifiers are attached to the gold particles.[34]

Accurate, rapid, and reproducible analysis using the CR sensors will require that the sensor materials be able to quickly adsorb and desorb analytes in sufficient quantities to produce the desired electrical response. This, in turn, is facilitated by high specific surface area of the deposit and the ability of a fabrication technique to tailor the sensor surface to achieve these characteristics. Lithographic techniques such as those used in conventional microelectronic device fabrication involve difficulties in tailoring the surface character of the films with organic modifiers and cannot create particulate deposits with high specific surface areas.

Patterning of nanoparticle thin films on substrates represents a critical step in manufacturing densely integrated arrays of sensors, and LBPD offers a significant promise as a suitable technique for this process. It should be possible to tailor the specificity, sensitivity, and response time of individual sensor elements by controlling the structures and compositions of the deposited nanoparticle layers. These structures will, in turn, be dependent upon the droplet sizes and densities in the deposition chamber, the laser intensity in the LBPD system, and the character and manipulation of the substrate.

Early experiments used the LBPD technique to pattern Au-thiolate nanoparticles having $CH_3(CH_2)_7SH$ and $C_6H_5C_2H_4SH$ monolayers onto glass and quartz. This class of Au-thiolate nanoparticles was successfully used for fabrication of a dual-chemiresistor vapor sensor array and showed to be a good candidate for a GC (gas chromatograph) detector.[35]

The patterns were uniform arrays of microstrips of less than 100-μm width, examples of which are shown in Figure 6. Au-$S(CH_2)_7CH_3$ microstrips less than 100-μm wide can be seen in Figure 6a. Figure 6b shows a fabricated ~100-μm wide strip of a few microns in thickness made of 7-nm Au-thiolate nanoparticles and deposited over an array of gold microelectrodes (15 μm wide and spaced by 15 μm) on quartz. The transverse profile of the deposited microstrips mirrored the radial distribution of the gold nanoparticles within the laser beam. These examples also demonstrate that the LBPD technique can produce repeated depositions with highly dense mesoscopic structures.

Another important feature of the LBPD is its ability to thermally process the deposited materials and structures. By increasing the power of the deposition laser, it is possible to melt particles during or subsequent to their deposition. As a result, selected parts of the patterned structure form solid gold electrodes as shown in Figure 6c.

The most critical issue for successful fabrication of CR sensors based on Au-thiolate nanoparticles is the preservation of the functionality of organothiol coatings on the surface of patterned gold nanoparticles. In the LBDP technique, the laser beam not only

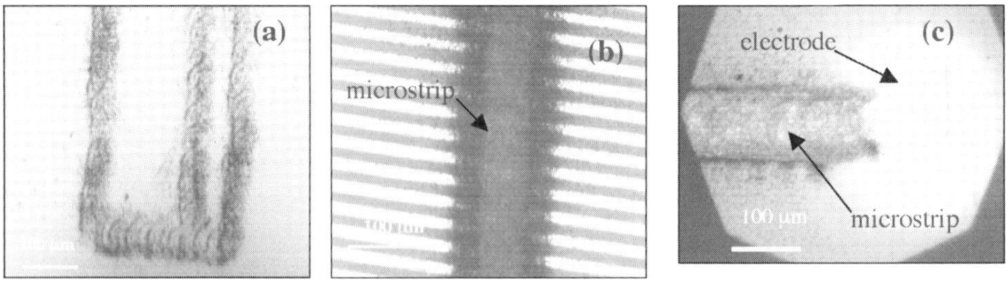

FIGURE 6 Optical images of ~100-µm-wide strips of gold-thiolate nanoparticles. (a) Parallel microstrips of #20-nm Au-S(CH_2)$_2$$C_6H_5$ nanoparticles deposited using 100-mW laser power. (b) A dense microstrip of the 7-nm Au-S(CH_2)$_7$$CH_3$ nanoparticles deposited on an array of gold microelectrodes using 200-mW laser power. (c) Microstrip made of 7-nm Au-S(CH_2)$_7$$CH_3$ nanoparticles with a gold electrode produced by in-situ melting. The Au-thiolate nanoparticles were deposited using 200 mW laser power. The electrode was melted using 500 mW laser power.

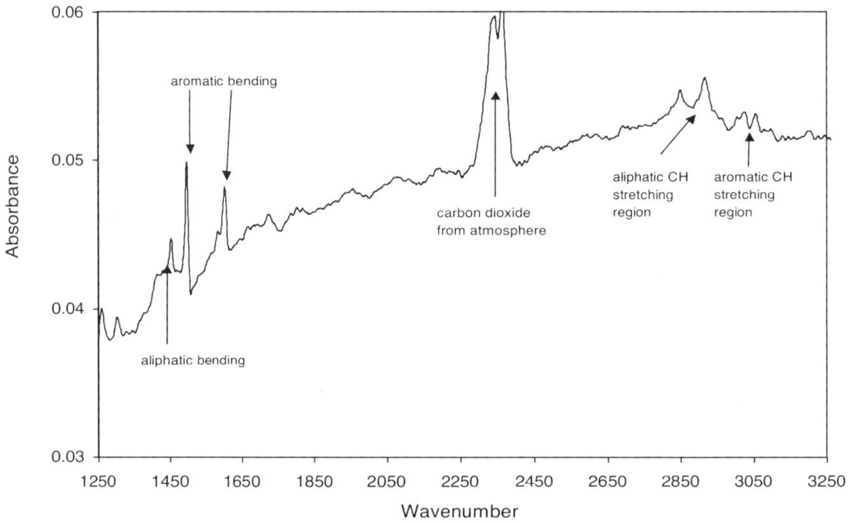

FIGURE 7 An FTIR spectrum from a $C_6H_5C_2H_4$-Au array deposited using a 532 nm laser at 100 mW power. The spectrum was acquired by UMA 500 IR microscope on a Digilab FTS-600 spectrometer using micro-ATR technique with a germanium element.

propels particles from the mist source to the substrate but also heats them in the process. For this reason, the laser intensity effects must be studied as a function of deposition conditions. Our preliminary experiments indicate that a laser intensity as low as 100 mW can guide the gold nanoparticles to the substrate for patterning. At the same time, such intensity is low enough to preserve the organic functionality of the deposited particles. This is illustrated by the FTIR spectrum of Figure 7: original aliphatic and aromatic bending of the $C_6H_5C_2H_4SH$ on gold nanoparticles is retained after deposition.

Finally, preliminary resistivity measurements were performed on dense meso-strips made of Au-S(CH_2)$_7$$CH_3$ nanoparticles and deposited between microelectrodes such as those in Figure 6b. The resistivity of arrays changed over a broad range of values, from 50 Ω to >20 MΩ, depending on the conditions of deposition such as laser power, deposition time, colloidal suspension concentration (and other unknown factors). This suggests that distribution and packing of nanoparticles in fabricated mesostructures, and perhaps some other factors, require better control, which will be the main topic of future research.

ACKNOWLEDGMENTS

This work was supported by the Engineering Research Centers Program of the National Science Foundation under Award Number EEC-9986866. The authors would like to express appreciation to Mr. Elvin Beach from Dow Chemical Company for imaging gold-thiolate nanoparticles using the field emission scanning electron microscopy, and to Dr. Qing-Yun Cai and Mr. Willie Steinecker for providing nanocluster samples.

REFERENCES

1. A. Pique and D.B. Chrisey (Eds), Direct-Write Technologies for Rapid Prototyping Applications: Sensors, Electronics, and Integrated Power Sources, Academic Press, San Diego, CA (2002).
2. S.I. Stupp, V. LeBonheur, K. Walker, L.S. Li, K.E. Huggins, M. Keser, and A. Amstutz, Science 276, 384–9 (1997).
3. B.T. Holland, Ch.F. Blanford, and A. Stein, Science 281, 538–40 (1998).
4. S.A. Jenekhe and X.L. Chen, Science 283, 372–5 (1999).
5. S. Sun, C.B. Murray, D. Weller, L. Folks, and A. Moser, Science 287, 1989–92 (2000).
6. Y.A. Vlasov, X.-Z. Bo, J.C. Sturm, and D.J. Norris, Nature 414, 289–93 (2001).
7. Z.-Z. Gu, S. Kubo, W. Qian, Y. Einaga, D.A. Tryk, A. Fujishima, and O. Sato, Langmuir 17, 6751–3 (2001).
8. M. Srinivasarao, D. Collings, A. Philips, and S. Patel, Science 292, 79–83 (2001).
9. Y. Lu, Y. Yin, B. Gates, and Y. Xia, Langmuir 17, 6344–50 (2001).
10. F. Iskandar, Mikrajuddin, and K. Okuyama, NanoLetters 1, 231–4 (2001).
11. M.J. Renn and R. Pastel, J. Vac. Sci. Technol. B 16, 3859–63 (1998).
12. M.J. Renn, R. Pastel, and H.J. Lewandowski, Phys. Rev. Lett. 82, 1574–7 (1999).
13. E.M. Nadgorny, R.L. Pastel, A.A. Struthers, and A. Miner, in Proc. International Conference on Mass and Charge Transport in Inorganic Materials (May 2000, Italy), Eds. P. Vincenzini and V. Buscaglia, Part B, 1107–1114 (2000).
14. J. Drelich, E. Nadgorny, J. Xu, E.T. Zellers, and Ch. Zhou, MTU Invention Disclosure #200209.00 (2002).
15. W.S. Moon, H.J. Chung, S.I. Woo, Ch.S. Hwang, M.Y. Lee, and S.B. Park, J. Aerosol Sci. 28 (Suppl. 1), 5525 (1997).
16. J. Xu, Ch. Zhou, S. Grant, E. Nadgorny, and J. Drelich, Proceedings of the Materials Research Society Symposium, 728, S7.6.1–S7.6.6 (2002).
17. J.W. Gardener and P.N. Bartlett, Electronic Noses: Principles and Applications, Oxford, Oxford University Press, UK 1999.
18. J. Park, W.A. Groves, and E.T. Zellers, Anal. Chem. 71, 3877–86 (1999).
19. V.M. Rotello, Materials Today 4, 24–29 (2001).
20. D.J. Schiffrin, MRS Bulletin 26, 1015–19 (2001).
21. C.C. Koch (Ed.), Nanostructured Materials: Processing, Properties and Potential Applications, Noyes Publications, Norwich, NY (2002).
22. H. Wohltjen, W.R. Barger, A.W. Snow, and N.L. Jarvis, IEEE Trans. Electron. Devices, ED 32, 1170–4 (1985).

23. J.W. Grate, M. Klusty, W.R. Barger, and A.W. Snow, Anal. Chem. 62, 1927–34 (1990).
24. J.M. Slater, J. Paynter, and E.J. Watt, Analyst 118, 379–84 (1993).
25. D.Y. Godovski, E.A. Koltypin, A.V. Volkov, and M.A. Moskvina, Analyst 118, 997–9 (1993).
26. E.J. Severin and N.S. Lewis, Anal. Chem. 72, 2008–15 (2000).
27. S.V. Patel, M.W. Jenkins, R.C. Hughes, W.G. Yelton, and A.J. Ricco, Anal. Chem. 72, 1532–42 (2000).
28. R.S. Ingram and R.W. Murray, Langmuir 14, 4115–21 (1998).
29. R. Chen, J. Phys. Chem. B 104, 663–7(2000).
30. B.I. Ipe, K.G. Thomas, S. Barazzouk, S. Hotchandani, and P.V. Kamat, J. Phys. Chem. B 106, 18–21 (2002).
31. T. Sato, H. Ahmed, D. Brown, and B.F.G. Johnson, J. Appl. Phys. 82, 696–701 (1997).
32. A.W. Snow and H. Wohltjen, Chem. Materials 10, 947–9 (1998).
33. P.W. Wuelfing, S.J. Green, J.J. Pietron, D.E. Cliffel, and R.W. Murray, J. Am. Chem. Soc. 122, 11465–72 (2000).
34. R. Parthasarathy, X.-M. Lin, and H.M. Jaeger, Condensed Matter (in press).
35. Q.-Y. Cai and E.T. Zellers, Anal. Chem. 74, 3533–3539 (2002).

Functionalization of a Synthetic Polymer to Create Inorganic/ Polymer Hybrid Materials

Christopher R. Lubeck*, **Michael C. Chin***, **and Fiona M. Doyle***

The fundamentals behind a novel approach for creating inorganic/polymer hybrids are described. The binding of +2 charged ions, specifically Cu^{2+} and Ca^{2+}, to ionic functional groups is explored through UV-Vis spectroscopy and zeta potential measurements. The novel approach uses an electric field to drive ions into a functionalized polymeric matrix, thereby inducing the precipitation of calcium phosphate species. Scanning electron microscopy images of the resulting composite illustrate the results of these precipitations with two distinct morphologies observed.

INTRODUCTION

When polymers are formed into structural materials there is significant void space within the material. Decreasing this space has been shown to improve the mechanical properties of the polymeric material, such as fracture strength or modulus. Fillers have long been used to decrease this void space. Numerous organic and inorganic materials have been used as fillers for polymeric materials (Bréchet et al., 2001; Hwu et al., 2002; Privalko et al., 2002; Shinzato et al., 2002; van Zyl et al., 2002). While it is not necessary for the filler to interact with the polymeric matrix to provide beneficial results, such an interaction dramatically improves the mechanical properties, even when the filler is present at small volume percentages (Liu et al., 1997; Ignjatovic et al., 2000). However, it is difficult to engineer a uniform microstructure in which the filler does interact with the polymeric matrix, especially when the disperse filler is only present at small volume percentages.

We are exploring a novel method to prepare polymeric composites with this increased strengthening, by precipitating sparingly soluble ceramic particles, such as calcium phosphates, directly into a polymeric matrix consisting of poly(methyl methacrylate) (PMMA) ($(-CH_2-C(CH_3)COOCH_3)-_n$) or a functionalized PMMA derivative. PMMA functionalized to create ionic or polar groups capable of interacting with inorganic ions

* Dept. of Materials Science and Engineering, Berkeley, Calif.

is a particularly promising matrix for this process, because the number and distribution of functional groups, and hence nucleation sites, can be tailored by controlling the degree of functionalization. Furthermore, functionalization also increases the hydrophilicity of the polymer, thereby allowing aqueous solutions to wet the three dimension porous structure of the polymeric matrix, and hence allowing ions to enter the matrix.

Unfortunately, the pores within polymeric materials useful for engineering applications are exceedingly small, with tortuous pathways between them. Consequently, diffusion of reactive ions into the functionalized matrix would be exceedingly slow, and it would be difficult to obtain a uniform dispersion of precipitates within such a polymeric matrix if diffusion were the principal ionic transport mechanism. Consequently, we are investigating the feasibility of using electrical migration to accelerate the transport of ions to reactive functional groups that can serve as nucleation sites for precipitates. For the precipitation of calcium phosphate ceramics, calcium and phosphate ions can be driven into the polymeric matrix with an appropriately applied electric field.

More specifically, the concept being tested is to start with anionic functional groups on the polymer. Mann (1996) has observed that polymers with phosphate functional groups can serve as the site of hydroxyapatite nucleation in a calcium and phosphate solution. Spanos and Koutsoukos (2001) recently investigated the use of carboxylate functional groups to nucleate calcium phosphates, most notably hydroxyapatite. An applied electric field is then used to draw reactive ions into pores within the polymeric matrix, causing the solubility product of sparingly soluble salts to be exceeded locally within the pores. In this way, it should be possible to precipitate calcium phosphate or other sparingly soluble salts throughout a polymeric matrix.

Here we report our initial studies in examining the feasibility of this novel concept for strengthening polymers through the in-situ precipitation of sparingly soluble salts into PMMA functionalized with carboxylate groups. UV-visible spectroscopy compares the affinity of commercially purchased PMMA for Cu^{2+} ions with that of partially functionalized PMMA. Zeta potential measurements provide insight into the interaction of PMMA and its functionalized derivatives with calcium at different pH. Finally, the results of preliminary in-situ precipitation tests are reported.

EXPERIMENTAL PROCEDURE

Materials

PMMA (molecular weight 120,000) was used as provided by Aldrich in the form of pellets smaller than $8mm^3$. A 1% (wt/vol) solution of unmodified PMMA in acetone was created by diluting 7ml of a 7% wt/vol PMMA stock solution to 1%. 1% (wt/vol) solutions of functionalized PMMA in acetone were created by diluting 7ml of a 7% PMMA stock solution with acetone and then adding 2mL of 0.01M aqueous NaOH in water. The base hydrolyzes the methacrylate groups, forming carboxylate groups and methanol:

$$\begin{gathered}(\text{-}CH_2\text{-}C(CH_3)COOCH_3)\text{-}_n + m\,OH^- = \\ (\text{-}CH_2\text{-}C(CH_3)COOCH_3)\text{-}_{n-m}(\text{-}CH_2\text{-}C(CH_3)COO^-)\text{-}_m + m\,CH_3OH \qquad \textbf{(EQ 1)}\end{gathered}$$

Electrostatic repulsion between ionized carboxylate groups would be expected to ensure that the carboxylate groups are dispersed rather uniformly throughout the polymer. A white precipitate formed upon initial addition of the NaOH, then redissolved upon shaking. A copper ion source was created by dissolving 0.1g of $CuCl_2*2H_2O$ in 10g of acetone.

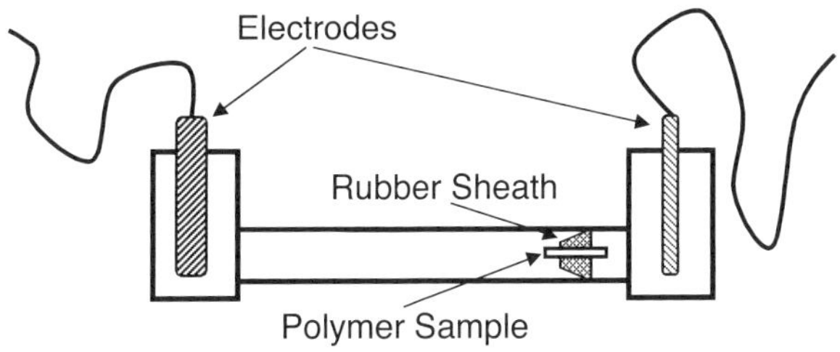

FIGURE 1 Schematic diagram of precipitation experiment design in zeta-meter cell

Interaction of Polymer Samples with Ions

The ability of unmodified and functionalized PMMA to interact with divalent metals was assessed initially using copper (II), to allow spectroscopic examination in the UV-visible range. Different amounts of copper ion solution were added to 5mL samples of the 1% solutions of unmodified and functionalized PMMA in acetone. After equilibration for 24 hours, the samples were analyzed using a Hewlett Packard 8452A diode array UV-visible spectrometer. A peak at 476nm (corresponding to a shift in the d–d transition of the copper due to binding with the polymer) was used to assess the degree of binding (Kramer and Dunsch, 1998). The absorbance intensity at 476nm was plotted as a function of the copper concentration.

Zeta Potential Measurements

Five grams of unmodified PMMA particles was dispersed in 500mL of double distilled water. 20mL of this suspension was pipetted into scintillation vials. Functionalized PMMA samples were made by reacting 1mL of 8.3% wt/vol solution of PMMA in acetone with 0.2mL of 0.01M NaOH in acetone. After a white precipitate appeared, the samples were evaporated to dryness and 20mL of double distilled water was added. The suspensions were then sonicated. Different amounts of calcium chloride were added to the unmodified or functionalized PMMA, then the pH was adjusted using sodium hydroxide and hydrochloric acid. The zeta potentials of the polymer samples were measured using a Zeta-Meter 3.0 with a Type GT2 cell, molybdenum anode and platinum cathode.

In-situ Precipitation of Calcium Phosphate in Polymeric Matrices

Solid samples of functionalized PMMA were produced by hydrolyzing 5mL of 8.3% (wt/vol) PMMA in acetone with 0.5mL of 0.01M aqueous NaOH. The solutions were then evaporated in Pasteur pipettes whose tips had been sealed. The samples were then heated to 120°C, to drive off any residual acetone and water. The samples were removed from the glass pipettes and placed into an aqueous solution containing 10,000ppm Ca^{2+}, as $CaCl_2$, for between one and five days. The samples were then removed from the solution and covered with a rubber sheath to prevent permeation of solution into the sides of the sample. Samples were then inserted into the zeta meter sample holder, as show in Figure 1. 0.0063M aqueous $CaCl_2$ and 0.026M aqueous Na_2HPO_4 were placed on opposite sides of the sample holder such that the Ca^{2+} and PO_4^{3-} ions would flow toward one another, into the sample, when an electric field was

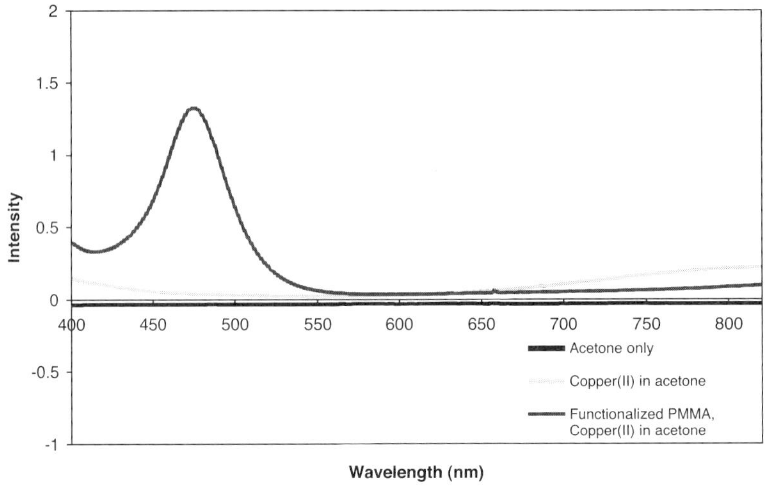

FIGURE 2 UV-visible spectra of acetone baseline, copper (II) ions in acetone and functionalized PMMA with copper (II) ions in acetone

applied. A potential of 100V (which created an electric field of about 870 V/m) was then applied for four days, after which the polymer samples were removed from the sample holder, rinsed in double distilled water, and fractured to reveal a fracture surface. The fracture surfaces were then mounted, sputtered with gold, and examined with a JEOL 6300 scanning electron microscope.

RESULTS AND DISCUSSION

Interaction of Polymer Samples with Ions

Figure 2 shows the UV-visible absorbance spectra of acetone, copper ions in acetone and PMMA with copper ions in acetone. The peak at 476nm on the PMMA/copper spectrum seems to correspond to the PMMA copper association. This peak is used to determine the amount of copper that can be bound to both the unmodified and the functionalized PMMA. Figure 3 plots the absorbance at 476 nm as a function of copper concentration; it can be seen that the absorbance for both polymers initially increases linearly with increasing copper concentration, then reaches a plateau, corresponding to reaching the binding capacity of the polymer for Cu^{2+} ions. The capacity of the functionalized polymer for Cu^{2+} ions (about 4.5×10^{-4} moles Cu^{2+}/gram polymer) is greater than that of the unmodified PMMA (about 6.4×10^{-4} moles/gram). This indicates that the ability to bind Cu^{2+} ions, and potentially other +2 charged ions, is enhanced by base-catalyzed hydrolysis of the ester groups to carboxylate groups. This is likely to increase the likelihood of precipitating inorganic particles within the polymeric matrix.

Zeta Potential Measurements

Additional evidence of the different binding ability of unmodified and functionalized PMMA is seen from the zeta potentials of polymer particles. Figure 4 shows the zeta potential as a function of pH in the absence of calcium. At low pH's, unmodified and functionalized PMMA particles possess a positive zeta potential, although the zeta potential of

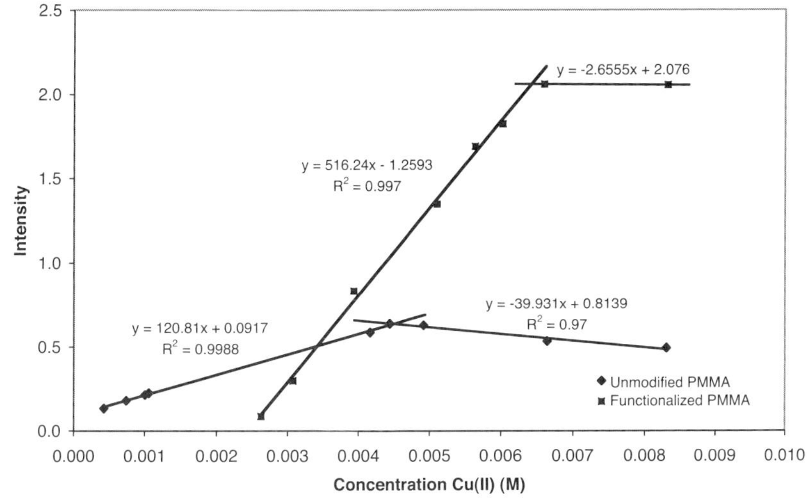

FIGURE 3 Comparison of binding capacity for 1% (wt/vol) functionalized and 1% (wt/vol) unmodified PMMA in acetone

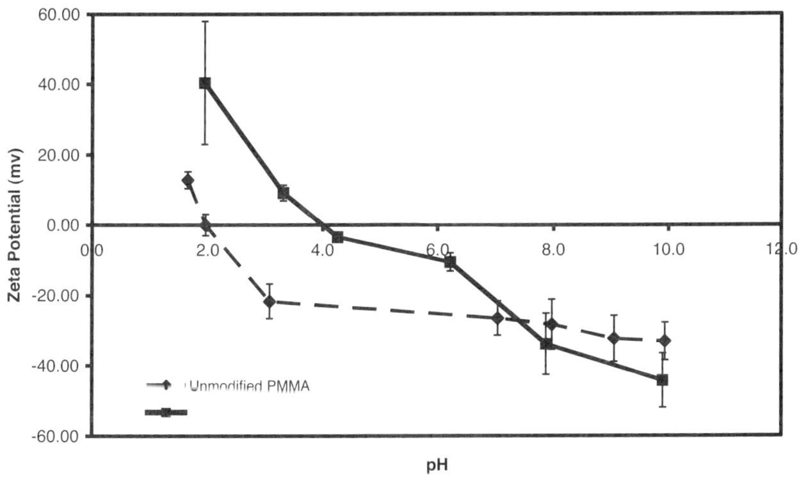

FIGURE 4 Zeta potential of unmodified and functionalized PMMA as a function of pH with no added calcium (II) ions

the unmodified PMMA was significantly greater than that of the functionalized PMMA. The isoelectric point of the unmodified PMMA lies at about pH 2, whereas that of functionalized PMMA is higher, at around pH 4. The latter is not unexpected, as the pK_a of many carboxylic acids is roughly 4. It appears that the origin of the surface charge differs for the unmodified and functionalized PMMA, because the zeta potential of the latter decreases more sharply with increasing pH than does the zeta potential of the former.

Figure 5 shows the zeta potential of unmodified PMMA as a function of Ca^{2+} concentration, for various pH values. At all pH values, increasing Ca^{2+} concentrations increase the zeta potential; except for pH 2, there is a charge reversal from negative to positive. At higher pH values (8 and 9) the zeta potential appears to converge on a constant value

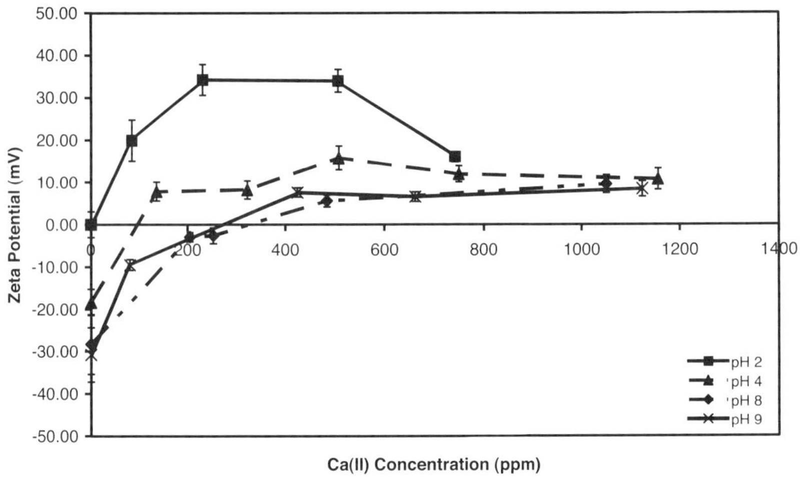

FIGURE 5 Zeta potential of unmodified PMMA particles as a function of calcium concentration for different pH values

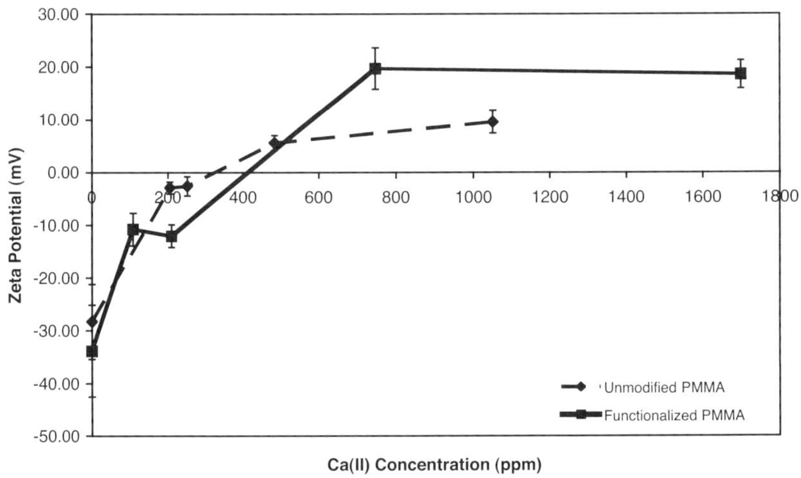

FIGURE 6 Comparison of zeta potential between unmodified PMMA and functionalized PMMA as a function of calcium concentration (pH = 8)

at higher calcium concentrations. The apparent maximum in the zeta potential for pH 2 at around 200 to 500 ppm Ca^{2+} probably occurs because of compression of the double layer at the higher Ca^{2+} concentrations; the ionic strength was higher at pH 2 because of the hydrochloric acid used for acidification.

Figure 6 compares the zeta potentials of unmodified and functionalized PMMA as a function of Ca^{2+} concentration, at pH 8. At high Ca^{2+} concentrations, the zeta potential of the functionalized PMMA reaches a steady, positive value that is significantly higher than that for unmodified PMMA, indicating that functionalized PMMA does appear to associate more strongly with calcium ions. While the presence of base can cause the conversion of ester groups to acid groups as mentioned previously, any hydrolysis would

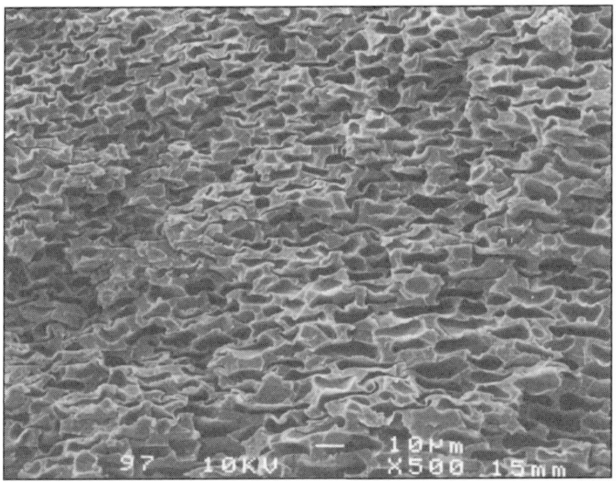

FIGURE 7 Morphology of functionalized PMMA sample without precipitation

probably be restricted to the solid polymer/aqueous solution interface when the polymer is suspended in an aqueous environment and hence would be much less extensive that the hydrolysis that occurred when PMMA was dissolved in acetone.

Within our experimental process, it would be desirable for the polymer to nucleate calcium phosphate through the sequestering or binding of calcium ions. Not only will the functionalized groups serve to concentrate the calcium ions within the polymer, but the resulting positively charged sites would then attract phosphate anions. Hence the functional groups are likely to be specific sites for the nucleation of calcium phosphate precipitates. One can envisage controlling the location of functional groups where the precipitation could occur, by such methods as creating a self assembled polymeric structure, thereby controlling the mechanical properties to maximize given properties according to the intended application.

In-situ Precipitation of Calcium Phosphate in Polymeric Matrices

SEM micrographs of the composites prepared by in-situ precipitation reveal the presence of precipitated material within the functionalized polymeric matrix. Figures 7 and 8 show the morphology of a functionalized PMMA sample before precipitation. The porous structure of the polymeric matrix is evident in both the low and high magnification images. After applying an electric field of 870 V/m to promote migration of ions, two distinctive precipitate morphologies were observed. Figure 9 shows an irregular, agglomerated precipitate morphology, whereas Figure 10 shows more discrete spheres, with diameters between about 0.1 and 1 μm. Table 1 lists the solubility product for some common calcium phosphate phases. Under the precipitation conditions used, $[Ca^{2+}]$ = 0.0063M, $[PO_4^{3-}] = 1.0 \times 10^{-5}$ M and pH = 9, the ion product, IP, for hydroxyapatite is defined as IP = $[Ca^{2+}]^5 * [PO_4^{3-}] * [OH^-]$, which is $(0.0063)^5 * (1.0 \times 10^{-5})^3 * (10^{-5}) = 9.95 \times 10^{-38}$. As described by Spanos and Koutsoakos (2001), the apparent degree of supersaturation is equal to the ion product divided by the solubility product. For the precipitation conditions, this corresponds to $\Omega = 4.2 \times 10^{21}$. Thus there is a large supersaturation and therefore a large driving force to form hydroxyapatite. The degrees of supersaturation for the other calcium phosphate phases were calculated in a similar manner and also appear in Table 1. Preliminary X-ray diffraction data suggest that the precipitates are amorphous, or poorly crystalline.

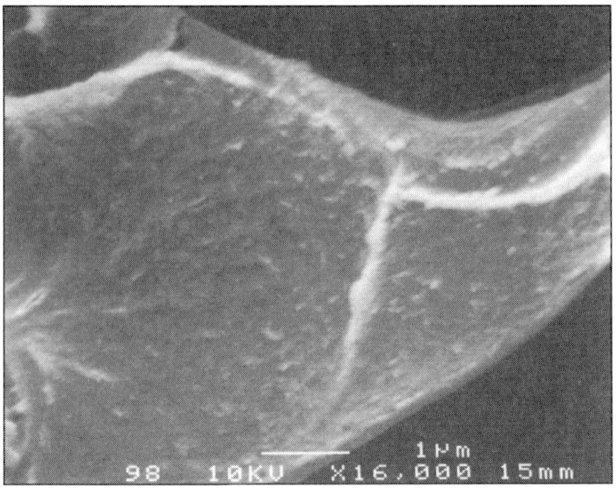

FIGURE 8 Close-up view of functionalized PMMA sample without precipitation

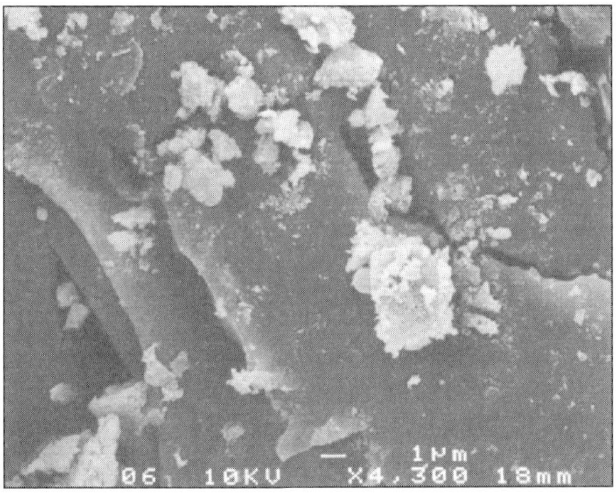

FIGURE 9 Irregular agglomerates of precipitate formed on functionalized PMMA sample. Solutions consist of 0.0063M $CaCl_2$*$2H_2O$ and 0.026M Na_2HPO_4*$7H_2O$. An electric field of roughly 870 V/m was applied.

Functionalized PMMA particles were prepared by hydrolyzing 5mL of 8.3% (wt/vol) PMMA in acetone with 0.5mL of 0.01M aqueous NaOH. The solution was then evaporated in a scintillation vial and dried at 50°C. Upon drying, the resulting solid polymer was broken into particles and contacted with a solution consisting of 4.2×10^{-4} M $CaCl_2$ and 2.5×10^{-4} M Na_2HPO_4 at pH 9, which is more dilute than the concentrations used to prepare the sample shown in Figures 9 and 10. The resulting precipitates, shown in Figure 11, were more uniform is size than those in Figures 9 and 10, close to 0.1μm, but more heavily aggregated (the tear visible in Figure 11 is an artifact due to beam damage). The apparent degree of supersaturation in this test also appears in Table 1. The similarity between the precipitation morphologies of these two different samples in different solutions reflect the fact that the actual degree of supersaturation at the nucleation sites may have been similar, despite the differences in apparent degrees of supersaturation, because ions had to migrate into the pores of the bulk sample shown in Figures 9 and 10.

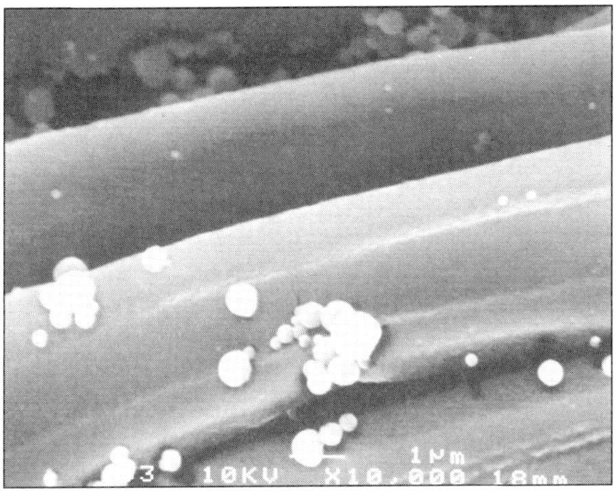

FIGURE 10 Spherical precipitates formed on functionalized PMMA sample. Solutions consisted of 0.0063M $CaCl_2*2H_2O$ and 0.026M $Na_2HPO_4*7H_2O$. An electric field of roughly 870 V/m was applied.

TABLE 1 Solubility products of some common calcium phosphate phases, along with the degrees of supersaturation during the precipitation of the samples shown in Figures 9 and 10, and the sample shown in Figure 11

Phase	Formula	K_{sp}	$\Omega = IP/K_{sp}$ (Figures 9 and 10)	$\Omega = IP/K_{sp}$ (Figure 11)
Octacalcium phosphate (OCP)	$Ca_4H(PO_4)_3$	2×10^{-49}	1.2×10^{12}	4.3×10^{-6}
Tricalcium phosphate (TCP)	$Ca_3(PO_4)_2$	2.83×10^{-30}	8.8×10^{8}	1.3×10^{-4}
Hydroxyapatite (HA)	$Ca_5(PO_4)_3OH$	2.34×10^{-59}	4.2×10^{21}	65

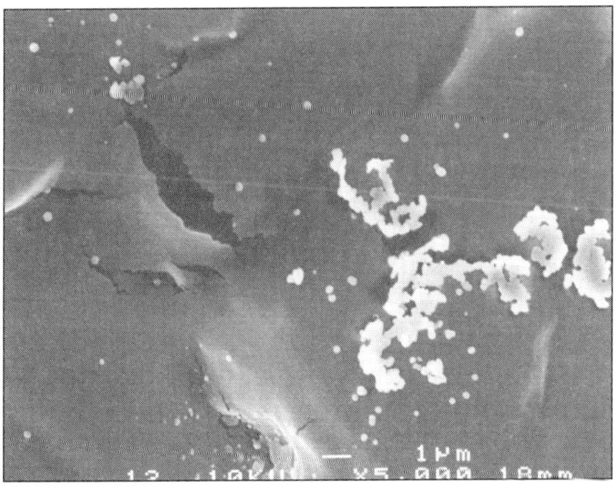

FIGURE 11 Spherical precipitates formed on the surface of functionalized PMMA in a solution consisting of 4.2×10^{-4}M $CaCl_2*2H_2O$ and 2.5×10^{-4}M $Na_2HPO_4*7H_2O$

Supersaturated concentrations, such as those used in the precipitation experiments, would be expected to lead to precipitation of calcium phosphate particles. Hence it is possible that precipitation could also occur within a non-functionalized matrix. However, a functionalized matrix is more likely to result in a chemical bond between the inorganic particles and polymeric matrix, which would be expected to improve the mechanical properties of the resulting composite, even at low volume percentage of ceramic particles.

The composite microstructure revealed by SEM would not be expected to have good mechanical properties, because of the high pore volume, although it did have structural integrity. Work is in progress to address this difficulty, using heat-induced densification either before or after precipitation. Furthermore, the low-density material prepared by this novel route may have many useful properties.

CONCLUSIONS

It appears to be possible to precipitate sparingly soluble, inorganic salts within a polymeric matrix, using electrical migration to accelerate the transport of ions into pores of the matrix. Precipitation is aided by the presence on the polymeric matrix of functionalized groups capable of interacting with one of the ions of the precipitating salt. UV-visible spectroscopy and zeta potential measurements show that the functionalized matrix exhibits increased binding ability over unmodified PMMA for both Cu^{2+} and Ca^{2+} ions.

REFERENCES

Bréchet, Y, J.Y.Y. Cavaille, E. Chabert, L. Chazeau, R. Dendievel, L. Flandin, and C. Gauthier. 2001. Polymer Based Nanocomposites: Effect of Filler-Filler and Filler-Matrix Interactions. *Advanced Engineering Materials*. 3(8): 571–578.

Hwu, J.Y.H.M., G.J. Jiang, Z.M. Gao, W. Xie, and W.P. Pan. 2002. The Characterization of Organic Modified Clay and Clay-Filled PMMA Nanocomposite. *Journal of Applied Polymer Science*. 83: 1702–1710.

Ignjatovic, N., M. Plavsic, and D. Uskokovic. 2000. Hydroxyapatite/poly-L-lactide (collagen) biocomposite with poly-L-lactide of different molecular weights. *Advanced Engineering Materials*. 2: 511–514.

Kramer, C., and L. Dunsch. 1998. Voltammetric and spectroscopic investigation of binding in complexes of divalent metal ions with styrene maleic acid copolymer and monomeric analogues. *Electrochimica Acta*. 44, 819–829.

Liu, Q., J. de Wijn, and C. van Blitterswijk. 1997. Nano-apatite/polymer composites: Mechanical and physicochemical characteristics. *Biomaterials*. 18: 1263–1270.

Mann, S. 1996. *Biomimetic Materials Chemistry*. New York: VCH.

Privalko, V.P., V.M. Karaman, e.g., Privalko, R. Walter, K. Friedrich, M.Q. Zhang, and M.Z. Rong. 2002. Structure and thermoelasticity of irradiation grafted nano-inorganic particle filled polypropylene composites in the solid state. *Journal of Macromolecular Science–Physics*. B41: 487–505.

Shinzato, S.T. Nakamura, K. Ando, T. Kokubo, and Y. Kitamura. 2002. Mechanical properties and osteoconductivity of new bioactive composites consisting of partially crystallized glass beads and poly(methyl methacrylate). *Journal of Biomedical Materials Research*. 60: 556–563.

Spanos, N., and P.G. Koutsoukos. 2001. Hydroxyapatite precipitation on a carboxylated vinyl acetate copolymer. *Journal of Materials Science*. 36(3): 573–578.

van Zyl, W.E., M. Garcia, B.A.G. Schrauwen, B.J. Kooi, J.T.M. De Hosson, and H. Verweij. 2002. Hybrid polyamide/silica nanocomposites: Synthesis and mechanical testing. *Macromolecular Materials and Engineering*. 287: 106–110.

Method for Characterizing the Wettability of Filler Particles

S.C. Brown*, R.C.G. Oliveira†, and B.M. Moudgil*

Surface wettability, or contact angle, dictates a number of multiphase ensemble interactions that influence particulate dispersability and adhesion behavior (e.g., filler compatibility) in addition to a number of other phenomena including biocompatibility and bulk powder segregation. The wettability of particulate surfaces significantly impacts nearly all process industries; however, the available techniques to access this quantity are either relative, or tedious–and are application or industry specific. In this study, key features of existing techniques have been combined to design an improved method to delineate particle wettability that is simple, versatile and capable of providing reproducible measurements in an industrial environment.

INTRODUCTION

The performance of filler particles is influenced by their surface wettability, or the contact angle that the particulate surface makes with the surrounding medium. This fundamental parameter dictates both the dispersive and the adhesive behaviors of particulates within a system, thereby influencing ensemble parameters such as bulk modulus, opacity, rheology, thermal resistance and electrical conductivity. Although particle wettability has a significant impact in nearly all process industries, the experimental techniques for accessing this quantity are either relative, application specific, tedious or limited to well-defined geometries. Hence, particle contact angles are typically not readily accessible in the industrial environment. It is believed that availability of wettability data can lead to significant improvements in process performance and quality control.

When a filler particle is introduced into a liquid environment the effective contact angle made at that particle's surface will determine both the time that it will take to wet the particle's surface and the extent at which this will occur. The use of mineral filler

* Dept. of Materials Science and Engineering, Engineering Research Center for Particle Science and Technology, University of Florida, Gainesville, Fla.

† Petróleo Brasileiro S/A–CENPES/Pentrobras, Rio de Janeiro, Brazil.

particles often requires that moisture or air pockets be removed from the particle surfaces during the dispersion process (Solomon and Hawthorne 1991). Variations in filler particle wettability can lead to unsatisfactory product performance not only by inadequate dispersion but also through unsatisfactory adhesion within the melt, matrix or surrounding environment.

The significance of particle wettability has spurred the development of a number of techniques for its interpretation and measurement. The most common of these methods—the capillary penetration (Washburn 1921; Bartell and Osterhof 1927) and the tablet formation techniques (Zografi and Tam 1976)—attempt to constrain an ensemble of particles into either a porous or a solid-like structure, respectively, to which the test liquid is applied. The capillary penetration methods involve packing the powder of interest into a porous plug that is partially immersed into a test liquid of known surface tension and density. The capillary rise or depression of the test liquid is monitored and correlated to an average particle contact angle through a variation of the Laplace equation by modeling the packed bed of particles as a bundle of capillaries. Significant artifacts and a lack of reproducibility occurs due to ill-defined and varying capillary structures that are formed by this technique along with further complications that arise from the mobilization and the reorganization of the particles with the advancing and the receding liquid fronts. The tablet formation technique avoids these issues by consolidating the powder of interest into a tablet onto which the contact angle is measured via a macroscopic flat plate routine—usually the sessile drop or captive bubble methods. However, equally severe errors may arise since the morphology, roughness, and chemical composition of the particles surfaces are often modified during tablet preparation, causing these surfaces to be non-representative of the primary particles. In both of these methods major errors arise from inadequacies of the multi-particle immobilization process.

Due to the inadequacies of the direct techniques above, a number of indirect methods have been devised. The film flotation technique (Fuerstenau and Williams 1987; Marmur et. al 1986) has been extensively used to rank particulate products by their relative hydrophillicities. In this method a known mass of particles are spread at the liquid-vapor interface and the number or the mass of particles that becomes engulfed by the bulk liquid as the surface tension is decreased is recorded. The particle's wettability is ranked either by critical surface tension (the surface tension at which the largest mass of particles sinks into the liquid phase) or by the maximum surface tension at which the complete wetting of all particles occurs (no particles present at the interface). Normally water is titrated with either methanol or ethanol to achieve the desired surface tension intervals. The results are frequently reported in terms of the percent titrant concentration required for the critical or necessary surface tension. Although this method is reproducible the values are only relative in nature and the method is limited to particles with apparent solid-vapor surface tensions in the approximate range of 20–72 mN/m. The sedimentation volume technique (Vargha-Butler et al. 1985; Omenyi et. al 1981) employs a similar approach. In this case the state of aggregation of particles dispersed in a series of solutions is investigated. The method assumes purely van der Waals forces between identical particles in which the interactions are governed by the free energy of cohesion. When the sedimentation volume is at an extremum the particle's surface tension is assumed equal to that of the suspending medium. Theoretically this method is only valid for pure liquids; however, in practice liquid mixtures are used to access a larger range of surface tensions. The results from this technique can be quantitative; however, artifacts arising from non van der Waal interactions or the presence of unlike particles (particle mixtures) can give misleading results.

Oil and water vapor absorption/adsorption studies have also been employed to indirectly determine particle wetting characteristics (Solomon and Hawthorne 1991; Buckton, Beezer and Newton 1986). In these measurements the particles are ranked based on their ability to absorb/adsorb a nonpolar organic compound or water vapor, respectively, on a per mass basis. The extent at which this occurs largely depends on the particles size, morphology and internal pore structure in addition to their surface energetics. Therefore the relative wettability of these particles can only be found after additional tests are preformed. Similarly, heat of immersion studies (Good and Girifalco 1958) have also been used to indirectly measure particle contact angles. In these methods the heat evolved per square centimeter of powder immersed in a liquid is directly measured and theoretically related to the average particle contact angle. The method requires that the specific surface area of the powder is known, and no other sources of enthalpy must be present (e.g., contributions from the partial dissolution of the particles). Yet, due to the temperature dependence of contact angles, heat of immersion data will normally only provide relative and semi quantitative information (Li and Neumann 1996). In general, indirect techniques require a series of time-consuming tests to achieve meaningful data and the technique used typically varies with industry and with particle application.

Recently, a number of advanced techniques for directly measuring particle contact angles through the incorporation of devices such as Langmuir troughs (Clint and Taylor 1992; Clint and Quirke 1993), Sheludko cells (Hadjiiski et al. 1996), and atomic force microscopes (Ducker, Xu, and Israelachvili 1994) have been developed. Surface pressure versus trough area isotherms generated by Langmuir troughs have been used to determine advancing and receding contact angle on monodispersed, well-defined powders distributed at the liquid-vapor interface. In these experiments the trough area is incrementally decreased which subsequently causes the surface pressure to increase due to the closer packing of the particulates at the interface. At a critical trough area the layer of floating particles becomes close-packed and further reductions in the trough area result in the particles (one-by-one) moving out into either the vapor or liquid phase. By calculating the free energy required to squeeze out these particles (upon an assumed close-packed structure) the advancing or the receding particle contact angle is found depending on whether the particles move into the liquid or vapor phases, respectively. Unfortunately, this technique is limited to particulate systems of well-defined geometries and size distributions for accurate results. By injecting dilute dispersions of particles into a Scheludko cell and subsequently removing part of the liquid in order to trap the particles in a liquid film, Hadjiiski and coworkers have measured contact angles on single particles. This film trapping method uses the interference patterns formed from monochromatic reflected light to numerically reconstruct the meniscus formed around a particle–from which the contact angle is derived. However, since this is an optical method, it is limited to micron-sized particles of well-defined geometries. The atomic force microscope (AFM) has also become a useful tool for measuring the contact angle on single particles. Through the used of an inert adhesive, a representative particle is immobilized onto the tip of an AFM cantilever to form a colloid probe. This colloidal probe is then used in a liquid cell to interact with a millimeter-sized confined bubble. The equilibrium distance that the particle jumps into the bubble is used to geometrically calculate the receding angle; whereas the advancing angle is calculated from the maximum force required to pull the particle out of the bubble and back into the liquid phase. As with the film trapping technique, micron-sized particles of well-defined geometries are required also for these experiments. Unfortunately, these advance methods are tedious, limited to particles of well-known geometries and must be performed in more-or-less ideal environments for adequate precision—therefore they have not been widely adopted for industrial use.

In this work, key features of the existing techniques for particle wettablity have been combined in order to create an improved method to access apparent particle contact angles. As mentioned earlier, the complications that arise in the capillary penetration and tablet formation experiments are primarily due to the inadequacies of the particle immobilization process. This can be circumvented by utilizing inert adhesives—as found in the colloidal probe method—to form a uniform coating of particles around a macroscopic body. Hence, the direct measurement of contact angles on the particles of interest by one of the well-developed macroscopic surface contact angle techniques—such as the sessile drop, external capillary rise, or one of the axisymmetric drop shape analysis (ADSA) techniques—can be conducted. This study examines the applicability of measuring the maximum external capillary height found on an inert particle-coated rod partially immersed in a large liquid well as a simple method for precise particle contact angle measurements that can be performed in an industrial environment.

EXPERIMENTAL

Materials

The poly methyl methacrylate (PMMA) beads (~100 μm) and the glass rods were obtained from Polysciences Inc. and Fisher Scientific, respectively. To form irregular glass particle the glass rods were crushed by mortar and pestle and the resulting powders were classified by sieving. The macroscopic PMMA rod surfaces were created by carefully coating the glass rods with an even layer melt-phase PMMA formed from the beads discussed above. The sodium dodecyl sulfate used in this study was of 99% purity as obtained from the Aldrich Chemical Co, whereas the Aerosil R-972 particles were obtained from Degussa. The acrylate adhesive used in all experiments was obtained from the 3M Corporation. Water used in this study was produced by a Millipore purification system and had an electrical resistance greater than 18 mega ohms and a carbon content of less than 7 parts per billion.

Methods

Particle Immobilization Procedure. After thorough cleansing, a glass rod was coated with a planar acrylate adhesive to which the test particles were applied. Mechanically instable particulates were subsequently removed by agitation to leave a uniform bed of particulates coated onto the rod as depicted in Figure 1.

Measurement of Contact Angles. The image of the particle-coated rod as it is vertically immersed into the test liquid was captured with the setup shown in Figure 2. The height of the advancing meniscus with reference to the level of the liquid far from the surface was measured and correlated to the apparent contact angle through the solution of the Laplace equation for the height of an external meniscus on vertical rod partially immersed into an infinite liquid well (Huh and Scriven 1968). All measurements were preformed a minimum of three times with the same rod diameter, each with a freshly coated rod surface. Examples of the images obtained by this method are found in Figures 3a and 3b for hydrophobic and hydrophillic particle-coated rods, respectively.

FIGURE 1 Optical microscope image of a rod coated by a layer of crushed glass particles

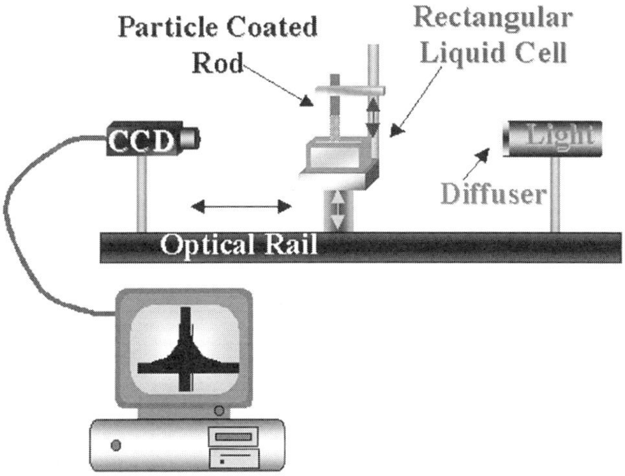

FIGURE 2 The experimental setup used to image the maximum height of the external meniscus formed between a test liquid and a rod

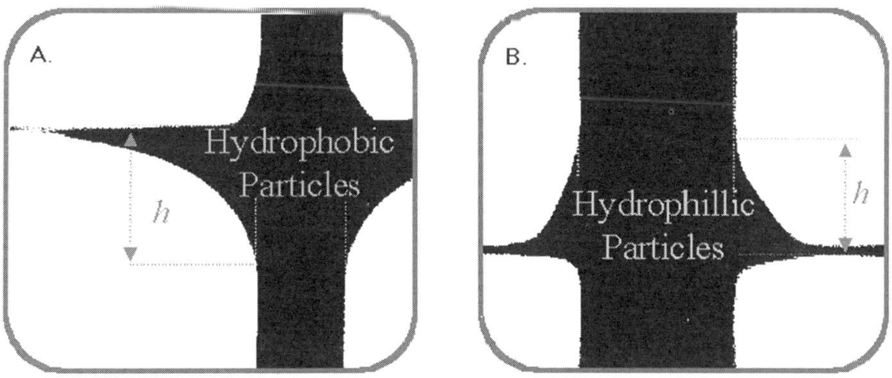

FIGURE 3 Example images captured by the experimental setup presented in Figure 2. A) A hydrophobic particle-coated rod. B) A hydrophilic particle-coated rod.

TABLE 1 Contact angle comparison between particle-coated rods and smooth rod surfaces of the same material with DI water as the test liquid

	Smooth Surface	Particle-coated Rod
Glass	23.8° ± 1.1°	23.5° ± 1.8°
PMMA	72.8° ± 0.7°	72.3° ± 1.1°
Tape adhesive (Scotch®)	77.2° ± 0.8°	—

TABLE 2 Effect of the particle size on the apparent contact angle of crushed glass particles

Sieve Cut(μm)	Apparent Contact Angle
<106	23.5° ± 1.8°
106–300	23.0° ± 1.9°
300–425	23.5° ± 3.25°
425–580	27.7° ± 4.9°

RESULTS AND DISCUSSION

The effect of chemically dissimilar domains and surface topography (roughness) has been studied extensively in the past. Through the works of Cassie (1948) and Wenzel (1936) the basic relations for accounting for chemical inhomogenities and surface roughness have been established. The coated rods in this study may be modeled as chemically and topographically heterogeneous surfaces. Hence, the apparent contact angles measured on these surfaces result from contributions from the particles surfaces, the exposed areas of adhesive, and the liquid or vapor filled voids that may be present at the surface. To test the influence of these contributions on the phenomenological contact angles obtained by this method, measurements were made with both particulate and macroscopic rod surfaces of PMMA and glass, using DI water as the test liquid. In both cases the deviation in the results from the different forms of the same material was within the error of the method–in the range of 1 to 2 degrees as shown in Table 1. From these preliminary experiments it appears that the apparent contact angles resulting from this technique almost exclusively represents the particles of interest, possibly due to their dominating the exposed surface areas. However, due to the irregularity of the particles used in this study this claim cannot be fully validated and further investigations with ideal particles are required to determine actual contributions from the underlying adhesive layers, surface morphology, and liquid or vapor voids.

The effect of particle size on the apparent contact angles measured by this technique was also investigated. Different size fractions of crushed glass particulates were immobilized on similar rods and the contact angles with DI water were acquired. The resulting data suggests a significant loss of precision as the particle size is increased from less than 106μm to approximately half a millimeter in size as shown in Table 2. This phenomena is thought to be attributed to the pinning of the contact line between the large voids that exist between the larger particles in addition to the added difficulty of accurately measuring the height of a thin meniscus on a highly corregated irregular surface (the surface features on the rod become evident for particles above 300μm). However, in all cases the measured contact angles were essentially the same.

The application of this technique to dynamic systems was also established. Figure 4 depicts the results from dynamic wetting studies on hydrophobic, irregular 16nm Degussa Aerosil particles exposed to various concentrations of SDS. In the absence of

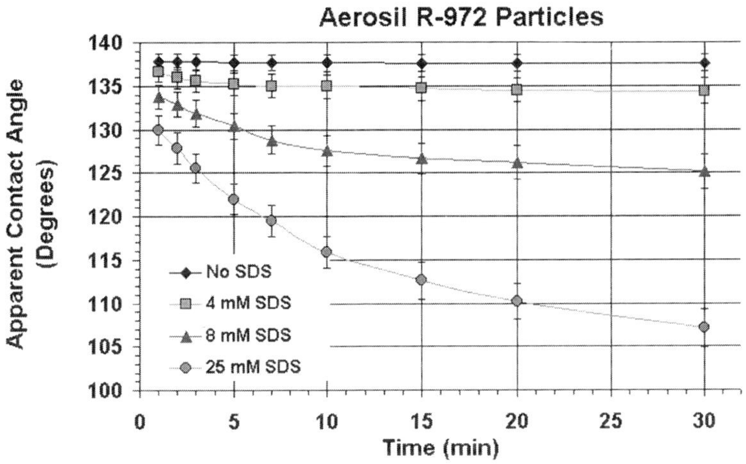

FIGURE 4 Dynamic contact angle experiments between a Aerosil R-972 particle coated rod and DI water with various concentrations of sodium dodecyl sulfate (SDS)

surfactant the contact angle is shown to be constant with time. This verifies that the acrylate adhesive does not dissolve or migrate to the particle surfaces within the time-frame of the experiments. As the SDS concentration is increased, the starting contact angle (t = 1min) is subsequently decreased as expected. The presence of the surfactant lowers the surface tension of liquid medium and allows for a larger quantity of monomer to hydrophobically adsorb at solid-liquid interface. The results of these studies show that this method is not only effective for dynamic measurements, but also precise and applicable to nanoparticle systems–with the largest error just exceeding ±2° in the absence of sophisticated thermal and vibration isolation systems.

CONCLUSIONS

It has been demonstrated that simple, precise measurements of apparent particle contact angles can be achieved through determining the maximum external meniscus height formed between a test liquid and a particle-coated rod. The method employed in this study can determine phenomenological static and dynamic contact angles on particles ranging from a few nanometers to several hundred microns in size. The technique is best suited for fine particles and could be employed in an industrial setting for quick measurements of particle contact angles.

ACKNOWLEDGMENTS

The authors acknowledge the financial support of the Engineering Research Center (ERC) for Particle Science and Technology at the University of Florida, The National Science Foundation (NSF) (Grant EEC-94-02989), and the Industrial Partners of the ERC for support of this research. One of the authors (R.C.G.O) also acknowledges the financial support of Petróleo Brasileiro S/A—CENPES/Petrobras.

REFRENCES

Bartell, F.E., and P.M. Osterhof, 1927. Determination of the wettability of a solid by a liquid. *Ind. Eng. Chem.* 19:1277.

Buckton, G., A.E. Beexer, and J.M. Newton, 1986. A vacuum microbalance technique for studies on the wettability of powders. *J. Pharm. Pharmacol.* 38:713.

Cassie, A.B.D., 1948. Contact Angles. *Disc. Faraday Soc.* 3:11.

Clint, J.H., and O. Quirke, 1993. Contact angle of particles from Langmuir trough studies. *Colloids and Surfaces*, 78:277.

Clint, J.H., and S.E. Taylor, 1992. Particle size and interparticle forces of overbased detergents. A Langmuir trough study. *Colloids and Surfaces*, 65:61.

Ducker, W.A., Z. Xu, and J.N. Israelachvili, 1994. Measurements of hydrophobic and DLVO forces in bubble-surface interactions in aqueous solutions. *Langmuir*, 10:3279.

Good, R.J., L.A. Girifalco, and G. Kraus, 1958. A theory for estimation of interfacial energies. Application to surface thermodynamics of teflon and graphite. *J. Phys. Chem.* 62:1418.

Hadjiiski, A., R. Dimova, N.D. Denkov, I.B. Ivanov, and R. Borwankar, 1996. Film trapping technique: Precise method for three-phase contact angle determination of solid and fluid particles of micrometer size. *Langmuir,* 12:6665.

Huh, C., and L.E. Scriven, 1968. Shapes of axisymmetric fluid interfaces of unbounded extent. *J. Colloid Interface Sci.* 3:323.

Li, D., and A.W. Neumann, 1996. *Applied Surface Thermodynamics.* Surfactant Science Series, Vol. 63. Edited by A.W. Neumann, New York: Marcel Dekkar, Inc. 516.

Omenyi, S.N., R.S. Snyder, D.R. Absolom, C.J. van Oss, and A.W. Neumann, 1981. Effects of zero van der Waals and zero electrostatic forces on droplet sedimentation *J. Colloid Interface Sci.* 81:402.

Mechanical Properties of Steric Dispersant Layers

Ivan U. Vakarelski*, Yakov I. Rabinovich*, and Brij M. Moudgil*

Polymer and surfactant dispersant are widely used to disperse filler particles within a medium. Surfaces active dispersants form self-assembled structures at the solid-liquid interface. The mechanical properties of the surfactant structures have been determined to strongly influence the dispersion behavior. In this study the atomic force microscope (AFM) is used to determine the elastic modulus and yield strength of a steric dispersant layer adsorbed to a rigid non-deformable surface. To interpret these values a model for elastic body contact based on a modified Hertz theory is applied. The results from this method may aid in the design of application-based dispersant systems.

INTRODUCTION

The state of dispersion of aqueous phase filler particles has a significant impact on the performance and quality of a number of composite slurries and suspensions, for instance paints, pigments, and paper coatings. Polymer and surfactant additives are often added to induce stability in these systems due to their ability to prevent aggregation under a wide range of solution and environmental conditions. It has been well established that both polymeric dispersants and surfactants mediate stability due to a steric force that results from the physical presence of surfactant aggregates or polymers at the interface. However, little work has been done on deciphering the mechanical properties of these steric dispersant layers, which may give rise to the fundamentals of how to control the extent, lifetime, and overall robustness of dispersions through nano- to molecular level modifications. Numerous studies have probed the forces between an atomic force microscope (AFM) tip (or colloidal probe) and adsorbed dispersant layers and have reported on the extent of the overall repulsion and the component interactions with respect to the adsorbed moieties and solutions conditions. Yet, rarely is the modulus of the adsorbed

* Dept. of Materials Science and Engineering, Engineering Research Center for Particle Science and Technology, University of Florida, Gainesville, Fla.

layer discussed, nor are methods presented to engineer these adsorbed layers to facilitate applications under high shear or high loading conditions.

Over the last decade the AFM has been established as a powerful tool for the direct measurement of interaction forces between surfaces at the nanoscale. In addition, there have been a number of advances in using the AFM for the measurement of the elastic response of deformable surfaces such as viscoelastic polymers (Aime et al. 1994; Domke and Radmacher 1998; Vakarelski et al. 2001), or biological samples (Tao, Lindsay and Lees 1992; Weisenhorn et al. 1993; Rotsch and Radmacher 1997). In these investigations the surfaces are compressed by indenting AFM tip into the sample and the resulting force curves are interpreted to investigate the substrate's elastic properties. A similar approach can be applied to surfaces with adsorbed surfactant aggregates, however, in this case, the deformable layer is only several nanometers thick and the indenting tip will "feel" the underlying substrate. An attempt has been made in the present study to use atomic force microscopy to investigate these interactions and these finite layer effects. In particular, a model system of an adsorbed self-assembled surfactant bilayer on top of an atomically smooth, rigid, non-deformable interface is investigated. To calculate the elastic modulus of the adsorbed surfactant layer, a modified Hertz theory is employed (Shull et al. 1998).

EXPERIMENTAL

Measurements were preformed with a Digital Instruments Nanoscope III AFM in a fused quartz liquid cell. The procedure for force measurements and typical force-distance curves in the presence of self-assembled surfactant aggregates have been published earlier (Adler et al. 2000). Forces were measured between a muscovite mica substrate and the AFM silicon nitride tip (NP-S DI) with a spring constant of 0.11 ± 0.1 N/m as given by the resonance frequency method (Cleveland et al. 1993). The radius of the tip used in the experiments was 15nm as determined by scanning electron microscopy (SEM). Octadecyl trimethyl ammonium bromide (C_{18}TAB) of 99% purity was obtained from Aldrich Chemical Co. The muscovite mica was freshly cleaved prior to each experiment and the AFM tips were subjected to UV treatment for approximately 30 minutes in order to remove any organic contaminants. The water used in the experiments was produced by a Millipore filtration system and had an internal specific resistance of 18.2 MΩ.

All measurements were repeated using several tips and the results exhibited similar interaction forces (±10%), which demonstrates the validity and the precision of the experiments. In addition, the force curves obtained in one experimental series (i.e., with the same tip and cantilever) for a given solution were found to have identical slopes with a variance of ±5% in the magnitude of the jump-in force.

THEORY

Figure 1 schematically illustrates the model employed to consider the interaction of the AFM tip and the adsorbed surfactant layer. To simplify the mathematical treatment of this phenomenon the geometry of the AFM tip is idealized as a rigid sphere, where as the surfactant bilayer is modeled as an initially planar homogeneous elastic layer on top of a rigid substrate. R represents the radius of the tip and h_0 is the undisturbed adsorbed layer thickness. As the AFM tip is pressed against the surfactant layer with a force F, it contacts an area of radius a and has a penetration depth δ. From the classical Hertz theory (Hertz 1896), the relationship between F and a for the case of a rigid sphere and an infinite half space of elastic (Young's) modulus E and Poisson' ratio ν is found to be:

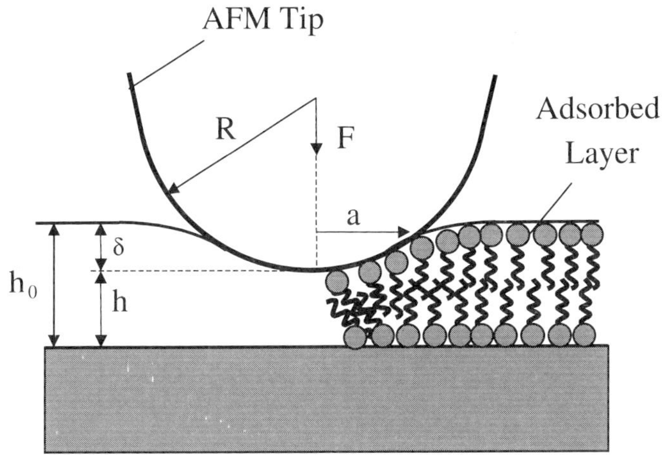

FIGURE 1 Model of interaction of a hard sphere (tip) with a soft flat layer (adsorbed surfactant bilayer)

$$F^H = \frac{Ka^3}{R} \qquad \textbf{(EQ 1)}$$

where $K = 3E/2(1 - \nu^2)$. Similarly, for δ and a:

$$\delta^H = \frac{a^2}{R} \qquad \textbf{(EQ 2)}$$

The exponent H is used to denote the use of Hertz theory. From Equation 1 and Equation 2, F^H can be derived to be

$$F^H = KR^{0.5}\delta^{H1.5} \qquad \textbf{(EQ 3)}$$

The normal stress distribution, p along the contact zone, as given by Hertz theory is:

$$p(r) = \frac{3Ka}{2\pi a_m}\left(1 - \left(\frac{r}{a}\right)^2\right)^{1/2} \qquad \textbf{(EQ 4)}$$

where r is the radial coordinate (Figure 1). From Equation 1, Equation 2 and Equation 3 the yield stress of the layer, Y, can be obtained. Taking $Y = p(0)$ for the point where the layer yields:

$$Y = \frac{3F_m}{2\pi a_m^2} \qquad \textbf{(EQ 5)}$$

where F_m and a_m are the force and the contact radius at the point of rupture of the layer.

For the case of a layer with finite thickness, the Hertz theory is valid only if $a/h_0 \ll 1$. When this condition is not satisfied the tip experiences the influence of the rigid bottom layer. To account for this phenomenon, Shull et al. (Shull, Ahn and Mowery 1997; Shull et al. 1998) proposed a semi-empirical correction term to the Hertz equation. The correction factor for the case of a thin elastomeric layer is given to be (Shull et al. 1998):

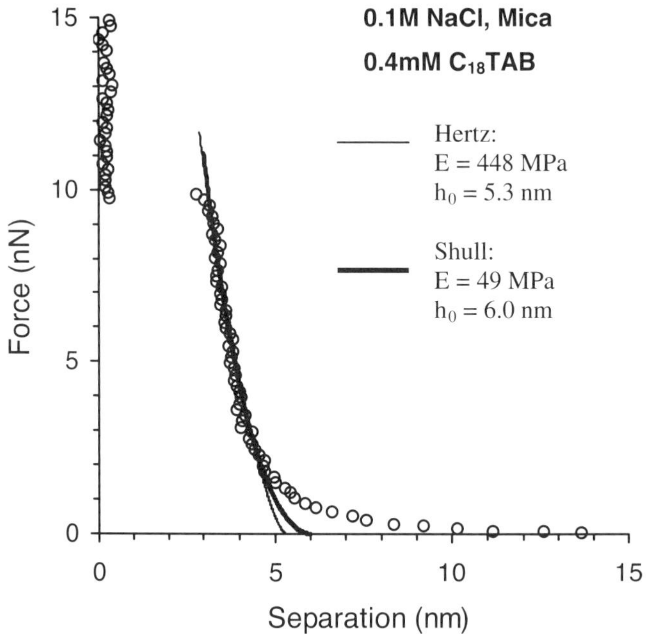

FIGURE 2 Experimental force/distance curve (circles) measured between an AFM tip and C_{18}TAB bilayer on silica substrate in 0.1 M NaCl solution of 0.4 mM C_{18}TAB. The thin solid curve is best fit of the data using Hertz model and the thick solid line is best fit with Shull's theory.

$$\delta = \delta^H(0.4 + 0.6\exp(-1.8a/h_0)) \quad \text{(EQ 6)}$$

And for the case of a frictionless contact, the force is given by:

$$F = F^H(1 + 0.15(a/h_0)^3) \quad \text{(EQ 7)}$$

where F^H and δ^H are as defined by Equation 1 and Equation 2 respectively. At small values of a/h_0, the Shull's equations converge to the Hertz theory. Through the use of a perturbation solution and finite element method simulation Hui et al. (2000) have shown that Shull's correction terms are accurate. They also found that Hertz stress distribution in the contact zone is nearly exact, even for moderately large contact radii. Therefore by using Shull's correction factors, the yield stress Y for higher values of a/h_0 can be calculated through Equation 5.

RESULTS AND DISCUSSION

Figure 2 represents a typical experimental force/distance curve (F vs. h) for the interaction between the AFM tip and a self-assembled C_{18}TAB bilayer on a mica substrate. This curve was generated at a concentration of 0.4 mM C_{18}TAB (two times the CMC of the bulk surfactant with no salt added) in order to ensure that a fully developed bilayer is present at the solid-liquid interface (Li et al. 1998). Additionally, 0.1M NaCl background electrolyte was used in order to shield the electrostatic interactions between the AFM tip and the surfactant aggregates. This allows for the interaction force/distance curves presented in Figure 2 to be taken to represent mechanical interactions of the surfactant bilayer in entirety. The break in the curve that occurs between 3–4 nm represents the disintegration of the surfactant structures as the AFM tip jumps into contact with the substrate. In this study the mechanical response of these structures prior to their destruction is of interest. To test the elasticity of the adsorbed bilayer, several experiments were preformed in which the bilayer was pressed and released at different rates (from 10 nm/s to 1000 nm/s). The generated force-distance curves showed no hysterisis (in the loading regime prior to layer destruction) between the approaching and retracting curves, and the rate of approach/retraction—which spanned three orders of magnitude—did not modify the results. These observations indications indicate that the bilayer may be correctly modeled as a purely elastic material in the timescale of these experiments.

The thick solid curves in Figure 2 represent the results obtained by Shull's method as discussed previously, whereas the thin curves are those found through the use of the Hertz theory. In the case of the Hertz model, the theoretical curve is plotted using Equation 3, whereas for Shull's method Equation 6 and Equation 7 are solved numerically by giving incremental values to the contact radius a. In both cases the best-fit theoretical curve yields an elastic modulus E and thickness h_0. However, the elastic modulus in the case of Hertz is one order of magnitude higher than that obtained using Shull's theory (448MPa and 49MPa respectively). The reason for this variation, as mentioned earlier, is that the Hertz fit represents the "effective" value corresponding to the elastic modulus of an infinite half space that gives the same elastic response as the existing two-layered system in this study. Because of the physical reality of the system the value obtained using Shull's method is a better estimate of the true elasticity of the aggregate layer. The elastic modulus and the layer thickness for the bilayer from each of these treatments are shown in Figure 2. In both cases the elastic modulus is calculated from K using $\nu = 0.5$, which is the average value for an elastomeric material (Sperling 1992). The yield stress values, $Y =$ 48MPa, was calculated from the Shull's model using Equation 5 in which F_m is the average experimental jump-in value, and a_m the corresponding value satisfying Equation 7. The elastic modulus and the yield strength obtained from Shull's method seems reasonable and comparable with rubber-like materials (E = (10 – 100) MPa (Sperling 1992)).

In a complementary study self-assembled micellar structures have also been investigated at the solid/liquid interface of both mica and silica substrates in the same manner. It was found in that study that the elastic modulus and yield strength of the surfactant structures linearly increased with carbon chain length of the surfactant tail and that the structure of the self-assembled aggregates could play a significant role in their mechanical properties. For spherical micelles it was found that the elastic modulus and yield strength were approximately half the value found for the same surfactant in a cylindrical micelle arrangement. It is therefore possible that modifying the dispersant structure and its morphology on a surface, one can control the mechanical properties of the steric structures. These effects can significantly affect the rheology and the efficacy of steric force induced dispersions.

CONCLUSIONS

Steric repulsive forces mediated by surfactants or polymers largely account for the stability of complex slurries and particulate suspensions also used as filler materials. The dispersion of filler particles can significantly impact the quality and performance of the resulting products; hence, a fundamental understanding of the mechanical interactions and the limits of these interactions is essential for the engineering of particulate systems for optimum performance under given process conditions. In this study, it has been shown that a surfactant bilayer can be modeled as a fully elastic material; however, the finite thickness of this aggregate layer must be taken in account. For this reason, a modified Hertz theory (Shull's method) has been applied to determine the elastic modulus (*E*) and the yield strength (*Y*) of the surfactant aggregate layer. The values derived using this procedure were an order of magnitude lower than those found by the Hertz theory alone. This shows that the surface onto which the steric layer is attached contributes significantly to force-distance profile and hence it is "felt" by interacting probe. The values for the elastic modulus and yield strength in the case of Shull's method are comparable to those of the highly elastomeric materials. The results of this study are expected to be helpful in engineering the state of dispersion for targeted performance.

ACKNOWLEDGMENTS

The authors acknowledge the financial support of the Engineering Research Center (ERC) for Particle Science and Technology at the University of Florida, The National Science Foundation (NSF Grant EEC-94-02989), and the Industrial Partners of the ERC for support of this research.

REFERENCES

Adler, J.J., P.K. Singh, A. Patist, Y.I. Rabinovich, D.O. Shah, and B.M. Moudgil. 2000. Correlation of Particulate Dispersion Stability with the Strength of Self-Assembled Surfactant Films. *Langmuir,* 16:7255.

Aime, J.P., Z. Elkaakour, C. Odin, T. Bouhacina, D. Michel, J. Curely, and A. Dautant. 1994. Comments on the use of the force mode in atomic force microscopy for polymer films. *J. Appl. Phys.,* 76(2):754.

Cleveland, J.P., S. Manne, D. Bockec, and P.K. Hansma. 1993. A nondestructive method for determining the spring constant of cantilevers for scanning force microscopy. *Rev. Sci. Instrum.*, 64:403.

Domke, J., and M. Radmacher. 1998. Measuring the Elastic Properties of Thin Polymer Films with the Atomic Force Microscope. *Langmuir,* 14:3320.

Hertz, H. 1896. In *Miscellanious papers*; London: McMillan & Co.

Hui, C.Y., Y.Y. Lin, J.M. Baney, and A. Jagota. 2000. The accuracy of the geometrical assumption in the JKR (Johnson-Kendall-Robertson) theory of adhesion. *J. Adhesion Sci. Technol.,* 14:1297.

Li, B., M. Fujii, K. Fukada, T. Kato, and T. Seimiya. 1998. In situ AFM observation of heterogeneous growth of adsorbed film on cleaved mica surface. *Thin Solid Films,* 312:20.

Tao, N.J., N.M. Lindsay, and S. Lees. 1992. Measuring the microelastic properties of biological material. *Biophys. J,* 63:1165.

Rotsch, C., and M. Radmacher. 1997. Mapping Local Electrostatic Forces with the Atomic Force Microscope. *Langmuir,* 13:2825.

Shull, K.R., D. Ahn, and C.L. Mowery. 1997. Finite-Size Corrections to the JKR Technique for Measuring Adhesion: Soft Spherical Caps Adhering to Flat, Rigid Surfaces. *Langmuir,* 13:1799.

Shull, K.R., A. Dongchan, W.C. Chen, C.M. Flanigan, and A.J. Crosby. 1998. Axisymmetric adhesion test of soft materials. *Macromol. Chem. Phy.,* 199:489.

Singh, P.K., J.J. Adler, Y.I. Rabinovich, and B.M. Moudgil. 2001. Investigation of Self-Assembled Surfactant Structures at the Solid-Liquid Interface Using FT-IR/ATR. *Langmuir,* 17:468.

Sperling, L.H. 1992. *Introduction to Physical Polymer Science*; New York: A Wiley-Interscience Publication. Chapter 8.

Vakarelski, I.U., A. Toritani, M. Nakayama, and K. Higashitani. 2001. Deformation and Adhesion of Elastomer Microparticles Evaluated by AFM. *Langmuir,* 17:4739.

Weisenhorn, A.L., M. Khorsandi, S. Kasas, V. Gotozos, M.R. Celio, and H.J. Butt. 1993. Deformation and height anomaly of soft surfaces studied with an AFM. *Nanotechnology,* 4:106.

SECTION 3

Filled Systems, Dispersion, and Compounding

Interfacial Effects in Polypropylene Block Copolymer/ Magnesium Hydroxide Composites Reactively Processed with 1,3-Phenylene Dimaleimide

C.M. Liauw*, V. Khunova†, G.C. Lees*, and R.N. Rothon*

The interphase modification activity of 1,3-phenylene dimaleimide (also known as m-phenylene bismaleimide (BMI)) has been studied in composites based on magnesium hydroxide (60% w/w) and a polypropylene block copolymer (PPBC) that featured an ethylene-rich elastomeric impact modifier phase. BMI was found to be a unique product that led to significant simultaneous improvements in composite strength and toughness, even at the high filler levels. BMI modification was found to promote encapsulation of filler particles with the elastomeric phase of the PPBC together with formation of a PP boundary layer of increased crystalline content.

INTRODUCTION

Over the past decade fillers have moved a long way from their traditional and historic role as cheap extenders and shrinkage reduction agents for polymeric materials. It is now widely recognised that fillers can be very useful functional additives; a good example being the use of metal hydroxides to impart halogen free flame retardancy (Hornsby and Watson 1986; Rothon 1996).

Highly flammable polymers, such as polyolefins and polystyrenes, can be made flame retardant by addition of organo-halogen compounds combined with antimony oxide. However, use of such flame retardant systems can result in evolution of thick smoke together with highly acidic hydrogen halide gases. Magnesium hydroxide, along

* Centre for Materials Science Research, The Manchester Metropolitan University, Manchester, United Kingdom.

† Dept. of Plastics and Rubber, The Slovak University of Technology, Bratislava, Slovak Republic.

with other materials such as aluminium hydroxide and mixtures of huntite and hydromagnesite, can be very effective halogen free flame retardants and smoke suppressants. As a result of these attractive properties, together with the environmental reasons, the use of these additives (known collectively as flame-retardant fillers) has been steadily increasing (Weber 2000).

The most significant disadvantage of flame retardant fillers is that large amounts (at least 60% w/w or ca. 40% v/v) of the additive are required to attain acceptable flame retardancy. In some instances these filler levels are close to the maximum packing fraction and composite melt viscosity can increase dramatically; this is often accompanied by an equally dramatic reduction in composite mechanical properties. At such high filler levels the interfacial region between the filler and matrix, from henceforth called the interphase, has a dominant effect on composite processing and mechanical properties. Therefore modification of the interphase structure is the key to attainment of the required composite properties (Rothon 1996).

Recent work conducted by the authors has shown the benefits of a single step reactive processing composite formation method in which the reactive interphase modifier is added directly to the premix of matrix and untreated filler. This so called in-situ addition method provides the compounder with much greater flexibility and control. Several multifunctional additives have been investigated for this role (Khunová 1995); however, 1,3-phenylene dimaleimide (or m-phenylene bismaleimide (BMI)) afforded the best results by a significant margin. Unlike the majority of surface treatment systems, particularly the silanes, it has shown that BMI is equally effective with both acidic (zeolites and bentonites (Khunová and Sain 1995)) and basic (calcium carbonate (Khunová and Sain 1995)) fillers. These studies demonstrated that processing temperature is the major determinant of the efficiency of BMI. The results obtained in these studies indicate the potential of BMI for use in highly filled systems such as magnesium hydroxide based polyolefin composites (Khunová, Hurst and Liauw 1999; Khunová et al. 1999; Khunová, Liauw and Alexy 1999; Liauw et al. 2000).

This paper investigates modification of polypropylene block copolymer (PPBC)/ magnesium hydroxide ($Mg(OH)_2$) based composites by reactive processing with BMI. The PPBC used in this investigation features a dispersed elastomeric impact modifier phase that is composed of a random copolymer containing mainly ethylene and propylene. Previous studies on BMI modified PP homopolymer (PPH)/$Mg(OH)_2$ composites have shown that the overall crystalline content of the PPH matrix is not significantly reduced despite noticeable increases in elongational viscosity due to chain branching. The branching is a result of addition of PP macroradicals to the maleimide alkenes. Very little if any true crosslinking occurred due to the fact that PP also undergoes chain scission when under radical attack. However, in BMI modified polyethylene (PE)/$Mg(OH)_2$ composites the crystallisation of the PE matrix was severely hindered by crosslinking. It was therefore considered that a PPBC would undergo reactions that form a very effective interphase as the ethylene rich dispersed elastomeric phase will undergo predominant crosslinking and the continuous PP phase will undergo simultaneous chain branching and chain scission. This combination of competing reactions may lead to crosslinked elastomeric encapsulation of the filler particles, that in turn may lead to composites with high strength and toughness.

In previous studies on BMI modified PP homopolymer/$Mg(OH)_2$ composites, judicious use of a lubricant, based on a blend of fatty acid esters and amides, successfully improved the processing characteristics whilst maintaining the interphase forming activity of BMI and hence much improved composite mechanical properties. The presence of the impact modifier phase is likely to complicate this situation and adsorption

TABLE 1 Coded values for three-factorial experimental design

	Factor	−1.681	−1	0	+1	+1.681	Step
X_1	Temperature (°C)	180	190	205	220	230	15
X_2	BMI (% w/w)	0.29	0.60	1.05	1.50	1.81	0.45
X_3	Lubricant (% w/w)	0.05	0.42	0.96	1.50	1.87	0.54

competition between the lubricant, elastomeric phase and BMI is likely to dominate the effect of the lubricant.

EXPERIMENTAL

Materials

The matrix selected was ICI GXM216 poly(propene) block copolymer (PPBC) based on propene and ethene (12–14% w/w) and contains a general-purpose stabiliser package [measured melt flow rate (MFR) was 7.3 dg min^{-1} (2.16 kg 230°C)]. GXM216 features a dispersed impact modifier phase that consists of ethylene rich poly(ethene-co-propene) and also sufficiently long sequences of poly(ethene) (PE) to enable formation of a crystalline PE phase. DSC analysis of GXM216 therefore yields two phase transitions; one at ca. 110°C corresponding to crystalline phase and the other at 160°C corresponding the predominantly isotactic poly(propene) continuous phase. The PP and PE crystalline contents were ca. 40% and 2%, respectively (based on the approximate masses of the respective comonomers in the PPBC and assuming values of 209 Jg^{-1} and 277 Jg^{-1} for the heats of fusion of perfectly crystalline PP and PE, respectively). The untreated magnesium hydroxide ($Mg(OH)_2$) used was Duhor N, from Duslo Sala [D_{50}, 1.0 μm, specific surface area, 7.8 m^2 g^{-1}] and the 1,3-phenylene dimaleimide (BMI) was obtained in a 90% active form from VUAS (Czech Republic).

Experimental Design

Initial investigations focused on variation of the BMI level (between 0 and 4% w/w), the composites contained no lubricant, only magnesium hydroxide (60% w/w), PPBC (40% w/w − % w/w of BMI) and BMI. Previous studies with BMI modified PP homopolymer based composites have shown that the fatty acid amide and ester based lubricant (Structol Polydis TRO-16) can relieve the problem of high melt viscosity without excessive compromise of composite mechanical properties.

A three factorial rotatable experimental design, described in Table 1, was therefore devised in order to investigate the simultaneous effect of processing temperature, BMI and lubricant concentrations. Five levels of coded variables were investigated and twenty composite formulations were devised by statistical means. Experimental data was evaluated on the basis of analysis of variance and full regression analysis. The composites investigated in this study were based on 60% w/w $Mg(OH)_2$. Experimental data was fitted to Equation 1. All equations determined only incorporate statistically significant coefficients with confidence levels of 95%.

$$Y = B_0 + \sum_{i=1}^{k} B_i x_i + \sum_{i=1}^{k-1} \sum_{j=i+1}^{k} B_{ij} x_i x_j + \sum_{i=1}^{k} B_{ii} x_i^2 \qquad \textbf{(EQ 1)}$$

Composite Preparation

All composites prepared contained 60% w/w $Mg(OH)_2$ which is considered to be an appropriate level for flame retardant applications (Hornsby P.R. 1986). The composite formulations were premixed before melt blending in a Brabender Plasticorder fitted with a W50 chamber and cam blades at a rotor speed of 70 r.p.m. For the first series of composites, where only the BMI level was varied, the mixing temperature was held constant at 220°C, and mixing continued for six minutes. For the second series of composites that were prepared in accordance with a factorial experimental design (Table 1), mixing temperature was varied between 180 and 230°C. Mixing time was held constant at six minutes. All composite samples were compression moulded at 220°C into plaques of 0.5 mm and 3 mm thickness for tensile and impact testing, respectively. The moulding (using frame moulds) was carried out using a pre-heating time of 3 minutes, pressure was applied for 2 minutes, and the mould was then transferred to a cooled press and pressure applied during cooling for 10 minutes.

Melt Flow Rate

Assessment of the melt flow behaviour of the composites, prepared as above, was carried out using an MFR apparatus with a set barrel temperature of 230°C and 10.0 kg load, in accordance with ASTM D 1238.

Mechanical Testing

ASTM D 638M Type M-III tensile dumbbell test pieces were cut from the compression moulded sheets and tensile strength was measured at ambient temperature using a crosshead speed of 50 mm min^{-1}. Unnotched impact strength was measured at ambient temperature using a Charpy impact testing instrument. Test piece dimensions were 3 mm × 10 mm × 70 mm and the span was set at 50 mm.

Analysis of Insoluble Matrix Fraction

In order to explore the possibility of encapsulation of filler particles with the impact modifier phase of the PPBC matrix, selected control composites were treated with refluxing decane in a Soxhlet apparatus to remove the soluble fraction of the matrix. Soxhlet extraction continued for 72 hours, followed by a 24 hour extraction with hexane (to remove the decane). The Soxhlet apparatus was thoroughly lagged such that a temperature of ca. 170°C was maintained in the extraction thimble. The resulting residues were oven dried to constant mass at 110°C and then a small amount was powdered by abrasion with silicon carbide paper.

The powdered composite residue was then diluted to ca. 5% w/w in KBr and analysed by diffuse reflectance Fourier transform infrared spectroscopy (DRIFTS) [Spectra-Tech cell in Nicolet 510P, 150 scans, resolution 4 cm^{-1}]. The presence of any crystalline PP was investigated by differential scanning calorimetry (DSC) [Perkin-Elmer DSC-7, heat-hold-cool-heat-cycle, 30–230°C with five minute hold after the first heat segment, heating rate 20°C min^{-1}, N_2 sample atmosphere]. The amount of insoluble matrix in the residues was then determined by measuring the mass loss on ignition at 1100°C (4 hours) after correcting for dehydration.

The interphase structure of the experimental design series of composites (Table 1) was also investigated by removal of the soluble matrix fraction followed by DRIFTS, DSC and thermal gravimetric analysis (TGA). For this series a more gentle extraction regime was used to minimise degradation and to preserve the interfacial region between the impact modifier phase (if encapsulating the filler) and the continuous PP phase. The composites were extracted by immersion in refluxing xylene for 24 hours. Prior to extraction, 0.5 mm thick compression moulded composite sheets were cut in to ca. 4 mm × 4 mm squares and ca. 1.0 g of these squares were placed in glass cloth bags. After extraction the envelopes containing the xylene insoluble fraction were dried to constant mass at 80°C.

The DSC analysis was then carried out as described above. The insoluble matrix content of the composites was then determined using a Netzch TG109 thermal gravimetric analyser (TGA) (with air purge) by first allowing the samples to equilibrate at 120°C for 5 minutes followed by heating to 900°C at 20°C min^{-1}, the samples were then allowed to equilibrate for 5 minutes at 900°C. The mass loss from the end of the 120°C equilibration period to the end of the 900°C equilibration period was used in calculation of insoluble matrix level, after taking account of mass loss associated with dehydration of pure $Mg(OH)_2$ under identical conditions. The insoluble matrix content was used for determination of crystalline content of the insoluble matrix by DSC analysis.

RESULTS AND DISCUSSION

Effect of BMI Level

The effect of increasing BMI level, whilst keeping mixing time constant at 6 minutes, on tensile strength, elongation to break, impact strength and MFR is presented in Figures 1 (a) to (d). From this data it is evident that tensile strength reaches a limiting value between 2 and 3% w/w BMI, this value is almost 60% greater than the unmodified composite and even exceeds the yield stress of the unfilled matrix. A maximum in elongation, 250% higher than the unmodified composite and greater than the yield strain (7.4%) of the unfilled matrix, is observed over the same BMI range. Impact strength was shown to increase with BMI level within the range investigated; at 4% BMI impact strength was increased by 200% relative to the unmodified composite. This increase in strength together with increase in toughness is likely to be associated with formation of an interphase that is elastomeric in nature but strongly bound to both the filler surface and the bulk matrix.

At this juncture it is worthwhile comparing these results with those obtained in a previous study (Liauw et al. 1998) on similar composites that had been modified by using a range of filler pre-treatments, including a proprietary vinylsilane based combination. In the latter study the best tensile results, at a filler loading of 60% w/w, were obtained using a vinylsilane based filler pre-treatment; tensile strength was 24 MPa and tensile elongation was 2.8%, a figure well below that afforded by BMI modification. $Mg(OH)_2$ based composites modified with maleanised PP (Jančář and Kučera 1990) also showed inferior relative tensile properties to those afforded by BMI modification. The latter two types of surface treatment are considered to be the most significant from a commercial standpoint; the results obtained in this study therefore highlight the superiority of BMI as an interphase modifier and the commercial potential of BMI and similar derivatives.

The MFR data (Figure 1(d)) shows that these noteworthy improvements in properties are counterbalanced by rapidly increasing melt viscosity. However, within the BMI range (2 to 3% w/w) at which the optimum mechanical properties were observed, the reduction in MFR had reached a plateau. Increasing BMI level to 4% w/w significantly

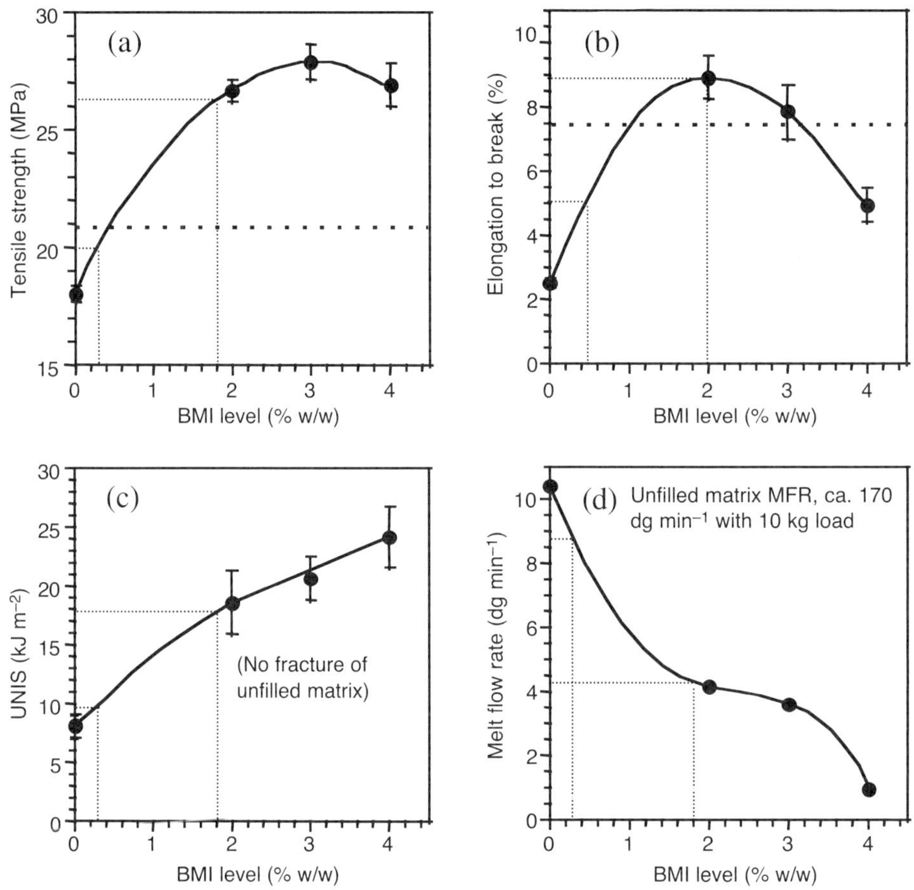

FIGURE 1 Composite properties as a function of BMI level: (a) tensile strength, (b) elongation to break, (c) unnotched impact strength (UNIS), and (d) melt flow rate (with 10 kg load). Long dashed line corresponds to tensile yield data for unfilled matrix. Short dashed lines show the BMI level range used in central composite design.

reduces MFR from 10.6 dg min^{-1} (for the unmodified composite) down to the unacceptable level of 0.96 dg min^{-1}. This data shows that from both rheological and mechanical property standpoints, it is evident that 3% w/w BMI is the optimum practical level in these composites. The rheological performance is clearly the weak point of BMI modification and is in need of improvement, this was attempted by incorporation of a fatty acid amide and fatty acid ester based lubricant (Structol TRO-16). In order to assess the simultaneous effects of the lubricant, processing temperature and BMI level, a central composite design approach was used (Table 1). The regression equation produced from the coefficients was used to construct response surfaces.

Effect of Lubricant Level Processing Temperature and BMI Level

From inspection of the regression coefficients for Equation 1 (Table 2), it is evident that all studied factors had a strong influence on the measured properties. It is clear that addition of lubricant (b_3) has a detrimental effect on tensile response as the values of the

TABLE 2 Statistically significant regression coefficients showing the effect of mixing temperature, BMI level and lubricant level on mechanical and rheological properties

	Tensile Strength	Elongation to Break	Impact Strength	MFR
b_0	19.34	2.79	6.68	9.01
b_1	1.30	0.35	0.87	–0.79
b_2	1.50	0.49	0.23	–5.04
b_3	-2.43	-0.67	-1.08	5.27
b_{11}	-0.74	-0.27	-0.97	-1.29
b_{12}	0.20	0.46	-0.19	-0.97
b_{13}	0.29	-0.14	-0.36	-0.28
b_{22}	-0.73	-0.08	-0.74	0.98
b_{23}	-0.77	-0.21	–0.96	-1.89
b_{33}	1.80	0.44	1.04	3.58

relevant coefficients are negative. Synergy between BMI and processing temperature is evident from Table 1 and is greater than the negative effect associated with lubricant addition. Predictably, the corresponding coefficients for melt flow rate (MFR) showed the reverse effect, i.e., addition of lubricant increased MFR and BMI and processing temperature decreased MFR. The interaction coefficients (b_{ij}, b_{ii}, b_{jj}) are relatively low and thus have little effect on the above trends. The more subtle effects will now be discussed with the aid of appropriate response surfaces generated using the coefficients in Table 2.

Effect on Mechanical Properties. The effect of these three variables on the tensile modulus is shown in Figure 2. It is evident that the modulus drops as the lubricant level is increased. This is principally due to a reduction in filler-matrix interaction caused by preferential adsorption of the lubricant and possibly a reduction in macroradical production arising from reduced mechanical degradation of the matrix. It is the underlying trends, however, that are the most important. At the lowest lubricant level the modulus decreases with increasing processing temperature but increases as the BMI level is increased, at the same time however, the tensile strength increases, i.e., at the BMI level = 1.8% w/w, processing temperature = 229°C line (axis Z-Z) tensile strength is highest but the modulus is relatively low. At the highest lubricant level the modulus increases very slightly with increasing temperature and modulus decreases with increasing BMI level. The tensile strength at this point is also low. These aspects are summarised in Figure 3 that shows the variation in tensile strength and impact strength (at axis Z–Z) versus the lubricant level. The change in modulus across the BMI level = 1.8% w/w plane of the plots (i.e., $E_2 - E_1$ (see Figure 2)) is also shown. It is evident that the maximum strength and toughness coincide with the lowest modulus. This points to elastomeric encapsulation of the filler particles and interesting changes in the interphase structure that are related to the level of lubricant and the competitive adsorption of the lubricant and BMI.

Effect on Melt Flow Rate (MFR). Examination of the contour plots for melt flow rate shows that there are again interesting variations, particularly at the Z–Z axis (Figure 4). The composite MFR goes through a minimum, or rather there is a discontinuity where MFR is effectively zero over a range of lubricant levels. In fact the lubricant reduced the melt flow rate even at very low addition levels, however, the lubricant related effects revealed unique insight into the development of the interphase structure.

The increase in melt viscosity observed (i.e., reduction in MFR) between 0.05 and 0.2% w/w lubricant was likely to be due to movement of the locus of addition reactions

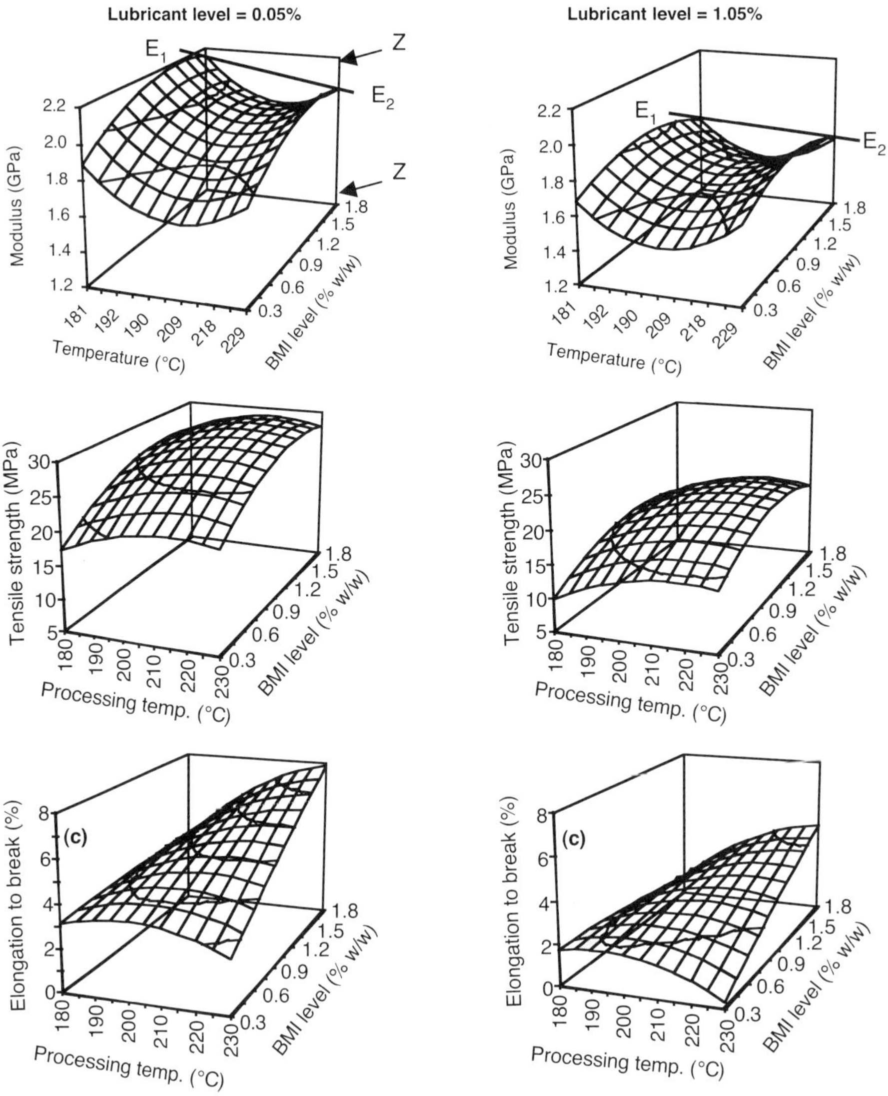

FIGURE 2 Contour plots showing effect of BMI level and processing temperature on tensile modulus, strength and elongation to break at the lowest and middle lubricant levels

to BMI from the interphase region to the bulk matrix. When the lubricant level is below 0.2% w/w, the BMI induced crosslinking of the elastomeric phase is confined mainly to the interphase region. It can therefore be envisaged that crosslinking of the interphase will slightly increase the effective size of the filler and increase filler-matrix interaction by entanglement with the chains of the bulk matrix. The latter effect may not significantly increase the viscosity of the composite melt because the bulk matrix chains can still move relative to one another with relative ease.

Preferential adsorption of the lubricant displaces both the BMI and the elastomeric phase from the filler surface, though addition reactions to BMI still occur both in the continuous PP phase and in the elastomeric phase. BMI reactions with the latter phase in

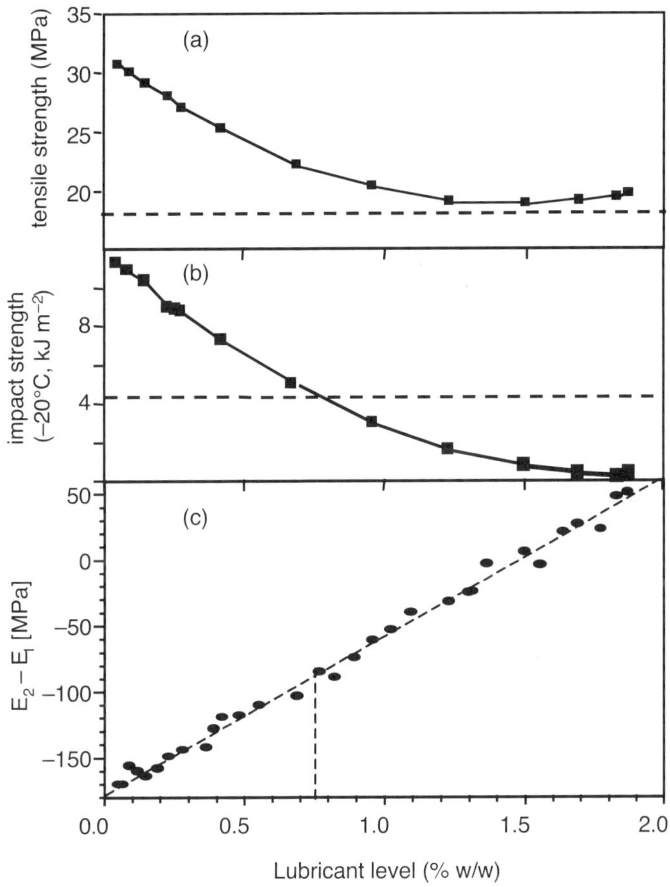

FIGURE 3 Effect of lubricant level on (a) tensile strength, (b) impact strength, and (c) the change in modulus ($E_2 - E_1$) across the BMI level = 1.8 plane

particular will result in extensive branching and some crosslinking of the bulk matrix, such effects will lead to an increase in composite melt viscosity. The author has also observed effects of this type in PP/calcium carbonate composites (Liauw 2002). Interfacial modification of these composites with trimethylolpropane triacrylate (TMPTA) alone increased filler–matrix interaction but only slightly increased the composite melt viscosity relative to the unmodified composite. Combined interfacial modification with TMPTA and isostearic acid (ISA) resulted in mechanical properties that were close to that when ISA was used alone. This was due to blockage of TMPTA adsorption sites by the more strongly adsorbing ISA. Macroradical addition to TMPTA, however, still occurred in the bulk matrix and resulted in a significant increase in the elongational viscosity of the composite melt relative to when the TMPTA was used alone.

Returning now to the BMI modified PPBC based composites; at lubricant levels greater than ca. 1.5% w/w, the composite melt viscosity reduced such that values of MFR were again obtainable; further additions beyond 1.5% w/w led to increased MFR. At such high levels of lubricant the mechanical properties were poor due to degradative effects that are discussed later.

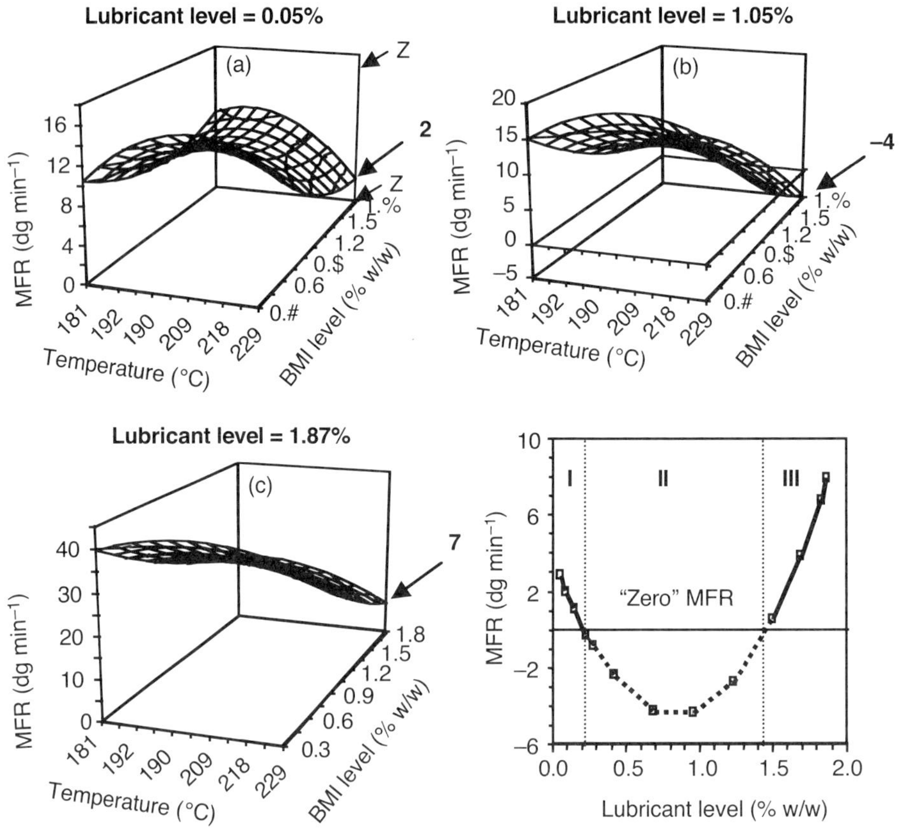

FIGURE 4 Effect of processing temperature, BMI level and lubricant level on melt flow rate; in (a), (b) and (c) the lubricant level is fixed at 0.05% w/w, 1.05% w/w and 1.87% w/w, respectively and (d) shows the variation in MFR at the intercept on the Z-Z axis as a function of lubricant level

Investigation of Interphase Structure

Confirmation of Elastomeric Encapsulation. Elastomeric encapsulation of the filler particles at high BMI level, high processing temperatures and low lubricant levels was confirmed by extracting the matrix from selected control composites with decane. The extraction was carried out in a thoroughly lagged Soxhlet apparatus and continued for 72 hours. The resulting composite residues, consisting of mainly the filler particles with the attached interphase and any insoluble fraction of the matrix, were analysed by DRIFTS and DSC. Relevant data is shown in Table 3. The amount of interphase/insoluble matrix in the PPBC based composite was far higher than that in an equivalent composite based on PP homopolymer. Furthermore the crystalline content of the PPBC based composite residue was much lower than that of the PPH based composite residue. This indicates that the filler particles are indeed encapsulated with the elastomeric phase of the PPBC. DRIFTS analysis confirmed this beyond doubt and it is apparent from Figure 5 that the methylene C-H stretching absorptions of the PBC based residue are more intense than those in the PPH based composite residue.

TABLE 3 Data relating to interphase structure

Composite	Bound Matrix Content (%)*	PP Crystalline Content of Bound Matrix (%)+
Unmodified	6.1	nd
BMI modified	19.5	3.4
BMI modified (PPH)	8.9	14.1

*,% by mass of filler in residue; +, crystalline content is based on crystallisable PP (there was no clear evidence of crystalline PE); nd, no peaks were resolved.

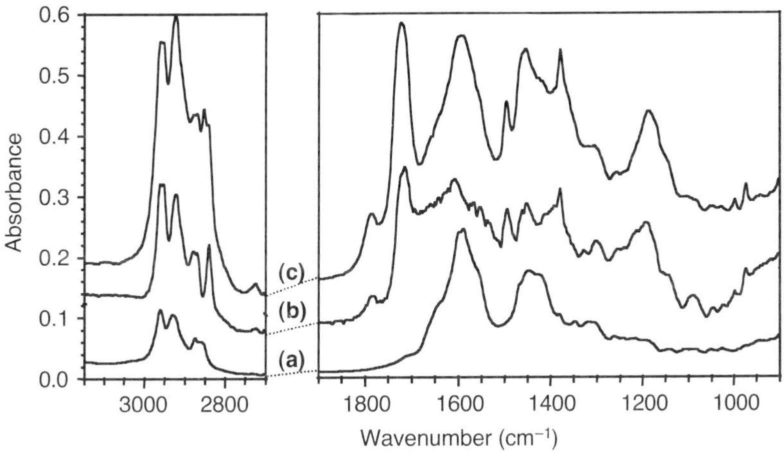

FIGURE 5 DRIFTS spectra of composite residues after decane extraction: (a) unmodified composite, (b) modified composite based on PPH, (c) modified composite based on PPBC

Simultaneous Effects of BMI Level Processing Temperature and Lubricant Level on Interphase Structure. Soxhlet extraction of the composites with decane was shown to be very rigorous as was manifested by the appearance of a carbonyl absorption (due to oxidative degradation) in the residue for the composite based on untreated filler. The crystalline content of this residue was also low despite the fact that it was propylene rich (Figure 5). It was decided therefore to use xylene for the remaining extractions and to immerse the samples (wrapped in glass cloth bags) in the refluxing xylene. The extraction time was reduced to 24 hours. This less rigorous extraction regime was expected to preserve the bulk matrix layer surrounding any elastomeric encapsulation and provide insight into the effect of interphase structure development on the phase separation.

Insoluble Matrix Content. Contour plots for insoluble matrix content as a function of processing temperature and BMI level at the lowest, middle and highest lubricant levels are presented in Figure 6. In these contour plots the role of the lubricant, in terms of influencing the location of BMI reactions, is clearly evident. Such effects are manifested as interesting changes in the temperature = 181°C plane of the contour plots; in this area the level of insoluble matrix increases with lubricant level. This observation points to a shift in the locus of addition reactions to BMI from the filler surface to the bulk matrix, this is most probably caused by preferential adsorption of the lubricant on the filler surface. The unmodified composite yielded an insoluble matrix content of 2.3%

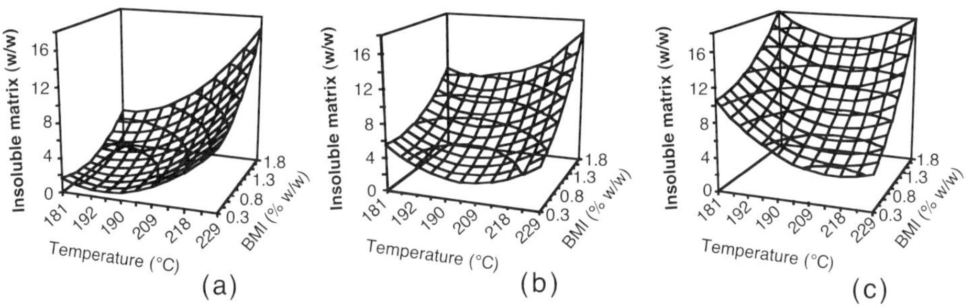

FIGURE 6 Response surfaces showing the effect of processing temperature and BMI level on the insoluble matrix content at lubricant levels of (a) 0.05%, (b) 0.96%, and (c) 1.87%

w/w, a figure consistent with strongly adsorbed matrix chains that are present in all unmodified particulate filled composites (Pukanszky 1995).

In the BMI = 0.3% w/w plane of the plots, the temperature coefficient of the insoluble matrix response (the 181–229°C secant) goes from positive (at the lowest lubricant level) through to zero at around the middle level of lubricant, and then becomes strongly negative at the highest level of lubricant. At low lubricant levels a significant proportion of the filler surface is available for reaction with BMI, therefore, as temperature increases the rate of BMI-filler and BMI-matrix coupling reactions increases, thus explaining the increase in insoluble matrix content. Once the level of lubricant exceeds monolayer coverage of the filler surface, any additional lubricant will reside in the bulk matrix. At low processing temperatures the level of insoluble matrix is high due to BMI addition reactions in the bulk matrix resulting in crosslinking, principally of the random ethylene-propylene copolymer impact modifier (EP) phase; though our earlier work has also shown that significant chain extension and branching of the PP phase can also occur (Liauw et al. 2000). As processing temperature increases the role of the lubricant as a pro-degradant becomes clear. Degradation of polyolefins associated with fatty acids and fatty acid salts/derivatives is an effect that has been observed by the authors in ethylene vinyl acetate/fatty acid treated magnesium hydroxide composites (Rothon et al. 1999), as the purity of the fatty acid was found to affect composite thermal stability with the acid of greater purity having a somewhat less detrimental effect. Therefore, the fatty acid amide/ester based lubricant used in this study may be involved in reactions associated with degradation, particularly if it is hydrolysed back to the fatty acid. It may therefore be envisaged that chain scission of the PP phase will occur more rapidly than addition reactions to BMI, resulting in an overall decrease in the molar mass and a possible increase in the level of chain branching. Such an effect may explain the observed decrease in insoluble matrix content arising from increasing processing temperature at low BMI levels. At higher BMI levels the lubricant associated effects are balanced by addition reactions to BMI.

These data generally confirm the MFR (at max. temp and max. BMI) versus lubricant level) data given in Figure 4, here the MFR decreased to zero then increased again at higher lubricant levels. These observations were explained by a change in the locus of BMI addition reactions from the filler surface, i.e., interphase structure development with minimal crosslinking of the bulk matrix (at the lowest lubricant levels), to the bulk matrix itself (at intermediate to moderately high lubricant levels). At the highest lubricant levels the MFR increased due to an increase in the level of degradation and the conventional effect of a lubricant.

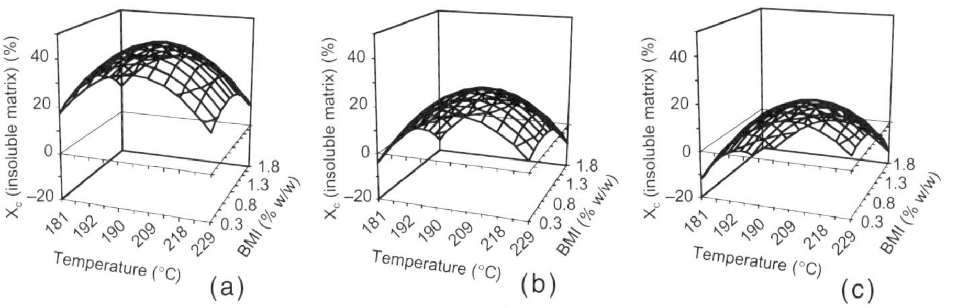

FIGURE 7 Response surfaces showing crystalline content of insoluble matrix fraction as a function of processing temperature and BMI level at lubricant levels of (a) 0.05%, (b) 0.96%, and (c) 1.87%

Crystalline Content. The crystalline content of the insoluble matrix in the composite residues was determined from the heats of fusion of PP determined by DSC and the insoluble matrix content determined from TGA. The insoluble matrix residue of the unmodified composite was 6% crystalline and the unfilled matrix was 35% crystalline. Response surfaces for the crystalline content as a function of processing temperature and BMI level at the three lubricant levels are given in Figure 7. The overall shape of the response surfaces was not significantly affected by the level of lubricant, though the position on the y-axis, i.e., overall crystalline content, was significantly affected by this parameter. Increasing lubricant level from the minimum to the middle level causes almost a step reduction in the crystalline content, whereas a further increase to the maximum lubricant level causes a rather more progressive decrease. On first examination these observations may seem a little unexpected, especially compared with the data obtained from samples that were solvent extracted under harsher conditions using decane. Decane extraction is likely to result in more thorough extraction of the PP component, partly due to the higher temperature (175°C) and due to more rapid degradation of the PP during extraction. Extraction with xylene is likely to leave behind more PP and possibly preserve a greater proportion of linkages between PP blocks and EP blocks. Therefore the increased crystalline content observed at the lower lubricant level may be explained by preferential BMI adsorption and degradation induced crosslinking of the EP phase at the filler surface and the attachment of this phase to the filler surface (via the adsorbed BMI), thus forming the elastomeric interphase. These effects are likely to further increase the degree of phase separation and leave the PP layer surrounding the EP covering of the filler more depleted in ethylene than at an equivalent distance from an EP domain in the unfilled matrix; therefore resulting in a higher crystalline content in the surrounding PP layer. Increasing the level of lubricant such that the BMI adsorption sites on the filler are blocked causes a greater proportion of addition reactions to BMI in the bulk matrix, the EP phase cannot encapsulate the filler so readily and the driving force for enhanced phase separation is lost. Additionally, reaction of PP macroradicals with BMI will increase the level of branching which, together with the increased rate of degradation associated with the fatty acid derived component of the lubricant, will ultimately result in mixed structures of limited structural regularity being formed. These factors explain the observed reduction in crystalline content.

DRIFTS of Composite Residues. A typical DRIFTS spectrum of a boiling xylene insoluble composite fraction displaying the quantified peaks and baselines is shown in Figure 8. The response surface shape for the relative level of BMI ($Abs_{(BMI)}/Abs_{(OH)}$) in

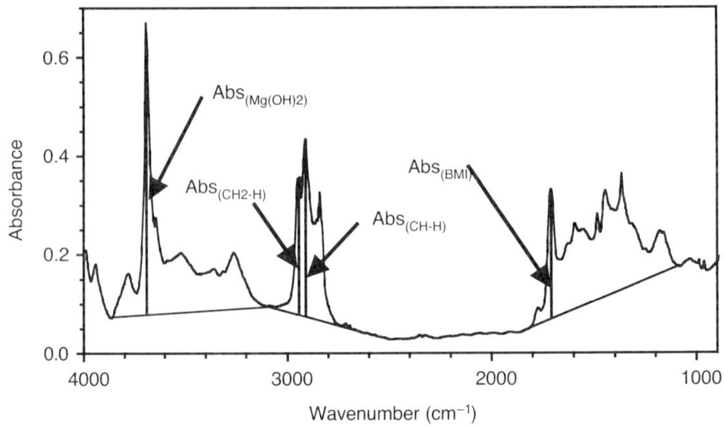

FIGURE 8 Example DRIFTS spectrum of composite residue showing measured absorptions

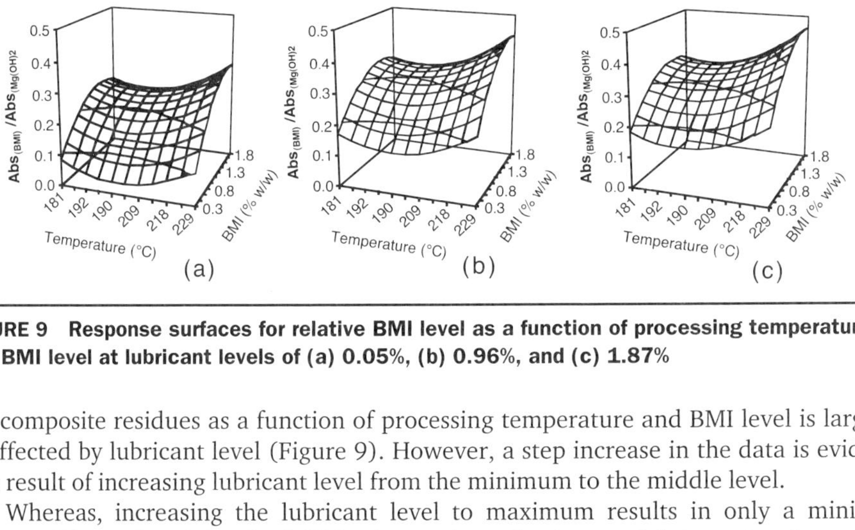

FIGURE 9 Response surfaces for relative BMI level as a function of processing temperature and BMI level at lubricant levels of (a) 0.05%, (b) 0.96%, and (c) 1.87%

the composite residues as a function of processing temperature and BMI level is largely unaffected by lubricant level (Figure 9). However, a step increase in the data is evident as a result of increasing lubricant level from the minimum to the middle level.

Whereas, increasing the lubricant level to maximum results in only a minimal change. The reduced level of BMI observed at the lowest lubricant level may be explained by its reaction with the filler surface; ring opening of the imide, to form the amide carboxylate, will result in a decrease in the carbonyl vibration frequency and intensity (Liauw et al. 2000). Close proximity to the filler surface may also result in the BMI being less resolvable through the relatively thick covering layer of polyolefin; most of the IR radiation will be absorbed in this layer particularly if the angle of incidence is low. Increasing the lubricant level to the extent that there is sufficient available to completely cover the filler surfaces results in an increase in the observed level of BMI in the composite residue, evidence that further verifies a shift in the locus of addition reactions to BMI. A sensible estimate of the level of added lubricant required to cover the filler is almost impossible to make, because the structures (and relative proportions) of the fatty acids on which the amide/ester blend is based are unknown, as is the proportion of lubricant able to reach the filler surface. However, the data shown in Figure 2 suggests lubricant levels in the range 0.6–0.8% w/w are required to render the filler surface inactive to BMI, at least as far as the impact strength response is concerned.

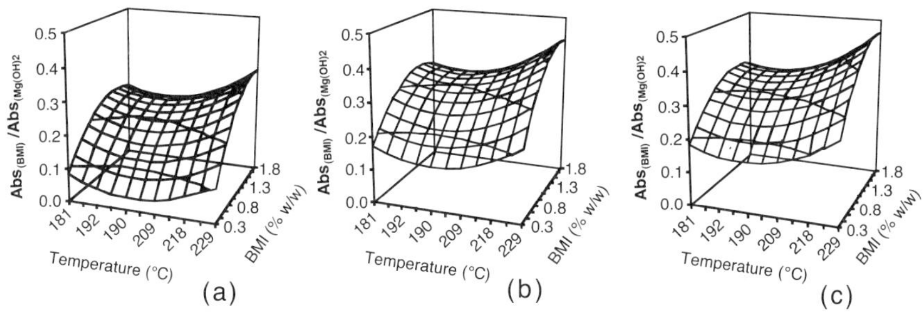

FIGURE 10 Contour plots showing the effect of processing temperature and BMI level on the relative ethylene content in the insoluble matrix at lubricant levels of (a) 0.05%, (b) 0.96%, and (c) 1.87%

The relative ethylene content response ($Abs_{(CH\text{-}H)}/Abs_{(CH2\text{-}H)}$) provides further evidence confirming elastomeric encapsulation of the filler. This can be seen in the temperature = 220°C plane of Figures 10(a) and (b), where at the lowest lubricant level the ethylene content is higher than at the middle level, this is particularly true at the lower BMI level. Furthermore, the ethylene content actually increases slightly with temperature at the lowest BMI and lubricant level. Under the latter conditions the layer of chain extended/crosslinked PP on top of EP layer is thin because there is just sufficient BMI available to interact with the filler and EP phase. Greater improvement in strength and toughness, however, are observed at higher BMI levels and processing temperatures. In this region the ethylene content decreases as a result of the increased thickness of the covering layer of chain extended/crosslinked PP. At the processing temperature = 181°C plane, however, the ethylene content is seen to increase progressively with increasing lubricant level. This may indicate that the EP phase (and BMI (Figure 9)) is displaced from the filler surface and therefore reacts in the bulk matrix. Increases in processing temperature (when lubricant level > 0.05% w/w) may enable the BMI to react more readily with the filler and thus displace the lubricant, this may restore the ability for the filler to undergo encapsulation with the EP phase. However, the level of degradation (associated with the lubricant) and re-arrangement caused by addition reactions of PP and EP macro-radicals to BMI has caused severe damage to the properties of the matrix in this area as reflected in the increased soluble matrix content and the reduced crystalline content.

CONCLUSIONS

This study has shown that PP block copolymer/magnesium hydroxide based composites of outstanding strength and toughness can be formed by reactive melt blending with BMI. This unique combination of properties is due to strongly bound encapsulation of the filler with the elastomeric phase of the matrix. The driving force for this encapsulation is considered to be adsorption of the BMI on the filler surface via ring opening and amide carboxylate formation. Macroradicals formed by mechanical degradation then add to the maleimide alkene. Due to the predominance of crosslinking of the ethylene rich impact modifier phase, it becomes preferentially attached to the filler and therefore forms the crosslinked elastomeric interphase region.

The problem with BMI modification of composites based on olefinic copolymers containing ethylene derived units, is that composite melt viscosity is increased relative to the unmodified composite. This occurs despite BMI being adsorbed onto the filler by diffusion through the matrix melt, during the course of this diffusion some matrix macroradicals will add to a fraction of the BMI. Therefore it is highly unlikely that all chain extension and branching reactions are exclusively confined to the interphase region.

Whilst addition of a fatty acid amide and fatty acid ester-based lubricant led to generally increased composite melt viscosity, three factorial response surface methods revealed interesting behaviour that successfully unravelled the interplay between BMI and lubricant adsorption on the filler and the resultant effect on the locus of addition reactions to BMI. Effective simultaneous improvement of tensile strength and toughness related properties requires a degree of chain extension/branching, and perhaps crosslinking, of the PP phase around the EP phase that encapsulates the filler. Such a condition cannot be met when the lubricant covers a significant proportion of the filler surface area. Under these conditions, adsorption of BMI and the EP phase is inhibited, resulting in a shift in the locus of BMI reactions from the interfacial region to the bulk matrix and thus preventing formation of the latterly described interphase structure.

REFERENCES

Hornsby, P.R., and C.L. Watson. 1986. Magnesium hydroxide–a combined flame retardant and smoke suppressant filler for thermoplastics. *Plastics and Rubber Processing and Applications*. 6: 169.

Khunová, V. 1995. Influence of Reactive Processing on the Properties of Particulate Polypropylene Composites. *Proceedings Euro-Fillers '95*, Centre de Recherches-CNRS, Mulhouse, France. 215.

Khunová, V., and C.M. Liauw. 2000. The role of m-phenylene bismaleimide (BMI) in reactive processing of polypropylene/magnesium hydroxide composites 3: analyses of interphase structure development, *Macromolecular Materials and Engineering*. 279: 34–41.

Khunová, V., and M.M. Sain. 1995. Optimization of mechanical strength of reinforced composites: 1. Study on the effect of reactive bismaleimide on the mechanical performance of $CaCO_3$-filled polypropylene. *Die Angewandte Makromolekulare Chemie*. 224: 9.

Khunová, V., and M.M. Sain. 1995. Optimization of mechanical strength of reinforced composites: 2. Reactive bismaleimide-modified polypropylene composites filled with talc and zeolite. *Die Angewandte Makromolekulare Chemie*. 225: 11–20.

Khunová, V., C.M. Liauw, and P. Alexy. The role of m-phenylene bismaleimide (BMI) in reactive processing of polypropylene/magnesium hydroxide composites: 2. The effect of processing temperature and composite formulation on rheological properties. *Die Angewandte Makromolekulare Chemie*, 269: 84–89.

Khunová, V., C.M. Liauw, P. Alexy, and M.M. Sain. 1999. The role of m-phenylene bismaleimide (BMI) in reactive processing of polypropylene/magnesium hydroxide composites: 1. The effect of processing temperature and composite formulation on mechanical properties. *Die Angewandte Makromolekulare Chemie*. 269: 78–83.

Khunová, V., S.J. Hurst, and C.M. Liauw. 1999. Reactive processing of particulate filled polymers: m-phenylene bismaleimide modified polyethylene/magnesium hydroxide composites. *Polymer Bulletin*. 42: 457–463.

Liauw, C.M. 2002. Filler-Matrix Interphase–The Hidden Resource. *Pre-Conference Seminar (Proceedings) Functional Fillers for Plastics 2002*, Intertech Conferences, Toronto.

Liauw, C.M., G.C. Lees, S.J. Hurst, R.N. Rothon, and S. Ali. 1998. Effect of silane based filler surface treatment formulation on the interfacial properties of impact modified polypropylene/magnesium hydroxide composites. *Composites Part A. Applied Science and Manufacture*. 29A: 1313–1318.

Pukánszky, B. 1995. Particulate Filled Polypropylene: Structure and Properties. In *Polypropylene Structure Blends and Composites.*, Vol. 3. Edited by J. Karger-Kocsis, Chapman and Hall, London.

Rothon, R.N. 1996. Effects of Particulate Fillers on Flame Retardant Properties of Composites (Chapter 6) in Particulate-Filled Polymer Composites, Edited by R.N. Rothon, Longman Scientific, London, 207.

Rothon, R.N., G.C. Lees, W.C.E. Schofield, and C.M. Liauw. 1999. Stearic Acid Modification of Magnesium Hydroxide Filler: The Effect of Acid Purity on the Mechanical and Fire Retardant Performance Before and After Aging. *Proceedings EuroFillers '99*, Lyon, France.

Weber, M. 2000. Markets for Mineral Flame Retardants–New Products based on Aluminium Hydroxide (ATH) and Magnesium Hydroxide Offer New Opportunities. *Proceedings Functional Effect Fillers 2000,* Intertech Conferences.

Influence of Surfactants on Interaction Forces Between Polyethylene Surfaces in a Hydrocarbon Solvent

Robert Aranowski*, Jakub Nalaskowski†, Jan Hupka*, and Jan D. Miller†

The polyolefins, mainly polyethylene and polypropylene particles, are used as an active filler material in polymer–polymer blends and are being used for the modification of asphalts to provide for high-load, low-maintenance road construction. The stability of prepared blends depends on the interaction forces between polyolefin filler and the matrix and/or between two polyolefin filler surfaces in the polymeric matrix. Various additives, especially surfactants, can be used to modify these interactions and improve the stability of polymer blends.

In current research, direct force measurements between polyethylene (PE) surfaces in n-tetradecane, a nonaqueous solvent imitating the polymer matrix, were performed in the presence of lauric acid and dodecylamine using the AFM colloidal probe technique. Additionally, the zeta potential of polyethylene particles in the presence of surfactants in n-tetradecane was measured. The interaction forces between the polyethylene surfaces can be described as a combination of electrical double layer repulsion and steric repulsion between polymer chains extending into the n-tetradecane solvent. The addition of surfactant to the n-tetradecane was found to increase the repulsive force between the PE surfaces. The strongest repulsion was observed for dodecylamine. Additionally, the temperature effect on the interactions in n-tetradecane was studied. It was found that steric repulsion between the PE surfaces increases with an increase in temperature.

INTRODUCTION

Preparation and stability of nonaqueous dispersions are of great technological importance. Many important products such as inks, paints, lubrication oils, pharmaceutical

* Dept. of Chemical Technology, Gdansk University of Technology, Gdansk, Poland.

† Dept. of Metallurgical Engineering, University of Utah, Salt Lake City, Utah.

formulations and cosmetics are examples where the stability of the nonaqueous dispersion is crucial for their application. Nonaqueous suspensions of mineral particles are also used in manufacturing of advanced ceramic materials (Bergstroem, 1996). Other systems where the stability of a dispersion in nonaqueous medium is critical for successful application include polymer-filler composites. Mineral particles are used as an active filler in polymer blends (Ahsan and Taylor, 1998; Guerrica-Echevarria, Eguiazabal, and Nazabal, 1999; Pastorini and Nunes, 1999) for mechanical strength and flame-retardant properties. Polymer particles are also used as active fillers in polymer blends, and polyethylene (PE) is most extensively used as a dispersed filler phase in polymer matrices (Arnal and Muller, 1999; Huneault et al., 2001; Lerdwijitjarud, Sirivat, and Larson, 2002). Polyolefin particles, especially PE, are also extensively used for modification of asphalts considered for high-load, low-maintenance road construction (Liang and Hesp, 1993; Giavarini et al., 1996). Addition of dispersed polyolefins to the asphalt blends helps to retard the low-temperature thermal stress cracking as well as rutting of the pavement at elevated temperatures. Other benefits include greater fatigue resistance, reduced temperature susceptibility, increased tire traction and greater adhesion to the aggregate. Polyolefin polymers such as PE do not contain double bounds and therefore have good thermal and ageing resistance. Different methods of stabilization of such dispersions are employed including modification of the filler surface, addition of surfactants, and/or steric stabilization by extending polymer chains (Morrison, Hedmark, and Hesp, 1994; Zeiler, Kellermann, and Munstedt, 2000). However, the major obstacles to widespread application of polyolefins for asphalt modification is still the low stability of the PE dispersion in asphalt at elevated temperature, which results in coalescence of polyolefin droplets and separation of polymer phase from the asphalt blend (Hesp and Woodhams, 1991).

Due to the technological significance, the stability of nonaqueous dispersions has been studied in detail using bulk observations of coagulation and settling and electrophoretic measurements (Romo, 1963; McGown, Parfitt, and Willis, 1965; McGown and Parfitt, 1966, 1967). To complement these measurements it seems appropriate that surface forces be measured to gain a more fundamental understanding of the phenomena involved. Unfortunately, despite numerous contributions which report direct interaction force measurements in aqueous systems, nonaqueous systems have received much less attention in this field. Only several papers pertaining to the direct measurement of interaction force between solid surfaces in nonaqueous surfactant solutions have been published (Gee and Israelachvili, 1990; Chen, Xu, and Israelachvili, 1992; Bergstrom and Blomberg, 2000; Briscoe and Horn, 2002). In these studies generally the surface force apparatus (SFA) (Israelachvili, and Tabor, 1972) was used with mica surfaces. Recent advances in force measurements using atomic force microscopy (AFM) (Binnig, Quate, and Gerber, 1986) in conjunction with the colloidal probe technique (Ducker, Senden, and Pashley, 1991) have made possible the measurement of interaction forces between a spherical particle (colloidal probe) and almost any other surface. With the AFM colloidal probe technique surfaces composed of almost any material of interest can be studied.

In this paper, direct force measurements between a PE spherical particle and a PE surface were made in n-tetradecane (nonaqueous solvent imitating polymer matrix) using the AFM colloidal probe technique. The effect of surfactant type (lauric acid and dodecylamine) on the interaction force and the effect of temperature on the interactions have been determined. Additionally, the electrophoretic mobility/zeta potential of polyethylene particles in n-tetradecane surfactant solutions was measured. According to our knowledge this is the first paper where interfacial forces between polyolefin surfaces

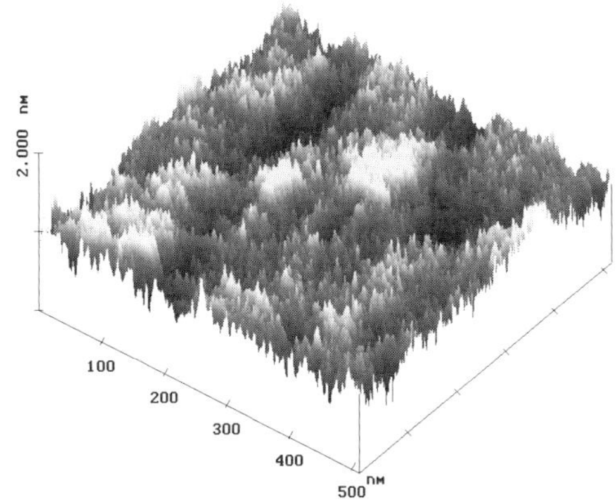

FIGURE 1 AFM image of flat PE surface (mean roughness 0.13 nm)

have been measured in hydrocarbon solutions of surfactants using AFM. The understanding of interaction forces existing between the polymeric particles in nonaqueous medium has not only significant fundamental implications but also may help in further modification of these forces in order to improve stability of the polymer-filler blends.

EXPERIMENTAL

Materials

The polyethylene (PE) powder used was a low-density PE (Scientific Polymer Product, Inc.) with a molecular weight MW = 1800 and a melting point mp = 117°C. Reagent grade n-tetradecane (Sigma), dodecylamine (DDA) (Aldrich) and lauric acid (Aldrich) were used during the experiments.

Methods

Preparation of Flat PE Surface. The PE surface was prepared by melting powdered material in a vacuum and pouring on a freshly cleaved mica surface. After cooling the PE disc was separated from the mica substrate under water. The PE surface obtained in this way had a very smooth surface as was determined by contact mode atomic force microscopy imaging, see Figure 1. The roughness measured from AFM images was determined to be 0.13 nm.

Colloidal Probe Preparation. Spherical polyethylene (PE) particles were obtained using a procedure, which involves suspending a powder of polymeric thermoplastic materials, such as PE, in glycerol, heating the suspension above the melting point of the polymer, and then solidification of the dispersed polymeric droplets at a reduced temperature. After appropriate filtration and drying, this procedure was found not to change the surface properties of PE particles, which retained a high degree of hydrophobicity. These particles had a relatively smooth surface and were particularly useful for investigating interaction forces using the AFM colloidal probe technique (Nalaskowski et al., 1999).

The prepared spherical particles were glued to the AFM cantilever with a small amount of epoxy resin using a procedure described elsewhere (Ducker, Senden, and

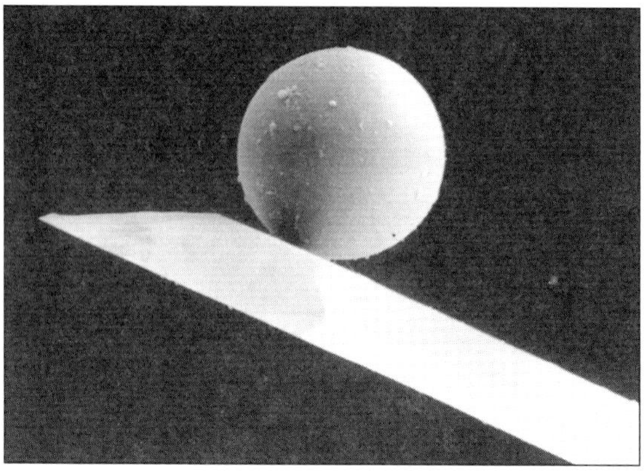

FIGURE 2 SEM image of 18 μm spherical PE particle glued to the end of AFM cantilever

Pashley, 1991). Tipless rectangular silicon cantilevers (Digital Instruments, Inc.) were used. The cantilever was placed under the CCD camera fitted with long distance lens giving a 500× magnification. A small amount of epoxy resin was transferred on the tip of the cantilever using a 20 μm diameter tungsten wire attached to a micromanipulator. Subsequently, a selected sphere was picked up using a new tungsten wire and carefully placed on the cantilever using the micromanipulator. It has been established that there is no contamination of spherical particles with epoxy resin during gluing. Using this procedure, particles in diameter from 1 μm to 200 μm can be precisely glued to the AFM cantilever. Cantilevers were used for measurements after at least 24 hours of drying. An example of the cantilever with attached PE spherical particle is shown in Figure 2.

Interaction Force Measurement. A Nanoscope IIIa atomic force microscope (Digital Instruments, Inc.) equipped with a fluid cell was used for the measurements of interaction force between PE sphere and PE flat surface in pure n-tetradecane and in 1×10^{-3} M n-tetradecane solutions of lauric acid and DDA. During the AFM surface force measurement the laser beam is focused on the back of the cantilever to detect the cantilever's deflection as it interacts with the surface beneath it. The reflected light is directed onto a split photodiode detector, which produces a voltage signal proportional to the cantilever deflection. If the spring constant is known, the deflection of the spring can be converted to force using the Hooke's law. Using such a setup the force acting on the cantilever can be determined with sensitivity greater than 0.1 nN. The sample beneath the cantilever is moved using a piezoelectric transducer. In the force measurements, motion in the x and y directions is disabled and the piezoelectric tube is used to move the surface in the z direction and the cantilever deflection is continuously measured. The surface is first moved toward the cantilever until the particle contacts the surface and then is retracted from the cantilever until the particle snaps off from the surface. The interaction force is obtained from measuring the deflection of the cantilever during the approach. For measurements at different temperatures, the AFM setup was placed in a modified solid-state cooling-heating chamber (ECHOtherm, Torrey Pines Scientific) which allows for operating AFM while in the chamber. The temperature was monitored and maintained by the internal controller. The spring constant of the cantilever was given by the manufacturer to be 0.12 N/m, a value which has been confirmed using the Cleveland method (Cleveland et al., 1993). After measurement, the particle diameter was measured using scanning electron microscopy (SEM). At least 10 consecutive

TABLE 1 Electrophoretic mobility and zeta potential of PE particles in n-tetradecane in the absence and presence of surfactants

System	Electrophoretic Mobility $\left(\frac{\mu m \cdot cm}{s \cdot V}\right)$	Zeta Potential (mV)
PE/n-tetradecane	2.14×10^{-3}	2.5
PE/1×10^{-3} M lauric acid solution in n-tetradecane	-4.01×10^{-3}	−4.78
PE/1×10^{-3} M DDA solution in n-tetradecane	1.04×10^{-2}	12.4

measurements were taken, values of the cantilever's deflection converted to force, normalized with respect to the PE sphere radius, and averaged.

Electrophoretic Mobility Measurements. Electrophoretic mobility of the spherical PE colloidal probe particles was measured using the Brookhaven ZetaPALS apparatus. This instrument employs phase analysis light scattering (PALS) method which has been found suitable for the measurement of low electrophoretic mobilities in nonaqueous media (McNeil-Watson, Tscharnuter, and Miller, 1998). After 15 minutes of equilibration of PE particles in the n-tetradecane solution, six consecutive measurements were taken at room temperature and an applied electric field of 357.14 V/cm. The measurements were averaged and the zeta potential was calculated using Smoluchowski equation with bulk properties of the pure n-tetradecane as the medium.

RESULTS AND DISCUSSION

Generally speaking, a suspension of particles in nonpolar medium can be stabilized by electrical double layer repulsion (Overbeek, 1966) or by entropic repulsion of the interacting polymer chains (Luckham and Manimaaran, 1997), especially in the studied case where PE can swell in the hydrocarbon liquid and where additional charging of the PE surface can occur due to the presence of surfactants. Both mechanisms may coexist and to differentiate between them may be quite difficult.

In many cases it is assumed that electrical double layer interactions in nonpolar liquids are of little significance. This assumption is usually explained by the fact that ions have extremely low solubility in nonpolar liquids. However, potential at the interface can still develop in a nonpolar solvent due to the differences in electronic properties of the two phases (Morrison, 1993). In addition the role of water dissolved in the nonpolar solvent cannot be neglected. For example the n-tetradecane used in this study was not dried before use and can be assumed to contain water originating from the laboratory atmosphere. It is expected that the level of dissolved water could amount to as much as 40 ppm which corresponds to a molar concentration of 2.2×10^{-3} M. Such an amount of water cannot be ignored and may allow for accommodation of ions and formation of the electrical double layer. Additionally, water molecules can be introduced to the tetradecane with surfactant. It has been generally accepted that the presence of even traces of water allow for increased dissociation of molecules and charging of particles in nonpolar media as has been shown by electrophoretic measurements (McGown, Parfitt, and Willis, 1965). Indeed the electrophoretic mobility and zeta potential of particles in nonpolar liquids has been successfully measured as reported in the literature (McNeil-Watson, Tscharnuter, and Miller, 1998; Kosmulski, 1999). The electrophoretic mobility and zeta potential of PE particles in n-tetradecane is given in Table 1. The addition of surfactants changes the measured zeta potential significantly (Kitahara et al., 1977), which is visible also in our system, especially in the presence of

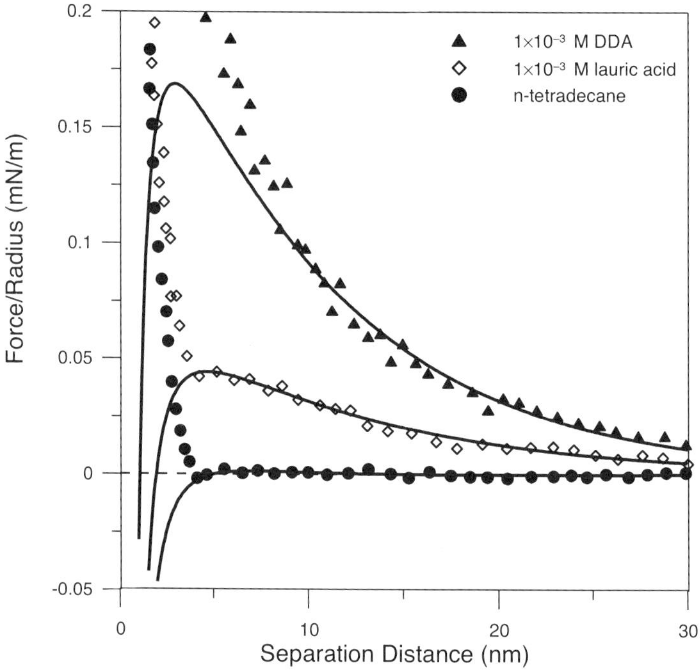

FIGURE 3 Interaction force between PE surfaces in n-tetradecane and in 1×10^{-3} M solution of DDA and lauric acid in n-tetradecane. Every fifth experimental point is shown for clarity. Continuous lines indicate constant charge DLVO fit. Hamaker constant calculated using combining relations is $A = 1.65 \times 10^{-21}$ J.

DDA. It has to be noted that the error of measurements reported in Table 1 is quite significant and was found to be in range of ±2 mV. Nevertheless the observed change in the value of zeta potential in the presence of surfactants is evident.

Interaction force curves measured between PE surfaces in n-tetradecane in the absence and the presence of surfactants are shown in Figure 3. Although the appropriateness of the use of the Derjaguin/Landau/Verwey/Overbeek (DLVO) theory for describing the behavior of particles in nonaqueous liquids is still discussed in the literature, many authors agree that DLVO theory can be applied to such systems quite successfully (Romo, 1963; McGown, Parfitt, and Willis, 1965; Briscoe and Horn, 2002). The experimental data from the force measurements have been fitted with the constant charge model for electrical double layer repulsion combined with van der Waals attraction, where force normalized with respect to particle radius is a function of separation distance as given by Equation 1:

$$\frac{F(D)}{R} = -\frac{A}{6D^2} + \frac{4\pi\sigma^2}{\varepsilon\varepsilon_0\kappa}e^{-\kappa D} \quad \textbf{(EQ 1)}$$

where A = Hamaker constant, σ = surface charge, ε = dielectric constant of medium (tetradecane) and κ = Debye constant (reverse Debye length).

It has been shown that such a simple model can be successfully used for describing interaction forces in nonpolar liquids especially when a significant amount of water is

TABLE 2 Parameters for fitted constant charge electrical double layer interactions between PE surfaces in n-tetradecane in the absence and presence of surfactants

System	Debye Length κ^{-1} (nm)	Surface Charge ($\mu C/m^2$)
PE/n-tetradecane	907.5	0.0017
PE/1 × 10^{-3} M lauric acid solution in n-tetradecane	10.5	0.109
PE/1 × 10^{-3} M DDA solution in n-tetradecane	9.36	0.204

present. The Hamaker constant for van der Waals interaction between PE surfaces across n-tetradecane has been calculated from Hamaker constants for the crystalline hydrocarbon PE ($A = 7.1 \times 10^{-20}$ J) and for n-tetradecane ($A = 5.1 \times 10^{-20}$ J) in vacuum (Israelachvili, 1991), and was found from combining relations to be $A = 1.65 \times 10^{-21}$ J. The parameters for the fitted DLVO curves (Debye length and surface charge) are given in Table 2.

In pure n-tetradecane the electrostatic repulsion is negligible as can be seen from experimental points. This is also reflected in the extremely small charge density value obtained from the theoretical fit. The thickness of the electrical double layer is extremely large, which is typical for a medium of low dielectric constant and with a very low concentration of ions in solution. At a separation distance around 5 nm van der Waals attraction is clearly visible and can be quite successfully fitted with the theoretical value for the Hamaker constant. The total interaction force between PE surfaces in n-tetradecane is consistent with the observed low stability of such systems, which are governed by the van der Waals forces. It has to be also noted that a long-range repulsive force related to interactions between extended polymer chains in solvent was not observed here, which can indicate that swelling of the PE under such conditions is not significant.

The addition of either surfactants (lauric acid or DDA) into the system dramatically changes the character of the interaction forces between PE surfaces. Strong repulsive components of the interaction force can be observed. The measured repulsive force is much stronger in the presence of DDA as is also demonstrated by the value of the surface charge that is almost two times greater than in the case of lauric acid. It has to be noted that the Debye length obtained from the theoretical fit is significantly smaller than for the pure hydrocarbon (see Table 2) that can indicate the presence of ions and water close to the interface. In the case of lauric acid the interaction force curve can be quite successfully fitted to the DLVO curve and even the small van der Waals component can be observed at separation distance close to 5 nm. However, in the presence of DDA the additional repulsive component of interaction force cannot be fitted to the DLVO curve at separations below 10 nm.

This additional repulsion can be explained by the presence of surfactant molecules, layers or micellar aggregates between the interacting PE surfaces. At distances below 14 nm in the presence of DDA significant oscillations in the force curve can be observed on a magnified part of the force curve as shown in Figure 4. The average value of the peak-to-peak distance for these interactions has been found to be 0.8 ± 0.07 nm. Such oscillations can suggest presence of surfactant layers or reversed micelles between interacting surfaces (Gee and Israelachvili, 1990; Chen, Xu, and Israelachvili, 1992). It has to be noted, however, that the observed value is significantly smaller than the size of the expected DDA micelle.

Although the swelling of PE was not observed in pure tetradecane, it cannot be completely ruled out for surfactant solutions. In order to characterize possible steric repulsion between polymeric chains extending into the solvent, a scaling theory framework developed by de Gennes (de Gennes, 1987) can be used. According to this theory the repulsive force between two surfaces due to the interaction of protruding polymer chains can be given by Alexander-de Gennes equation (Equation 2):

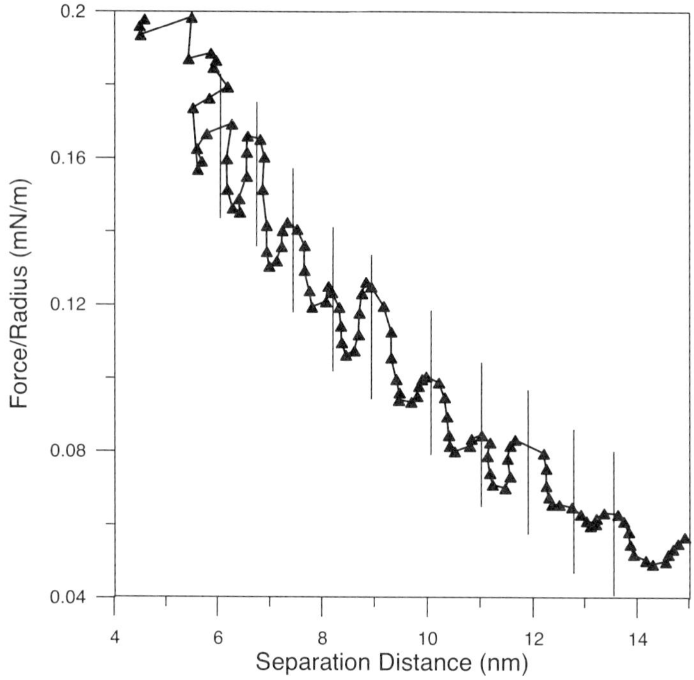

FIGURE 4 Interaction force between PE surfaces in 1 × 10^{-3} M solution of DDA in n-tetradecane. Vertical lines indicate peaks of the oscillations of force curve. Average peak-to-peak distance is 0.8 nm.

$$\frac{F(D)}{R} = \frac{16\pi kTL}{35s^3}\left[7\left(\frac{2L}{D+2\delta}\right) + 5\left(\frac{D+2\delta}{2L}\right) - 12\right] \quad \textbf{(EQ 2)}$$

where L is the thickness of brush layer extending from the surface, s is the distance between the chain anchoring points and δ is the thickness of the highly compressed polymer layer. This theoretical approach was developed originally for the grafted polymer chains at surfaces but was chosen here due to its simplicity and small number of parameters describing the state of the polymer layer.

The experimental values of interaction force between PE surfaces in pure tetradecane and with surfactants added, together with corresponding fits from the Alexander-de Gennes equation are shown in Figure 5. The Alexander-de Gennes equation was fitted to the experimental values of force at larger distances (larger than 5 nm) in order to see how well experimentally measured interactions correlate with values from the scaling theory. It can be shown that especially for the interactions in pure tetradecane and in the presence of lauric acid significant deviations from the theoretical fit are present. A much better fit was obtained for the interaction forces between PE surfaces in the DDA solution. The corresponding parameters of the Alexander-de Gennes equation are given in Table 3. The real significance of the values of L and s obtained for the pure nonpolar solvent and for the lauric acid solution may be, however, small in view of the poor fit to the theoretical equation. The steric component of repulsion with layer thickness L = 13.3 nm and distance between

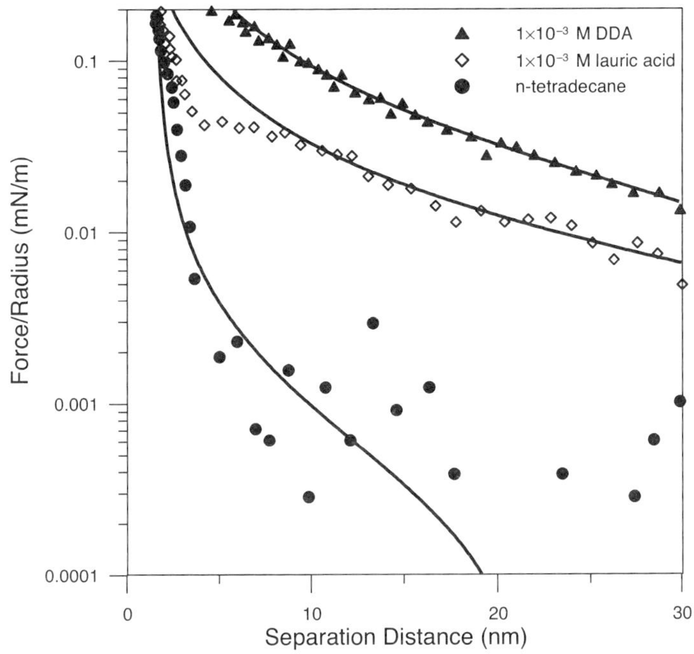

FIGURE 5 Interaction force between PE surfaces in n-tetradecane and in 1×10^{-3} M solution of DDA and lauric acid in n-tetradecane. Every fifth experimental point is shown for clarity. Continuous lines indicate fit to the Alexander-de Gennes equation.

TABLE 3 Parameters for the theoretical fit of the Alexander-de Gennes equation for PE surfaces in n-tetradecane in the absence and presence of surfactants

System	Brush Layer Thickness L (nm)	Distance Between Anchoring Points s (nm)
PE/n-tetradecane	13.3	102.8
PE/1×10^{-3} M lauric acid solution in n-tetradecane	41.5	82.9
PE/1×10^{-3} M DDA solution in n-tetradecane	13.6	63.7

anchoring points s = 102.8 nm cannot be ruled out, however, for the PE surfaces in the DDA solution.

Finally, the effect of temperature on the interaction forces between PE in nonpolar solvent has been studied. Interaction forces were measured in the lauric acid solution at 30°C and at 40°C. Experimental values have been fitted to the Alexander-de Gennes equation at a distance close to the contact of surfaces (below 20 nm) as can be seen in Figure 6. It can be observed that repulsive forces significantly increase with an increase of temperature. These results can be attributed to the more pronounced swelling of the polymer at the surface at higher temperatures and increased repulsion between polymer chains. The parameters for the theoretical equation as given in Table 4 show an increase in the brush layer thickness L and an increase in the compressed polymer layer thickness δ at 40°C when compared to results at 30°C.

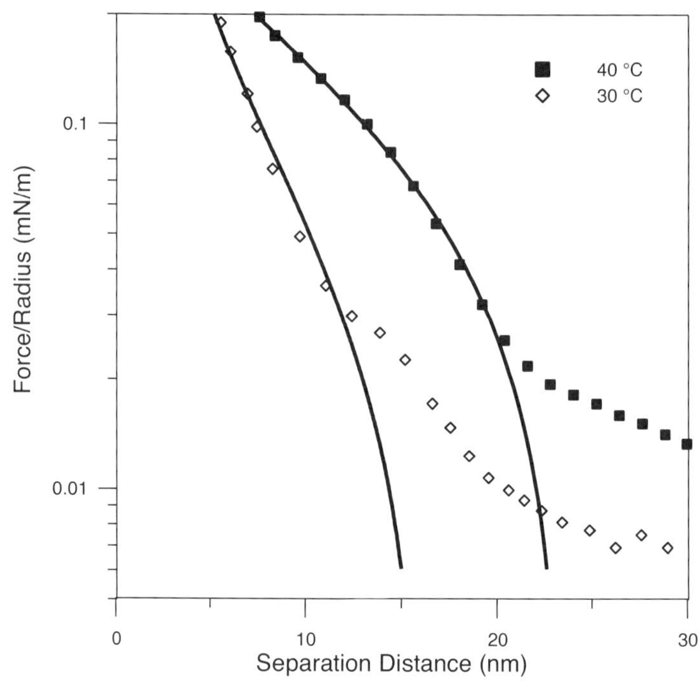

FIGURE 6 Interaction force between PE surfaces in 1 × 10^{-3} M solution of lauric acid in n-tetradecane at 30°C and 40°C. Continuous lines indicate fit to the Alexander-de Gennes equation.

TABLE 4 Parameters for the theoretical fit of the Alexander-de Gennes equation for PE surfaces in 10^{-3} M lauric acid solution in n-tetradecane at different temperature

System	Brush Layer Thickness L (nm)	Distance Between Anchoring Points s (nm)	Compressed Polymer Layer Thickness δ (nm)
PE/1 × 10^{-3} M lauric acid solution in n-tetradecane at 30°C	12.4	23.9	0
PE/1 × 10^{-3} M lauric acid solution in n-tetradecane at 40°C	29.8	22.1	7.6

CONCLUSIONS

The interaction forces between PE surfaces in pure n-tetradecane were found to be governed mostly by the attractive van der Waals component. After addition of surfactant into the n-tetradecane, a strong repulsive component appears, especially strong for the DDA solution. The mechanism of this repulsion can be attributed to the combined effect of electrical double layer repulsion due to the presence of traces of water in the system, steric repulsion related to the possible presence of the surfactant aggregates between the interacting surfaces, and possible steric repulsion between polymer chains extending into the solvent. Although strong swelling of PE was not evident from measurements of interactions between PE in pure tetradecane, the repulsive force increases with an increase in temperature which can suggest a slight swelling of PE surface in the n-tetradecane/lauric acid solution at elevated temperature. It can be concluded that addition of DDA into the PE dispersion in nonpolar medium should stabilize PE dispersions and prevent coagulation/coalescence of PE droplets/particles.

ACKNOWLEDGMENTS

Financial support provided by the Department of Energy (DoE), Basic Science Division Grant No. DE-FG-03-93ER14315 is gratefully acknowledged. Although the research described in this article has been funded partially by DoE, it has not been subjected to the agency's required peer and policy review and therefore does not necessarily reflect the views of these agencies and no official endorsement should be inferred.

REFERENCES

Ahsan, T. and Taylor, D.A. 1998. The influence of surface energetics of calcium carbonate minerals on mineral-polymer interaction in polyolefin composites. *J. Adhes.* 67:69.

Arnal, M.L. and Muller, A.J. 1999. Fractionated crystallisation of polyethylene and ethylene/.Alpha.-olefin copolymers dispersed in immiscible polystyrene matrices. *Macromol. Chem. Phys.* 200:2559.

Bergstroem, L. 1996. Rheological properties of concentrated, nonaqueous silicon nitride suspensions. *J. Am. Ceram. Soc.* 79:3033.

Bergstrom, L. and Blomberg, E. 2000. Probing polymeric stabilization in nonaqueous media by direct measurements. *J. Am. Ceram. Soc.* 83:217.

Binnig, G., Quate, C., and Gerber, C. 1986. Atomic force microscope. *Phys. Rev. Lett.* 56:930.

Briscoe, W.H. and Horn, R.G. 2002. Direct measurement of surface forces due to charging of solids immersed in a nonpolar liquid. *Langmuir.* 18:3945.

Chen, Y.L., Xu, Z., and Israelachvili, J. 1992. Structure and interactions of surfactant-covered surfaces in nonaqueous (oil-surfactant-water) media. *Langmuir.* 8:2966.

Cleveland, J.P., Manne, S., Bocek, D., and Hansma, P.K. 1993. A nondestructive method for determining the spring constant of cantilevers for scanning force microscopy. *Rev. Sci. Instrum.* 64:403.

de Gennes, P.G. 1987. Polymers at an interface; a simplified view. *Adv. Colloid Interface Sci.* 27:189.

Ducker, W.A., Senden, T.J., and Pashley, R.M. 1991. Direct measurement of colloidal forces using an atomic force microscope. *Nature.* 353:239.

Gee, M.L. and Israelachvili, J.N 1990. Interactions of surfactant monolayers across hydrocarbon liquids. *J. Chem. Soc., Faraday Trans.* 86:4049.

Giavarini, C., De Filippis, P., Santarelli, M.L., and Scarsella, M. 1996. Production of stable polypropylene-modified bitumens. *Fuel.* 75:681.

Guerrica-Echevarria, G., Eguiazabal, J.I., and Nazabal, J. 1999. Morphology and physical properties of injection-molded, mineral-filled polyamide 6/poly(hydroxy ether of bisphenol a) blends. *J. Appl. Polym. Sci.* 73:1805.

Hesp, S.A.M. and Woodhams, R.T. 1991. Asphalt-polyolefin emulsion breakdown. *Colloid Polym. Sci.* 269:825.

Huneault, M.A., Mighri, F., Ko, G.H., and Watanabe, F. 2001. Dispersion in high viscosity ratio polyolefin blends. *Polym. Eng. Sci.* 41:672.

Israelachvili, J.N. 1991. *Intermolecular and surface forces.* New York: Academic Press.

Israelachvili, J.N. and Tabor, D. 1972. Measurement of van der waals dispersion forces in the range 1.5 to 130 nm. *Proc. Roy. Soc. London, Ser. A.* 331:19.

Kitahara, A., Amano, M., Kawasaki, S., and Konno, K. 1977. The concentration effect of surfactants on zeta-potential in non-aqueous dispersions. *Colloid Polym. Sci.* 255:1118.

Kosmulski, M. 1999. Zeta potentials in nonaqueous media. How to measure and control them. *Colloids Surf., A.* 159:277.

Lerdwijitjarud, W., Sirivat, A., and Larson, R.G. 2002. Influence of elasticity on dispersed-phase droplet size in immiscible polymer blends in simple shearing flow. *Polym. Eng. Sci.* 42:798.

Liang, Z. and Hesp, S.A.M. 1993. In situ steric stabilization of polyethylene emulsions in asphalt binders for hot-mix pavement applications. *Colloids Surf., A.* 81:239.

Luckham, P.F. and Manimaaran, S. 1997. Investigating adsorbed polymer layer behavior using dynamic surface force apparatuses–a review. *Advances in Colloid and Interface Science.* 73:1.

McGown, D.N.L. and Parfitt, G.D. 1966. Stability of nonaqueous dispersions. IV. Rate of coagulation of rutile in aerosol ot + p-xylene solutions. *Discuss. Faraday Soc.* No. 42:225.

McGown, D.N.L. and Parfitt, G.D. 1967. Stability of nonaqueous dispersions. V. Theoretical predictions for dispersions in hydrocarbon media. *Kolloid Z. Z. Polym.* 219:48.

McGown, D.N.L., Parfitt, G.D., and Willis, E. 1965. Stability of nonaqueous dispersions. I. Relation between surface potential and stability in hydrocarbon media. *J. Colloid Sci.* 20:650.

McNeil-Watson, F., Tscharnuter, W., and Miller, J. 1998. A new instrument for the measurement of very small electrophoretic mobilities using phase analysis light scattering (pals). *Colloids Surf., A.* 140:53.

Morrison, G.R., Hedmark, H., and Hesp, S.A.M. 1994. Elastic steric stabilization of polyethylene-asphalt emulsions by using low molecular weight polybutadiene and devulcanized rubber tire. *Colloid Polym. Sci.* 272:375.

Morrison, I.D. 1993. Electrical charges in nonaqueous media. *Colloids Surf., A.* 71:1.

Nalaskowski, J., Drelich, J., Hupka, J., and Miller, J.D. 1999. Preparation of hydrophobic microspheres from low-temperature melting polymeric materials. *J. Adhes. Sci. Technol.* 13:1.

Overbeek, J.T.G. 1966. Colloid stability in aqueous and nonaqueous media. *Discuss. Faraday Soc.* No. 42:7.

Pastorini, M.T. and Nunes, R.C.R. 1999. Mica as a filler for abs/polycarbonate blends. *J. Appl. Polym. Sci.* 74:1361.

Romo, L.A. 1963. Stability of nonaqueous dispersions. *J. Phys. Chem.* 67:386.

Zeiler, T., Kellermann, S., and Munstedt, H. 2000. Different surface treatments to improve the adhesion of polypropylene. *J. Adhes. Sci. Technol.* 14:619.

Interaction Forces Between Silica Surfaces in Aqueous Solutions of Poly(Ethylene Oxide)s

V.K. Paruchuri*, J. Nalaskowski*, and J.D. Miller*

An understanding of forces responsible for the stability of dispersions of colloidal particles when interacting with each other in aqueous solutions is of great interest academically and industrially. When water-soluble polymers are adsorbed onto colloidal particles, various effects on the stability of their suspensions can be expected ranging from steric stabilization to bridging flocculation. The possibility of detailed characterization of such systems using direct measurements of interaction forces between particles can help in further improvement of existing technologies of solid-liquid separation.

An atomic force microscope (AFM) colloidal probe was used to measure the interaction forces between a silica sphere and a silica substrate in the presence of water-soluble polymers. Poly(ethylene oxide) (PEO) with three different molecular weights (7,000, 43,000, and 63,000), each with a very narrow molecular weight distribution, were used in this study. The interaction forces measured with the AFM over an extended time scale are discussed in terms of the adsorption kinetics measured through in-situ FTIR/IRS measurements.

INTRODUCTION

Polymers can be used as dispersants or flocculants either to stabilize or aggregate particulate suspensions (Israelachvili, 1991). This is of great importance in many industrial areas such as slip casting, filler preparation (Drelich and Miller, 1995), cosmetics (Gans, 1990), paints and inks (Schroeder, 1998), oil production, and water purification. Considering the complexity of the interactions between polymer-covered surfaces and the technological importance for the control of colloidal suspension stability, these interactions have been the subject of many studies (Klein et al., 1986, 1984; Wong et al., 1992; Biggs, 1995, Milling, 1996; Braithwaite et al., 1996). Different techniques

* Dept. of Metallurgical Engineering, University of Utah, Salt Lake City, Utah.

such as photon correlation spectrometry (Cohen Stuart et al., 1984), microcalorimetry (Korn et al., 1980), and infrared spectroscopy (Eisenlauer et al., 1980) have been used to characterize polymer adsorption in general.

In addition, a number of tools have been developed to measure directly the interaction forces between surfaces. The surface force apparatus (SFA) first introduced by Tabor and Winterton (Tabor, 1969) and developed further by Israelachvili and co-workers (Israelachvili, 1973; Israelachvili and Adams, 1978) has been used with good success. However, the SFA technique can only be applied to a limited number of surfaces (mainly mica) because of its operating principles. More recently, the use of the atomic force microscope (AFM) (Binning et al., 1986) for measuring inter-particle forces was demonstrated by Ducker et al. (1991). Since then the AFM has become a popular tool for studying the interactions between surfaces (Parsegian et al., 1996; Luckham, 1996; Butt et al., 1995). When a colloidal particle is attached to an AFM tip, the force between this "colloidal probe" and a macroscopic surface can be measured as a function of separation distance. With the AFM, it is possible to investigate any surface, including particulate and fibrous materials.

Poly(ethylene oxide) (PEO) is an uncharged homopolymer which has been successfully used as a dispersant for latex (Cowell and Vincent, 1983) and flocculant for clays. The dispersing or flocculating effect of PEO depends on its conformation at the surface. Many parameters such as surface charge density, surface chemistry, electrolyte concentration, molar mass and adsorption time influence the adsorption of the polymer and its conformation. PEOs are available in a wide molecular weight range with narrow molecular weight distributions. Its adsorption behavior on silica has been studied extensively (Kawaguchi et al., 1984; Killman et al., 1985; van der Beek and Cohen Stuart, 1988; Dijt et al., 1990, 1992; Scheutjens et al., 1986; Mathur and Moudgil, 1997). The interaction forces between PEO covered mica surfaces in 0.1 M KNO_3 were studied by Klein and Luckham (Klein et al., 1986, 1984) using the SFA. At full coverage of polymer the interaction was repulsive at all separations and could quantitatively be explained by scaling theory for the interactions between adsorbed polymers (de Gennes, 1982, 1987). It was found that the range of interaction scaled with the molecular weight of the adsorbed polymer and commenced at separation distances in the range of 5–6 times the radii of gyration (R_g) of the polymer in solution.

In the present work we used the AFM colloidal probe technique, mentioned earlier, to measure the interaction forces between a silica sphere and a silica flat surface in different PEO solutions for different times of adsorption. Three different PEO solutions of molecular weights 7,000, 43,000, and 630,000 were studied. The approaching and the retracting parts of the force curves were monitored over extended periods of time, up to 24 hours, for any changes in the interaction forces. The measurements reported here were recorded at a scan rate of 1 Hz, which for a scan size of 500 nm, means the surfaces are driven together at a rate of 1 $\mu m\ s^{-1}$, which roughly corresponds to the Brownian collisions in colloidal suspensions. Hence, the forces measured can be considered representative of the interaction between colloidal particles in solution. The changes in the force curves over time are compared with the adsorption kinetics measured by the FTIR/IRS technique.

EXPERIMENTAL

PEO polymers, with molecular weights of 7, 43, and 630 kg mol^{-1} (hereafter referred as 7k, 43k, and 630k, respectively) were obtained from Scientific Polymer Products, Inc. (Ontario, NY). The polymers were of very narrow molecular weight distribution ($M_w/M_n \sim 1$). Some characteristic properties of the polymers used are summarized in Table 1.

TABLE 1 Characteristics of the PEO molecules used in this study

PEO Sample	Molecular Weight (kg/mol)	Polydispersity (Mw/Mn)	Radius of Gyration (nm)
7k	7,000	1.01	2.4
43k	43,500	1.10	7.8
630k	630,000	1.05	36

Potassium nitrate (KNO_3) was obtained from Sigma Chemicals (St. Louis, MO). Polymer solutions, 0.05% by weight, were prepared as aqueous solutions with 0.25 M KNO_3 as background electrolyte for ionic strength control. Solutions were prepared a day before the experiments and allowed to stay overnight for equilibration. Deionized water (18 mΩ) from a Milli-Q system (Millipore) was used for preparing the solutions. Colloidal silica spheres (4.3 μm diameter) were from Bangs Labs, and the fused silica substrate was obtained from Harrick Scientific.

AFM Force Measurements

Force measurements were conducted using a Nanoscope IIIA atomic force microscope (Digital Instruments, Inc., Santa Barbara, CA). The colloid probe technique of measuring forces with the AFM has been described extensively elsewhere (Ducker et al., 1992) and will not be discussed here.

Standard silicon nitride cantilevers (Digital Instruments, Inc., Santa Barbara, CA) were used for all the force measurements. The spring constant for the cantilevers was reported to be 0.12 Nm^{-1} and the same value was used to convert the force versus deflection curves into force versus separation curves.

The colloidal silica spheres were mounted on the tipless AFM cantilevers using a speed bonder and activator (Loctite Corporation) by means of a micromanipulator and a CCD camera/monitor system. The cantilevers with silica spheres and the silica substrates were cleaned in Argon plasma cleaner just prior to the experiments.

After assembling the cantilever and wafer in the fluid cell, the fluid cell was filled with electrolyte solution and the repulsive silica-silica interaction was measured to check for the cleanliness of surfaces. Subsequently, polymer solution was pumped into the cell and measurements were taken at the stipulated times. All the measurements reported here were done in situ, i.e., in the presence of the polymer solutions. The experiments were done at the natural pH of the solution.

FTIR/IRS Adsorption Density Measurements

FTIR transmission spectra of PEO solutions were collected as a function of concentration in order to calculate the absorptivity of the PEO solution Beer's law. The FT-IR internal reflection experiments were done as a function of time using the closed cylindrical cell with a cylindrical silicon IRE and the Bio-Rad 6000 FTS spectrometer. The areas under the CH_2 peaks were integrated and the surface excess was calculated, according to the FTIR/IRS adsorption density equation, which was first derived by Sperline et al. (1987) and later developed by Miller and co-workers (Kellar et al., 1989, 1990; Free and Miller, 1996)

$$A = N\varepsilon b_{eff}C + N\varepsilon\frac{d_e}{d_p}1000\ \Gamma$$

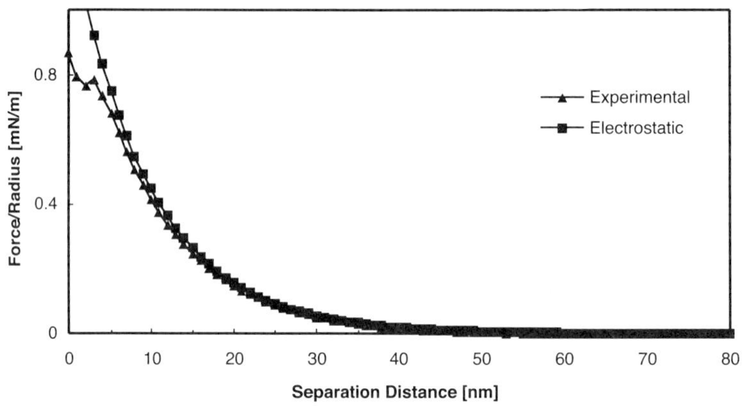

FIGURE 1 Interaction force between two silica surfaces in the presence of 10^{-3}M KCl solution. The experimental data is fitted to a theoretical electrostatic force predicted according to the DLVO theory.

Where: A is integrated absorbance (cm^{-1}), ε is absorptivity (liter/mole·cm^2), C is concentration (mole/liter), N is number of reflections, d_e is effective path length (cm), d_p is effective depth of penetration (cm) and Γ is the surface excess (moles/cm^2).

RESULTS

Figure 1 shows the interaction force curve between silica surfaces in the presence of 1×10^{-3} M KCl solution. The measured force curve shows complete repulsive behavior both on approach and on retraction. At larger separation distances the force decays exponentially, characteristic of an electrostatic double layer decay length. The results are in nice agreement with the Derjaguin/Landau/Verwey/Overbeek (DLVO) theory for electrical double layer repulsive interactions between silica surfaces, thereby standing evidence that the surfaces are clean. The theoretical curves shown in the figure were calculated by fitting the experimental data according to the Poisson Boltzmann equation assuming constant potential.

The interaction forces between silica surfaces in the presence of PEO solutions, with different molecular weights, are presented in Figures 2–4. The interaction forces are presented for three different adsorption times–5 minutes, 60 minutes and 24 hours. In each case the approaching part and the retracting part of the force curves are presented separately. The approach curve represents the electrostatic or steric forces experienced by the surfaces as they approach each other, whereas, the retraction curve gives an indication of the adhesion forces acting between the surfaces. As mentioned earlier, each force curve presented here is an average of 30 observations taken at different places on the flat silica surface.

Interaction forces between the silica surfaces in the presence of 7k molecular weight PEO solution are shown in Figure 2. Three different curves for adsorption times of 5 minutes, 60 minutes and 24 hours are presented. The interaction forces did not show any changes between 5 and 60 minutes but were significantly different for 24 hours. In the approaching part of the force curve the magnitude and range of attraction were almost identical for adsorption times of 5 and 60 minutes. However, the attraction decreased significantly for forces measured after 24 hours. The retracting force curves

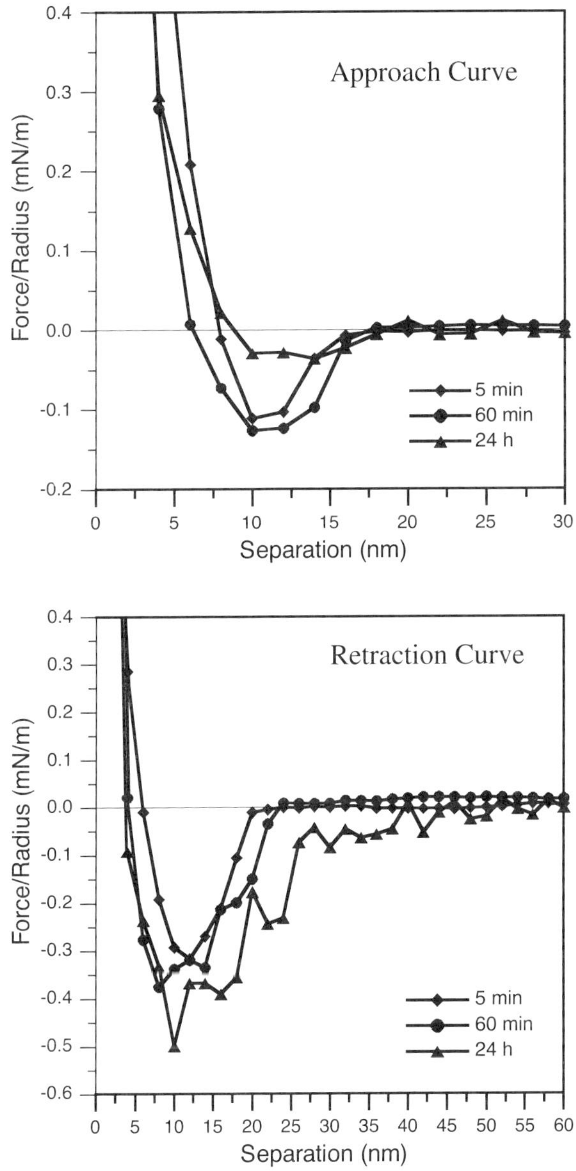

FIGURE 2 Interaction force curves between a silica sphere and a silica flat in aqueous solution of 0.05% PEO (7,000 molecular weight) and 0.25 M KNO_3, at natural pH. Approach and retraction curves are shown for three different adsorption times—5 minutes, 60 minutes and 24 hours.

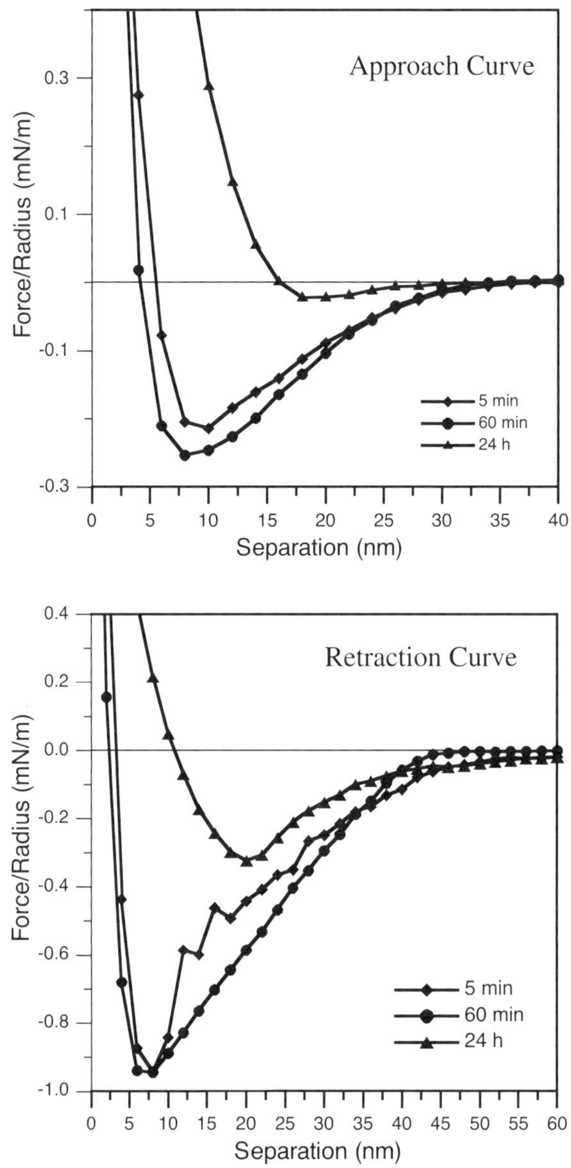

FIGURE 3 Interaction force curves between a silica sphere and a silica flat in aqueous solution of 0.05% PEO (43,000, molecular weight) and 0.25 M KNO_3, at natural pH. Approach and retraction curves are shown for three different adsorption times—5 minutes, 60 minutes and 24 hours.

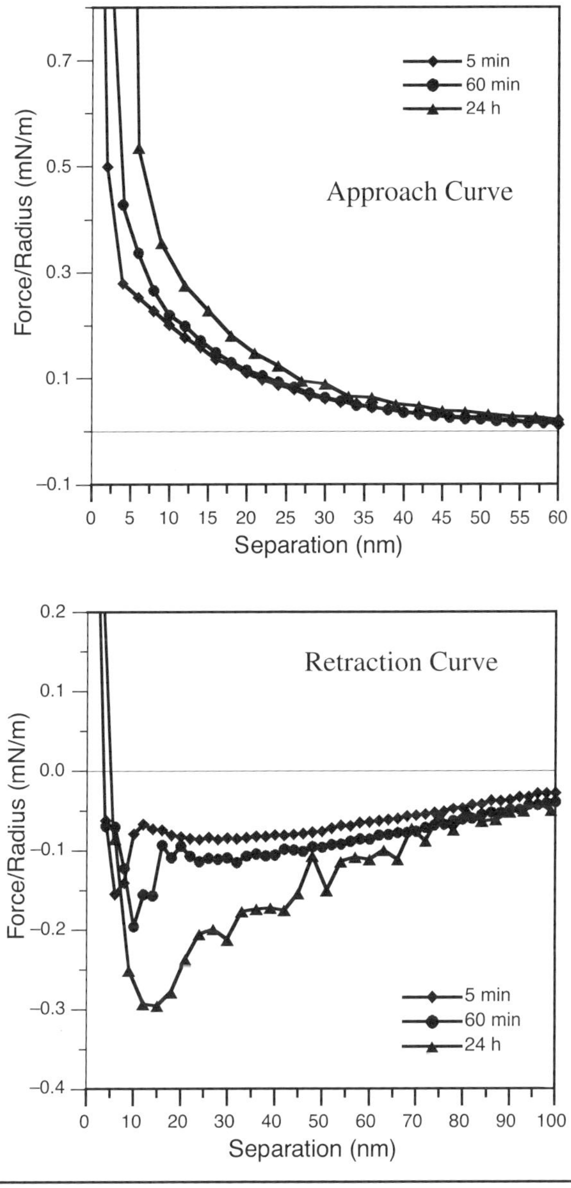

FIGURE 4 Interaction force curves between a silica sphere and a silica flat in aqueous solution of 0.05% PEO (630,000, molecular weight) and 0.25 M KNO_3, at natural pH. Approach and retraction curves are shown for three different adsorption times—5 minutes, 60 minutes and 24 hours.

showed some interesting changes as well. The range and magnitude of the pull off (adhesive) forces did not show significant changes between 5 and 60 minutes but differed significantly after 24 hours. Both the range and magnitude showed a significant increase.

The force curves obtained for the adsorption of 43k PEO solution are presented in Figure 3. Again, the approach part of the force curves was similar for 5 minutes and 60 minutes. The attractive force was identical in both the cases, both in range and in magnitude. But the magnitude of the attraction force was larger in this case than for the 7k PEO system. The retraction part of the force curves showed a different trend. While the range and magnitude of the adhesive forces were more or less same for 5 and 60 minutes, the magnitude of the adhesive force decreased after 24 hours. This is in sharp contrast with the 7k PEO case where the magnitude and range of adhesive forces increased after 24 hours.

Interaction forces observed for the 630k PEO solution are presented in Figure 4. The force curves are significantly different from the previous two cases. For the approach curve, a repulsive force was observed for all equilibration times. Even though the force curves exhibited the same behavior for different times of adsorption, there were some subtle differences. The distance at which the repulsive forces start, and the distance of closest approach for the two surfaces, was different for different times. The retraction force curve, however, differed significantly, with the range and magnitude of the adhesion increasing significantly with time.

DISCUSSION

The approaching part of the force curves showed a significant attraction for the PEO solutions with molecular weights 7k and 43k (Figures 2, 3) for smaller incubation times, i.e., 5 minutes and 60 minutes. This attractive component can be attributed to bridging. This may seem surprising, since usually bridging is only found for unsaturated surfaces (Fleer et al., 1993), but our experimental conditions are well within the plateau region of the adsorption isotherm of PEO on silica. The first reading here is reported for 5 minutes of adsorption but reflectometer experiments by Dijt et al. (1990) showed that under appropriate flow conditions, the maximum adsorption of PEO on silica is reached within 40 seconds.

Even with the PEO adsorption layers presumably saturated, there is no sign of steric repulsion before constant compliance is reached and it appears that it is still possible for the polymer chains adsorbed at one silica surface to attach to another silica surface. This might be explained as follows. Upon approach, the concentration of polymer segments between the surfaces increases. Due to the flexibility of the PEO chains, fast reconformation of the adsorbed layers can take place and additional adsorption of polymer segments is possible, with some of the newly formed PEO/silica contacts formed are bridging contacts. From the fact that attraction occurs on approach and the interaction forces are the same at different places, we can conclude that on the time scale of the experiments, the conformation of the polymers can indeed adapt to the changing conditions. Similar explanation can be applied to the long range adhesion observed in the retraction force curves. The bridging contacts will try to overcome the retracting force thereby stretching the polymer chains. So before the surfaces are totally retracted apart the polymer chains first need to be stretched to break the bridging contacts. However, for longer times of adsorption, although there is no necessary increase in the adsorbed amount of PEO, the polymer chains are given enough amount of time to organize and reorganize at the surfaces there by leaving little scope for any bridging contacts. Accordingly, after 24 hours

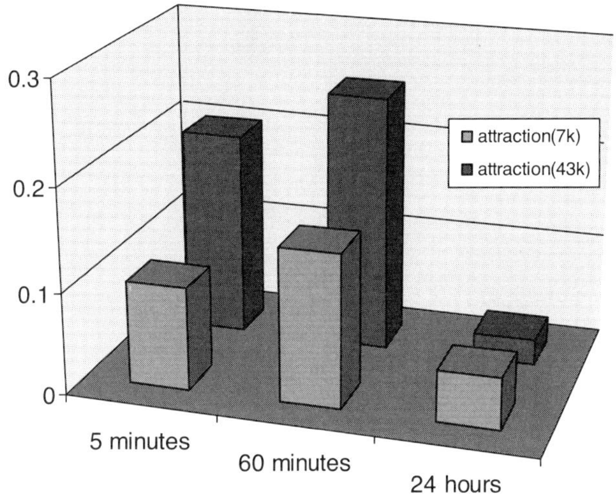

FIGURE 5 Maximum attractive force between silica surfaces for an aqueous solution of 0.05% PEO (7k and 43k), 0.25 M KNO_3solutions, at natural pH. The data are taken from the force curves shown in Figures 2 and 3.

of adsorption, the magnitude of attraction decreased in both the cases of 7k and 43k molecular weights PEO solutions, as shown in Figure 5.

In order to provide further evidence that the change in interaction forces shouldn't necessarily come from the change in the adsorption density of polymers, in-situ Fourier-transform infrared (FTIR) internal reflection experiments were done to study any changes in adsorption density that might occur during the equilibration time. The infrared spectra are shown in Figure 6. From these spectra the adsorption density of PEO was calculated as explained in the earlier section. The plot of adsorption density as function time is shown in Figure 7. No significant change in adsorption density with time was observed from FTIR experiments, which can suggest the rapid saturation of silica surface with PEO (plateau adsorption density of 0.3 mg/m^2). These results are consistent with the trends reported in the literature (Dijt et al., 1990, have observed a plateau adsorption density of 0.5 mg/m^2, for a PEO with molecular weight 56,300, from reflectometry experiments); however the absolute values of adsorption density are somewhat smaller. From FTIR measurements it was found that changes in force curves cannot be related to simple increase in adsorption density but also to the rearrangement of PEO molecules in time.

For the PEO system with very high molecular weight (Figure 3), force curves on approach show a steep repulsion starting at greater separation distances. The repulsive interaction increases monotonically at first, before changing the slope at closer distances. This can be explained as follows. The adsorption density of polymers shows a direct dependence on molecular weight. The adsorbed thickness of PEO on silica varies accordingly as $M^{0.56}$ (Kato et al., 1981), where M is the polymer weight. Therefore the thickness of the adsorbed PEO layer on silica is much higher in the case of the high molecular weight PEO (630k) than the lower molecular weight PEO (7k and 43k). This thick adsorbed layer makes any kind of bridging impossible. As the surfaces approach closer, the osmotic pressure between the adsorbed polymers increases steeply, giving rise to the greater steric repulsive forces. This fact is further corroborated by the evidence that the distance of closest approach is much higher for 630k PEO than either of 7k or 43k PEO solutions.

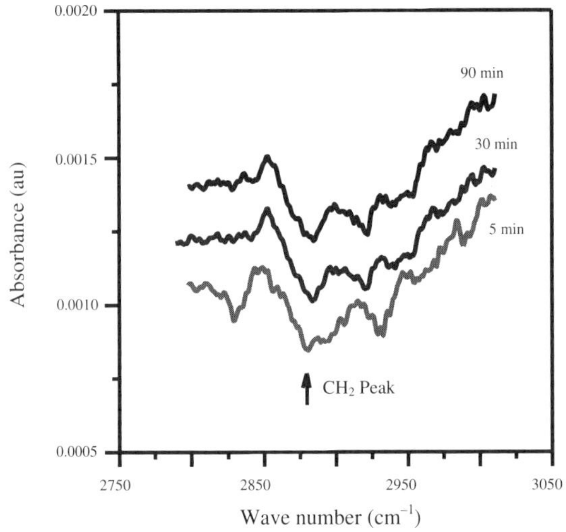

FIGURE 6 In-situ Fourier transformation infrared (FTIR/IRS) spectra taken at the surface of a silicon IRE for an aqueous solution of 0.05% 43k PEO in 0.25 M KNO_3 solution, at natural pH. The spectra, collected for three different adsorption times (5 minutes, 30 minutes and 90 minutes), are shown.

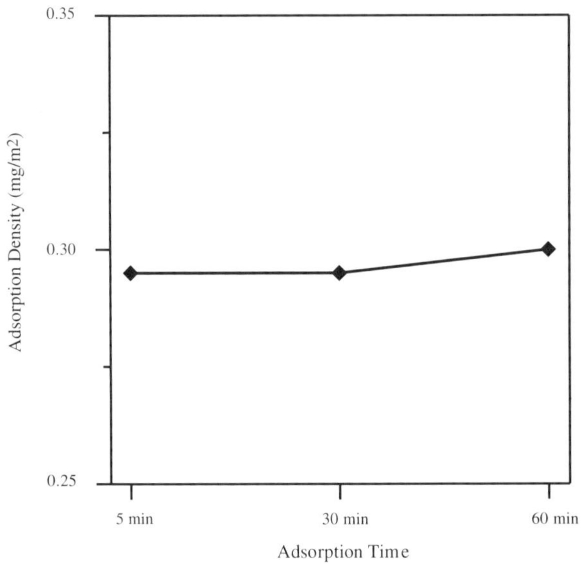

FIGURE 7 The calculated adsorption density values adsorption from an aqueous solution of 0.05% 43k PEO, 0.25 M KNO_3, at natural pH, at the surface of a silicon IRE. The values were calculated integrating the areas under the CH_2 peak of spectra shown in Figure 6.

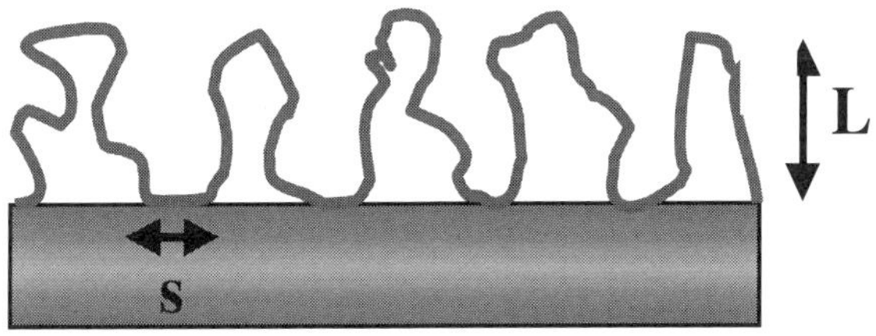

FIGURE 8 Schematic representation of surface polymer configuration assumed in the Alexander-de Gennes' scaling theory

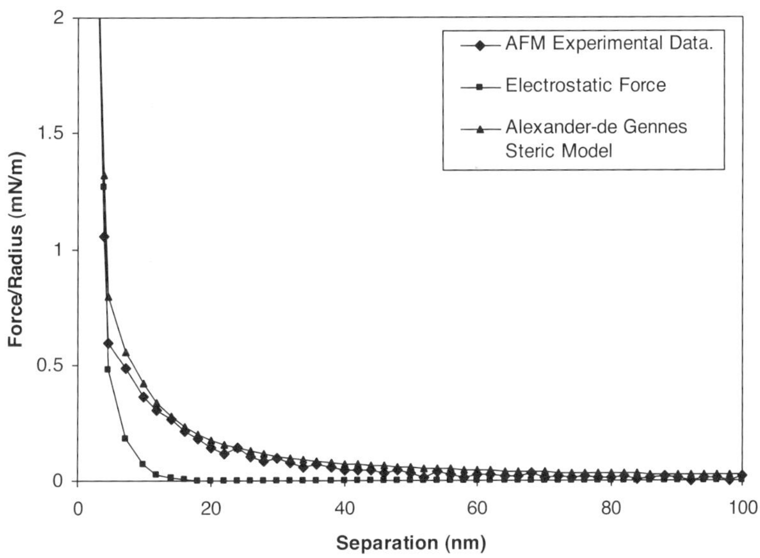

FIGURE 9 The experimental approach curve in Figure 4 fitted with the Alexander-de Gennes' scaling theory. There is an excellent fit at greater separation distances, although a slight deviation is observed at closer separation distances. The parameters used for fitting the curves are summarized in Table 1.

Several theories have been developed to describe the steric forces experienced by surfaces in the presence of adsorbed polymer layers, both qualitatively and quantitatively. Without going into detail, for an adsorbed polymer conformation as shown in Figure 8, the Alexander-de Gennes' scaling theory (de Gennes, 1982, 1987) can be given as

$$\frac{F(D)}{R} = \frac{16\pi kTL}{35s^3}\left[7\left(\frac{2L}{D}\right)^{5/4} + 5\left(\frac{D}{2L}\right)^{7/4} - 12\right]$$

where, k = Boltzmann Constant (J/K), T = temperature (K), L = layer thickness (m), S = mean separation distance between adsorbed polymers (m), and F = force as a function of separation distance 'D' between the surfaces.

In Figure 9, we show the experimental force curves fitted with the theoretical values predicted according to the Alexander-de Gennes' scaling theory. Also, just to bring out the difference the calculated electrostatic force curve based on the DLVO theory is

TABLE 2 The values of the mean separation distance between the adsorbed polymers at the surface 'S', and the thickness of the adsorbed polymer 'L', obtained from fitting the approach curve in Figure 4 with the Alexander-de Gennes' steric model. The geometrical interpretation of 'S' and 'L' values are shown in Figure 8.

Time	L (nm)	S (nm)
5 min	65.7	9.3
60 min	71.4	8.3
24 hours	82.8	6.5

shown in the figure. Here it should be mentioned that as the scaling theory is only applicable to surfaces experiencing steric forces, attempts were not made to fit the force curves obtained in the presence of lower molecular weight solutions, due to the presence of an attractive component in the forces.

It can be seen that the fit to the scaling theory curve is very remarkable indeed, thereby suggesting that the forces being seen are due to steric repulsion. The values for the layer thickness and the mean separation distance obtained through fitting for different times of adsorption are summarized in Table 2. The layer thickness for the polymer adsorbed on two surfaces is close to twice the radius of gyration of the molecule given in Table 1. Once again it can be seen that there are some small reconformations (changes in 'S' and 'L' values) with time.

The retraction force curves in Figure 4, for the 630k PEO system, show great adhesion forces both in range and in magnitude, although there is no attraction seen in the approach curves. In the absence of attractive force it is expected that there is no bridging interaction between the surfaces (i.e., the polymer adsorbed on one surface interacting with the other surface). However, that doesn't necessarily mean that there is no interaction between the adsorbed polymer molecules on both surfaces. Hence, although steric repulsion is experienced between the two approaching surfaces, once they overcome the repulsive barrier and enter the constant compliance region, there is some interaction between the polymer molecules adsorbed on the two surfaces as the dangling tails interact with each other. The adhesion force seen from the retraction curves is a result of these contacts being broken as the surfaces are being pulled apart. The polymers will be stretched completely before eventually giving away, leading to the great range of the adhesion forces. Again, the changes in the range and magnitude of adhesion with time can be attributed to the reconformation of the polymer at the surfaces, as can also be evidenced by the changes in the fitting parameters, summarized in Table 1, with time.

CONCLUSIONS

From the AFM results, it can be concluded that interaction forces between silica surfaces depend on the molecular weight of the polymer and the time of adsorption. For the low molecular weight PEO attractive bridging interactions are observed. The attractive component of the surface force decreases with time, which can be related to adsorption and rearrangement of the PEO layer. The range of the force during approach and retraction increases with molecular weight of PEO, which might be expected assuming some stable conformation of polymer at the silica surfacc. Increase in the range and magnitude of the pull-off force and steps observed during retraction can be explained by bridging between silica surfaces.

In the case of the very high molecular weight PEO, steric repulsive forces were observed in the approach curves at all times of adsorption. The steric forces corresponded very well with those calculated based on the Alexander-de Gennes' scaling theory. Adhesion forces of greater magnitude and acting at larger separation distances were seen in the retraction curves.

Any changes in the force curves are attributed to the reconformation of polymers at the surfaces and not necessarily due to the changes in adsorption density values, as is evident from adsorption density measurements by FTIR/IRS.

ACKNOWLEDGMENTS

The financial support provided for this work by the NSF Engineering Research Center (ERC) for Particle Science and Technology at University of Florida and by the Department of Energy, Basic Science Division Grant No. DE-FG-03-93ER14315, is gratefully appreciated. Discussions with Simon Biggs, University of Leeds, were most helpful in preparing this manuscript.

REFERENCES

Biggs, S., 1995, *Langmuir* 11, 156.
Binnig, G., Quate, C.F. and Gerber, C., 1986, *Phys. Rev. Lett.*, 56, 930.
Braithwaite, G.J.C., Howe, A. and Luckham, P.F., *Langmuir,* 12, 4224.
Butt, H.J., Jaschke, M. and Ducker, W.A., 1995, *Bioelectrochem. Bioenerg.*, 38, 191.
Cohen Stuart, M.A., Waajen, F.H.W.H., Cosgrove, T., Crowley, T.L. and Vincent, B., 1984, *Macromolecules,* 17, 1825.
Cowell, C. and Vincent, B., 1983, *J. Colloid Interface Sci.*, 95, 573.
Israelachvili, J.N., 1991, *Intermolecular and Surface Forces,* 2nd ed., Academic Press: London.
de Gennes, P.G., 1982, *Macromolecules,* 15, 4942.
de Gennes, P.G., 1987, *Adv. Colloid Interface Sci.*, 27, 189.
Dijt, J.C., Cohen Stuart, M.A., Hofman, J.E. and Fleer, G.J., 1990, *Colloids Surfaces*, 51, 141.
Dijt, J.C., Cohen Stuart, M.A. and Fleer, G.J., 1992, *Macromolecules,* 25, 5416.
Drelich, J. and Miller, J.D., 1995, *Minerals & Metallurgical Processing,* 12, 197.
Ducker, W.A., Senden, T.J. and Pashley, R.M., 1991, *Nature,* 353, 239.
Ducker, W.A., Senden, T.J. and Pashley, R.M., 1992, *Langmuir,* 8, 1831.
Eisenlauer, J., Killmann, E. and Korn, M., 1980, *J. Colloid Interface Sci.*, 74, 120.
Fleer, G.J., Scheutjens, J.M.H.M. Cohen Stuart, M.A., Cosgrove, T. and Vincent, B., 1993, *Polymers at Interfaces,* Chapman and Hall, London.
Free, M.L. and Miller, J.D., 1996, *Int. J. Miner. Process.*, 48, 197.
Gans, E.H., 1990, *Polymeric Materials Science and Engineering, Proceedings of the ACS Division of Polymeric Materials Science and Engineering,* 63, 225.
Israelachvili, J.N., 1973, *J. Colloid Interface Sci.,* 44, 255.
Israelachvili, J.N. and Adams, G.E., *J. Chem. Soc., Faraday Trans.*, 74, 1975.
Kato, T., Nakamura, K., Kawaguchi, K. and Takahashi, A., 1981, *Polym. J.*, 13, 1037.
Kawaguchi, M., Mikura, M. and Takahashi, A., 1984, *Macromolecules,* 17, 2063.
Kellar, J.J., Cross, W.M. and J.D. Miller, 1989, *Appl. Spectrosc.,* 43, 1456.
Kellar, J.J., Cross, W.M. and J.D. Miller, 1990, *Separation Sci. and Tech.*, 25, 2133.
Killman, E., Maier, H., Kaniut, P. and Gutling, N., 1985, *Colloids Surfaces*, 15, 261.
Klein, J. and Luckham, P.F., 1986, *Macromolecules,* 19, 2007.
Klein, J. and Luckham, P.F., 1984, *Macromolecules,* 17, 1041.
Korn, M., Killmann, E. and Eisenlauer, J., 1980, *J. Colloid Interface Sci.*, 76, 7.
Lafuma, F., Lindner, P., Aguerrre, O. and Cabane, *J. Colloid Interface Sci.,* 153, 55.
Luckham, P.F., 1996, *Curr. Opinion Colloid Interface Sci.*, 1, 39.
Mathur, S. and Moudgil, B.M., 1997, *J. Colloid Interface Sci.,* 196, 92.

Parsegian, V.A. and Evans, E.A., 1996, *Curr. Opinion Colloid Interface Sci.*, 1, 53.
Scheutjens, J.M.H.M., Fleer, G.J. and Cohen Stuart, M.A., 1986, *Colloids Surfaces*, 21, 285.
Schroeder, J., 1998, *European Coatings Journal*, 5, 334.
Sperline, R.P., Muralidharan, S. and Freiser, H., 1987, *Langmuir*, 3, 198.
Tabor, D. and Winterton, R.H.S., 1969, *Proc. R. Soc. London Ser. A*, 312, 435.
van der Beek, G.P. and Cohen Stuart, M.A., 1988, *J. Phys. Paris*, 49, 1449.

Influence of Dispersants and Solution Conditions on Kaolin Stability Mechanisms

Abbas A. Zaman*, Brij M. Moudgil*, and Hassan El-Shall*

The addition of organic and inorganic dispersants to aqueous suspensions of heterogeneous particles (kaolin) was found to significantly affect the solubility of metal ions from kaolin particles and stability of the dispersion. In most of the industrial applications of kaolin, highly concentrated slurries of controlled stability and fluidity are required. The solubility of kaolin particles in terms of aluminum and silicon ions and its effect on the stability of the slurry was investigated. The main objectives of this study were (1) to study the effect of the extracted metals on the properties and performance of the slurry, (2) to understand the reaction mechanism of dissolution of metals from kaolin, and (3) to understand the interaction mechanism of dispersants with kaolin particles. The controlling variables were dispersant dosage, slurry pH, solids loading, and aging.

INTRODUCTION AND BACKGROUND

In most of the applications of kaolin such as ceramics, paper coatings, paints, pigments, and other industrially important materials, highly concentrated stable dispersions of controlled fluidity and stability are needed which can be achieved through the addition of inorganic and organic dispersing agents to the system. There are several important factors such as concentration of the dispersing agents, solubility of kaolin particles, soluble salts, and metal ions concentration arising from the solubility of the particles in the suspending media that may have a significant role in governing the rheological behavior and stability of the dispersion (Konta, 1995). Therefore, a basic understanding of mechanism of adsorption, interaction of dispersing agents with the active sites of the particles and their effect on solubility of kaolin, and also reaction mechanism of dispersants with soluble ions from the particles becomes of significant importance.

* Dept. of Materials Science and Engineering, Engineering Research Center for Particle Science and Technology, University of Florida, Gainesville, Fla.

Kaolinite crystal consists of altering layers of silica tetrahedra and alumina octahedra and has the unit cell chemical composition of $Si_2Al_2O_5(OH)_4$. It has a complicated surface chemistry due to the presence of heterogeneously charged edges and faces constituting the particle (Brady et al., 1996). Kaolin edges contain both silica and alumina-like charge sites and the faces contain only silica-like charge sites, which remain negatively charged across the pH range. The edges are positively charged at low pH, but progress through an isoelectric point to posses a negative charge at high pH (Johnson et al., 1998, 2000). In the absence of dispersing agents, electrostatic edge-to-face attraction causes agglomeration of the particles in the form of a card-house type structure (even at very low volume fractions) depending on the pH of the dispersion (Herrington et al., 1992). In some applications, the solids content of kaolin dispersions can be as high as 60–70 %solids (wt) and the slurry is required to be stable with acceptable viscosity for processibility (Sjoberg et al., 1999). Increasing the pH of the suspension above the iso-electric point of alumina (IEP) will cause an increase in electrostatic repulsion between the particles, which may result in a well-dispersed suspension (Herrington et al., 1992). However, this may not be effective at high solids loadings and, also, high pH may not be desirable due to high corrosion rate, safety, and high operational cost. Due to hydrophilic nature, kaolinite can be dispersed in water using anionic inorganics, polyelectrolytes, and surfactants. Effects of pH, ionic strength, and polymer charge, on the adsorption of polymers and surfactants on the surface of kaolin has been studied in the past (e.g., Sastry et al., 1995; Lee et al., 1996). Since the concentration of the ions resulting from the solubility of the particles in the suspending media may affect stability and rheology of kaolin dispersions, our main objective in this study is to investigate the effect of dispersing agents on solubility of kaolin particles and to understand the reaction mechanism between different dispersants and the dissolved metal ions in supernatant. Results are presented and discussed in terms of the effect of the dosage of the dispersing agents, and pH on the concentration of aluminum and silicon ions in supernatant.

MATERIALS AND METHODS

A high purity-grade East Georgia crude kaolin supplied by Engelhard Corporation (without any pretreatment of the mineral surfaces) was used in this investigation. The BET nitrogen-specific surface area and density of the particles were measured to be 20.32 $m^2 \cdot g^{-1}$ and 2.72 $g \cdot cm^{-3}$ respectively. Chemical analysis of the powder indicated a relatively pure sample with insignificant amounts of Mg^{2+}, Ca^{2+}, Na^+ and other impurities. Sodium salt of a low molecular weight polyacrylic acid (NaPAA, MW = 3,400, polydispersity index = 1.18), sodium hexametaphospate (NaHMP), and sodium silicate (Na-Silicate) were used as dispersing agents to prepare the suspensions in this investigation. Stock solutions of dispersing agents were prepared in the concentration of 15,000 ppm and diluted with ultra pure water to the desired concentration and used as the suspending fluid. The required mass of dry kaolin was then slowly added to the solution while mixing the sample by shaking and vibration. After adding all the particles, the pH of the suspension was adjusted to the desired value using Fisher brand HCl and Na_2CO_3 as pH modifiers (in one direction only). The samples were then aged for a given period of time while agitating on a Burrell Model 75 wrist action shaker. For chemical analysis, the samples were centrifuged at 15,000 rpm for 15 minutes, supernatant was carefully withdrawn and then analyzed for the dissolved metal ions, mainly aluminum and silicon, using a Perkin Elmer Plasma (ICP-AES) instrument. The metal ions concentrations are reported as mg/l (ppm) in the supernatant.

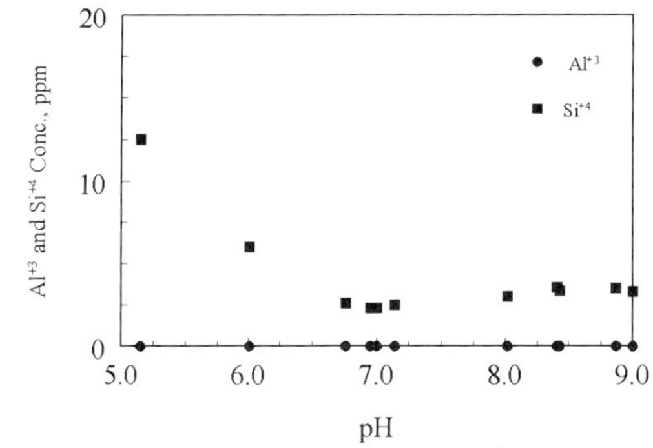

FIGURE 1 Effect of the pH of the dispersion on the concentration of aluminum and silicon ions in supernatant for dispersions of kaolin particles at 35 %wt solids in the absence of dispersing agents (25 °C)

RESULTS

Solubility of Kaolin in the Absence of Dispersing Agents

Solubility of kaolin in terms of the concentration of the dissolved metal ions in supernatant was investigated in the absence of dispersing agents as a function of pH of the slurry. Figure 1 represents the concentration of metal ions in supernatant as a function of pH of the dispersion at a fixed level of solids contents. As can be observed, in the absence of dispersing agents, while the concentration of silicon ions in supernatant decreases with increasing pH, aluminum ions do not appear in the solution over the entire pH range studied. This may be due to (1) formation and precipitation of $Al(OH)_3$ which its solubility in water is very small in the pH range of 4.5–8.5 (Wefers and Misra, 1987; Mcmurry and Fay, 1998), and (2) formation and precipitation of alumino-silicate gel (Yuan, 1998). Yuan et al. (1998) have explained the reduction in the concentration of silicon ions in terms of the formation of insoluble alumino-silicate gel that will be precipitated with solids during the centrifugation process. Also, the formation of alumino-silicate gel layer on the surface of kaolin may prohibit dissolution of the particles, which will result in a decrease in the concentration of the released ions in supernatant.

In addition to the possible production of alumino-silicate gel, formation of aluminum hydroxide with negligible solubility in the pH range of 4.5–8.5 that precipitates with solids may be another reason for the decrease in the concentration of Al^{+3} ions in supernatant.

Effect of the Dosage of Dispersants on Solubility of Kaolin

The dosage of the dispersing agents used to prepare kaolin dispersions has a significant role on the solubility of the particles as shown in Figures 2 and 3 which are plots of the concentration of aluminum and silicon ions in supernatant as a function of the dispersants dosage used to prepare the slurry. It appears from the data that NaPAA and NaHMP have similar effects on the solubility of kaolin particles for dosage levels up to 5 mg/(g solids). When the dosage of the dispersing agent is increased above 5 mg/(g solids),

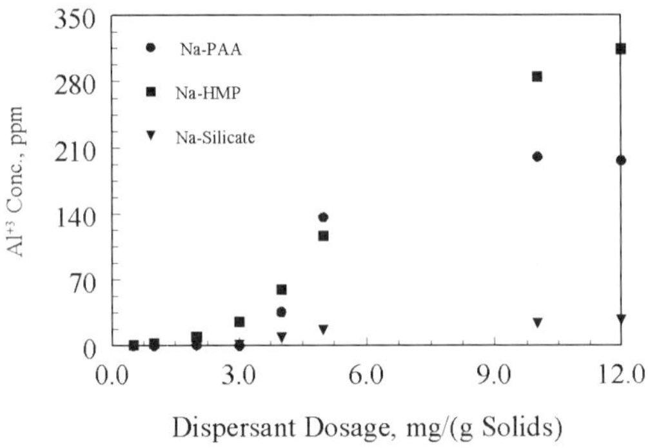

FIGURE 2 Effect of dosage of different dispersing agents on the concentration of aluminum ions in supernatant for dispersions of kaolin particles at 55 %wt solids (25 °C)

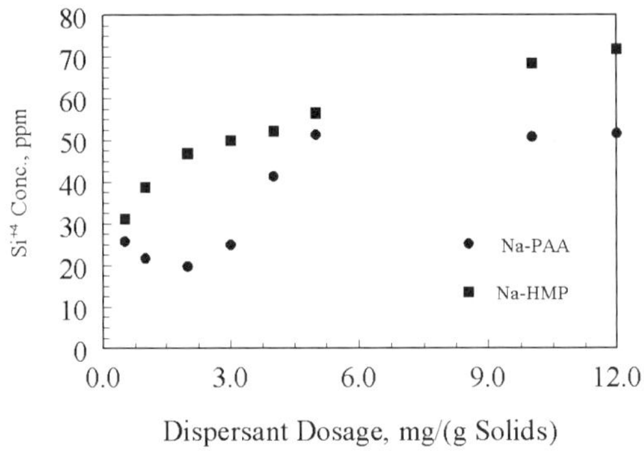

FIGURE 3 Effect of dosage of different dispersing agents on the concentration of silicon ions in supernatant for dispersions of kaolin particles at 55 %wt solids (25 °C)

the concentration of aluminum ions in supernatant is considerably higher in the presence of NaHMP. The phosphorous ions have strong affinity to aluminum ions and will compete with OH^- for aluminum ions to form water soluble complex ions. Similar behavior was observed by Yuan et al. (1998), who studied the influence of different dispersants on solubility of calcined kaolin and concluded that these complexes can displace the H_2O or OH^- ligand in the first coordination sphere of Al^{+3} inhibiting $Al(OH)_3$ crystallization and precipitation. It is shown by Hsu (1989) that the phosphates ability to form stable soluble complexes is so strong that crystalline $Al(OH)_3$ cannot develop even at pH levels above 9–10 and, therefore, almost all Al^{+3} will stay in the dissolved form.

When NaPAA was used as dispersing agent, the concentration of the released aluminum ions in supernatant increased significantly with increasing the polymer dosage in the system. PAA contains ionizable carboxylic acid groups, which are capable of dissociation depending on the pH of the solution. Kaolinite consists of silica like charge sites on the basal planes and alumina like and silica like charge sites on the edge. The results of our earlier work on the adsorption of PAA on the surface of alumina, silica, and kaolin (Zaman et al., 2002) show that negatively charged dispersing agents such as NaPAA and NaHMP adsorb on alumina sites of kaolin particles. Results presented in

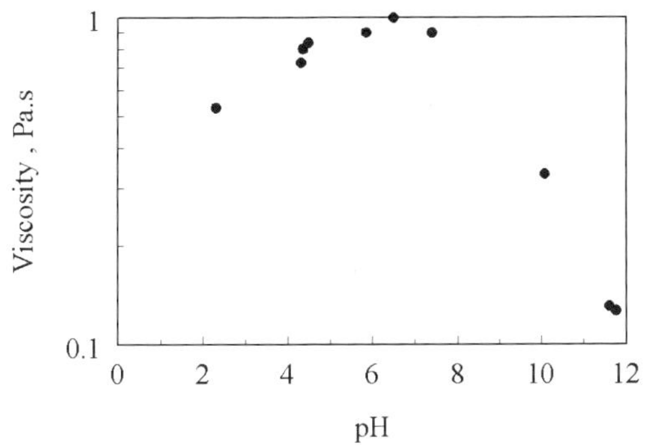

FIGURE 4 Effect of pH on the viscosity of dispersions of kaolin particles of 15 %vol solids content at a shear rate of 100 s^{-1} in the absence of dispersing agents (25 °C)

Figure 2 indicate that NaPAA and NaHMP may have similar reaction mechanisms with kaolin particles. The high concentration of aluminum ions in supernatant in the presence of these dispersing agents in the system suggests that aluminum ions form a water soluble complex with each of these dispersing agents which remains in supernatant and, therefore, can be detected by ICP measurements.

When Na-silicate was used as the dispersing agent, the concentration of aluminum ions in supernatant was very small as compared with the other two dispersing agents. Aluminum ions were not detectable in supernatant for Na-silicate dosage levels of up to 4 mg/(g solids). Above this level, an increase was observed in the concentration of aluminum ions in supernatant, which appeared to reach a plateau at higher levels of Na-silicate dosage. The low concentration of aluminum ions in supernatant may be attributed to the formation of insoluble alumino-silicate gel as was explained earlier. It is reported by Okamoto (1957) and Iler (1973) that as aluminum becomes available in the solution, it will inhibit the release of silicon from kaolin.

Steady Shear Flow Properties

The rheological properties of kaolin dispersions is a function of different factors including particle size and particle size distribution, particle shape, the volume fraction of particles, type of the stabilization method, and the amount and type of dispersing agents added to the dispersion. In the absence of dispersing agents, interparticle interactions between the edges and the basal planes of the kaolin particles promote aggregation of the particles into edge-to-edge, edge-to-face, or face-to-face structures in concentrated aqueous dispersions causing a significant increase in the viscosity of the dispersion. This can be observed in Figure 4, which is a plot of viscosity of kaolin dispersions of 15 %vol as a function of pH of the slurry and shear rate. The viscosity is maximum over a pH range, which is believed to be the isoelectric point of the edge of the particles. As the pH of the system is decreased, the overall charge on the edge of the particles becomes positive which will reduce edge-to-edge interaction between the particles leading to a better dispersed slurry of lower viscosity. At pH levels below the isoelectric point of the edge but above the point of zero charge of the basal planes, overall charge on the edge will be positive, but the basal planes will be negatively charged and therefore edge-to-face interactions will cause aggregation of the particles. When the pH is increased above the

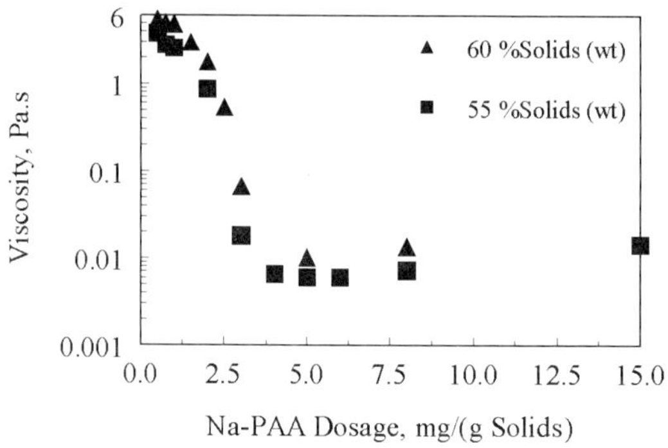

FIGURE 5 Effect of the dosage of NaPAA on the viscosity of dispersions of kaolin particles of 55 %wt solids and 60 %wt solids at a shear rate of 1000 s^{-1} (suspension pH = 7, 25 °C)

isoelectric point of the edge, both basal planes and the edge will be negatively charged which will reduce both edge-to-edge and edge-to-face interactions. At high pH levels the negative charge on the edge of the particles will be sufficiently high and the slurry will be well dispersed through electrostatic repulsion between the suspended particles, which is accompanied by a significant decrease in the viscosity of the system. Similar behavior has been reported in the literature about the yield stress of kaolin dispersions by several investigators (e.g, Johnson et al., 2000; Rand and Melton, 1977). Johnson et al. (2000) have explained the variation of the yield stress with pH of the slurry in terms of the relative face-to-face, edge-to-face, and edge-to-edge interaction forces which were calculated using the zeta potential of the edge and face in conjunction with appropriate Hamaker constants and employing the electrical double layer theory of Hogg et al. (1966). At low pH, many particles interact in an edge-to-face manner due to edge-to-face attraction. Face-to-face association drives the formation of discrete lamellar structured tactoids in kaolin dispersions, resulting in a decrease in shear yield stress. Results of Johnson et al. (2000) show that as the pH is increased, edge-to-face force becomes more attractive and face-to-face force less attractive and due to edge-to-face association, a three-dimensional house of card structure results causing an increase in the yield stress of the slurry. At high levels of pH, the interaction forces between all surfaces are either repulsive or approaching zero and under this condition the slurry is well dispersed and yield stress vanishes. These results indicate that a well dispersed slurry of low viscosity can be achieved by increasing the pH of the slurry above 10. However, this may not be desirable in practice and therefore, steric and/or electrosteric stabilization must be employed to prepare highly concentrated well-dispersed kaolin suspensions.

Steric and electrosteric stabilizations are widely used in mineral processing operations through the addition of polymers and polyelectrolytes to the system. These substances can adsorb on the surface of the particles and produce a barrier that may prevent adhesion and agglomeration of the particles, which will result in a well-dispersed dispersion of low viscosity. Effect of NaPAA on the viscosity of kaolin dispersions of 60 %wt solids is presented in Figure 5, which is a plot of viscosity as a function of the dosage of NaPAA at different shear rates. As can be observed, there is several orders of magnitude decrease in the viscosity of the dispersion as the dosage of NaPAA is increased to 5 mg/(g solids) in the system. The results of our earlier work (Zaman et al., 2002) indicate the PAA adsorbs on alumina sites present on the edge of kaolin particles and the dosage of 5 mg/(g solids) corresponds to the saturation adsorption of the polymer on the surface of the edge of the particles.

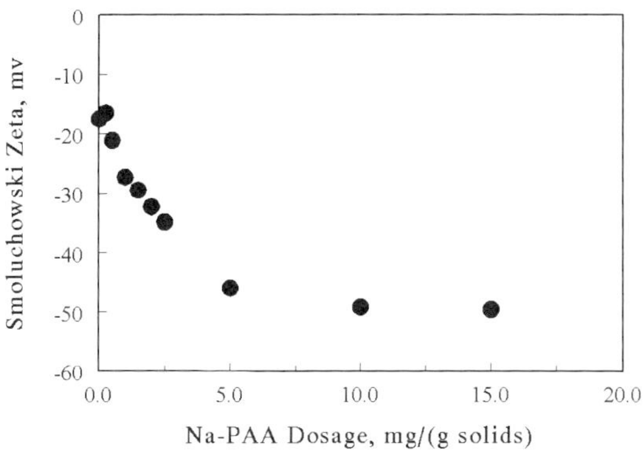

FIGURE 6 Effect of the dosage of NaPAA on zeta (Smoluchowski) potential of kaolin (25 °C, pH = 7, 2 %vol) [Zaman et al., 2002]

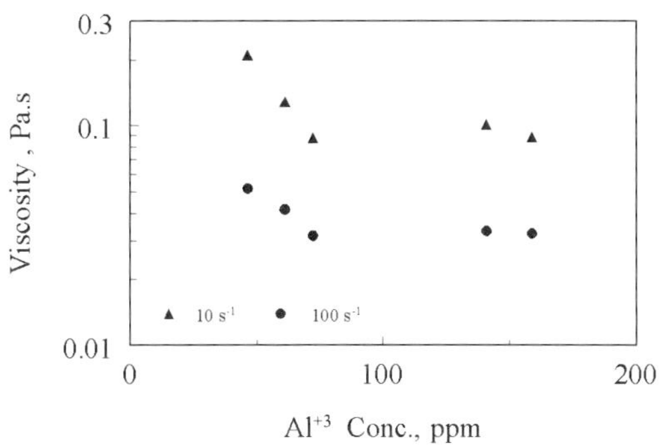

FIGURE 7 Effect of concentration of aluminum ions in supernatant on the viscosity of kaolin dispersions at 65 %wt solids at two different shear rates (25 °C)

Effect of the dosage of NaPAA on the Smoluchowski zeta potential of dispersions of kaolin particles (suspension pH = 7) is shown in Figure 6 (Zaman et al., 2002). As can be observed, over the entire range of polymer dosage, the zeta potential values indicate a strong negative charge on the surface of the particles. The magnitude of the Smoluchowski zeta potential increases significantly as the polymer gradually covers the surface of the particles and reaches a plateau corresponding to the value of saturation adsorption density. The Smoluchowski zeta potential of the bare kaolin surface at pH 7.0 is approximately equal to −16 mV. It can be observed that, as the amount of NaPAA added to the dispersion is increased to 5 mg/(g solids), the Smoluchowski zeta potential approaches a plateau value of approximately −48 mv, corresponding to the saturation adsorption and critical concentration of the polymer added to the dispersion for minimal viscosity.

Effect of the Dissolved Ions on the Viscosity of the Dispersion

Effect of the concentration of aluminum ions in supernatant on the viscosity of kaolin dispersions is shown in Figure 7, which is a plot of viscosity as a function of aluminum ions concentration at different shear rates. A decrease in viscosity of the system is observed as the dissolved ions concentration is increased in supernatant. Due to the complexity of the problem, more experiments has to be conducted to be able to have an accurate explanation of the phenomena in terms of the effect of different factors such as pH, dispersant type and dosage, and interaction of dissolved metal ions with dispersing agents, which is the subject of future research.

SUMMARY AND CONCLUSIONS

The objective of the present work was to investigate the effect of different organic and inorganic dispersing agents on solubility of kaolin particles and stability of the dispersion to develop more effective dispersing agents for engineered slurries of particles of heterogeneous charges. Three different anionic commercially available dispersing agents used in this work were (1) sodium salt of a low molecular weight polyacrylic acid (NaPAA), (2) sodium hexametaphosphate (NaHMP), and (3) sodium silicate (Na-Silicate).

It appears from the data that the concentration of the metal ions in supernatant is highly affected by the dosage of the dispersing agents and pH of the slurry. The concentration of aluminum ions in supernatant in the presence of both NaPAA and NaHMP is maximum over a pH range of 5.5 to 6.5. Results indicate that NaPAA and NaHMP have similar reaction mechanisms with kaolin particles. The high concentration of aluminum ions in supernatant in the presence of these dispersing agents in the system suggests that aluminum ions form a water soluble complex with each of these dispersing agents which remains in supernatant. In the presence of Na-silicate, aluminum ions were not detectable in supernatant for Na-silicate dosage levels of up to 4 mg/(g solids), which is attributed to the formation and precipitation of insoluble aluminum hydroxide and/or alumino-silicate gel.

Overall, the concentration of aluminum and silicon ions in supernatant, surface charge development and the viscosity of kaolin dispersions are highly affected by solution conditions. Preliminary results indicate that keeping the dissolved aluminum ions in the form of water-soluble species in supernatant may have a decreasing effect on the viscosity of the dispersion. More research has to be conducted to understand the effect of the solubility and reaction between the dispersing agents and the dissolved ions on the performance of the slurry in terms of stability and flow properties of the system.

ACKNOWLEDGMENTS

The authors are grateful for the financial support provided by Vinings Industries Inc. and Engelhard Corporation and also for the partial support by the University of Florida Engineering Research Center for Particle Science and Technology (NSF Grant No. EEC-94-02989) and the industrial partners of the ERC. Useful discussions with Mr. P. Macy from Vinings Industries Inc., Mr. T. Say from ERC, and Dr. E. Finch from Engelhard Corporation are greatly acknowledged. Help from F. Demir for performing experiments is also greatly acknowledged.

REFERENCES

Brady, P.V., Cygan, R.T., and Nagy, K.L. 1996. Molecular control on kaolinite surface charge, *J. Colloid Interface Sci.,* 183 (2): 356–364.

Herrington, T.M., Clarke, A.Q., and Watts, J.C. 1992. The surface, charge of kaolin, *J. Colloid and Surface Sci.,* 68: 161–169.

Hogg, R., Healy, T.W., and Fuersten, D.W. 1966. *Transactions of the Faraday Society,* 62 (522P): 1638.

Hsu, P.H. 1989. Aluminum hydroxides and oxyhydroxides. *Minerals in Soil Environments, 2nd edn.*, Edited by J.B. Dixon and S.B. Weed, Soil Science Society of America Book Series No. 1, pp. 354–356.

Iler, R.K. 1973. Effect of adsorbed alumina on the solubility of amorphous silica in water. *J. Colloid Interface Sci.,* 43: 399–408.

Johnson, S.B., Franks, G.V., Scales, P.J., Boger, D.V., and Healy, T.W. 2000. Surface chemistry–rheology relationships in concentrated mineral suspensions, *Int. J. Miner. Process.* 58: 267–304.

Johnson, S.B., Russell, A.S., and Scales, P.J. 1998. Volume fraction effects in shear rheology and electroacoustic studies of concentrated alumina and kaolin suspensions. *Colloids Surf. A.,* 141: 119–130.

Konta, J. 1995. Clay and man: Clay raw materials in the service of man. *Appl. Clay Sci.,* 10: 275–335.

Lee, D.H., Condrate, R.A., Sr., and Reed, J.S. 1996. Infrared spectral investigation of polyacrylate adsorption on alumina. *Journal of Material Science,* 31: 471–478.

Mcmurry, J., and Fay, R.C. 1998. *Chemistry*. 2nd ed., Prentice-Hall, Inc., New Jersey.

Okamoto, G., Okura, T., and Goto, K. 1957. Properties of silica in water, *Geochimi. Cosmochim. Acta.,* 12: pp. 123–132.

Rand, B., and Melton I.E. 1997. Particle interactions in aqueous kaolinite suspensions. 1. effect of pH and electrolyte upon mode of particle interaction in homoionic sodium kaolinite suspensions. *J. Colloid Interface Sci.,* 60 (2): 308–320.

Sastry, N.V., Sequaris J.-M., and Schwuger M.J. 1994. Adsorption of polyacrylic acid and sodium dodecylbenzenesulfonate on kaolinite, *J. Colloid Interface Sci.,* 171: 224–233.

Sjoberg, M., Bergstrom, L., Larsson, A., and Sjostrom, E. 1999. The effect of polymer and surfactant adsorption on the colloidal stability and rheology of kaolin dispersion, *J. Colloids and Surface Sci.,* 159: 197–208.

Wefers, K., and Misra, C. 1987. *Oxides and Hydroxides of Aluminum.* Alcoa Laboratories.

Yuan, J., Garforth, W.L., and Pructt, R.J. 1998. Influence of dispersants on the solubility of calcined kaolin, *Appl. Clay Sci.,* 13: 137–147.

Zaman, A.A., Tsuchiya, R., and Moudgil, B.M. 2002. Adsorption of a low molecular weight polyacrylic acid on kaolin: infrared spectroscopy and adsorption studies, in press, *J. Colloid Interface Sci.*

SECTION 4

Surface Modification, Matrix Interactions, Interfaces, and Adhesion

AFM Pull-off Force Measurements with Polystyrene (Deformable) Colloidal Probes

Elvin R. Beach III* **and Jaroslaw Drelich†**

Pull-off forces were measured for 8–9 μm diameter polystyrene (PS) particles on smooth spin-coated polystyrene films and hexadecanethiol (HDT) self-assembled monolayer surfaces using atomic force microscopy (AFM). Three parameters were controlled during experimentation: (i) the maximum applied load pressing the particle onto the substrate was varied from 50 to 4,000 nN, (ii) loading/unloading rate for the particle to attach and separate from the substrate varied from 0.01 to 9.3 Hz, and (iii) contact time for particle to reside on the substrate surface before retracting the particle to separate it from the surface changed from 1 μsecond to 200 seconds. The pull-off forces increased with increasing applied loading, increasing loading/unloading rate, and increasing contact time for the PS particle on PS film and on HDT monolayer. The PS particles experience elastic and plastic deformations during pull-off force measurements. Plastic deformation is primarily associated with flattening of surface irregularities. The results obtained in this research suggest that the particle-substrate contact areas of the magnitude comparable to those predicted by the JKR contact mechanical model can be reached in the AFM experiments by using high applied loads and long particle-substrate contact times. This result re-opens the prospect of using AFM for studying surface energies of deformable microparticles, even if the microparticles have small irregularities on the surface.

INTRODUCTION

The development of micro and nanotechnology has summoned the need for a technique to measure surface interactions at a sub-micron scale. Pull-off force measurements made by atomic force microscopy (AFM) are being developed for this application. In this technique, a microscopic or sub-microscopic probe is brought into contact with a surface and

* Dow Chemical Company, Midland, Mich.

† Dept. of Materials Science and Engineering, Michigan Technological University, Houghton, Mich.

then separated. The maximum force of separation (pull-off force) is determined from the maximum deflection of the cantilever with probe, multiplied by the probe's spring constant. The pull-off force results normalized by the size of the probe can be converted to the work of adhesion by applying one of several contact mechanics models that represent different situations of tip or surface deformation upon loading [1,2].

Force measurements involving polymeric components have received a large research effort since the introduction of scanning probe microscopy. Pull-off force studies are normally done with one polymeric component and one rigid material component to limit the deformation to the polymer component in the system [3,4]. Measurements between two polymer components are also reported [5–7]; however, analysis is more complex for such measurements.

In 1994 Amié et al. used the AFM to study adhesion of a Si_3N_4 cantilever tip to rigid poly-acetylene films and to soft polystyrene-co-poly-acetylene films [8]. This paper was the first to recognize that there are significant differences in the behavior of pull-off forces measured between a tip and a rigid polymer or a tip and compliant polymer. The problems associated with the viscous flow of polymeric materials (viscoelastic effects) were pointed out as an important factor for further research by this study [8].

The first direct measurement of the pull-off force between polymer surfaces to determine the solid surface free energy was performed in 1994 by Mangipudi et al. using the surface force apparatus (SFA)* [7]. The SFA is not discussed in detail here, the reader is instead referred to references [9,10]. Pull-off forces between poly (ethylene terepthalate) (PET), polyethylene (PE), and PET-PE films were reported in Mangipudi's paper. The authors noted that pull-off force increased with increasing contact time for the PET-PET and PE-PE systems. The use of a silicone-based adhesive tape to attach the thin polymer film samples to the SFA influenced the deformation observed in their experiments [7].

A study by Schmitt et al. employing the bimorph surface force apparatus (BSFA) with two polystyrene (PS) spheres of low surface roughness (less than 1 nm RMS roughness) showed that the measured pull-off force increased with increasing applied load [5]. This paper demonstrated that additional factors, including the instrumental applied loading, scan rate, and contact time, influence measured pull-off forces and complicate the interpretation for polymer-polymer systems as compared to pull-off force measurements for systems with rigid, elastic components.

Feldman et al. used a silicon dioxide (SiO_2) AFM tip to measure pull-off forces with isotactic polypropylene (iPP), polymethylmethacrylate (PMMA), PS, and a PS/PMMA polymer blend [11]. This study used very small tip-polymer loading, less than 5.5 nN, and reported no significant effect of the applied load on the measured pull-off force. Also, hysteresis in the contact regime was noted at different loading/unloading rates (scan rates) demonstrating that the scan rate selected during measurements is an important variable for consideration in viscoelastic systems.

Luengo et al. studied adhesion between poly (butyl methacrylate) (PBMA) surfaces at various contact times and temperatures [6]. This study proved that the pull-off force increased with applied load, contact time, and temperature (temperatures in this study were in the range of 15 to 35 °C). The work also showed a trend for the pull-off force to increase with the loading/unloading rate up to 5 mN/s then decrease in the 5 to 10 mN/s range [6]. Additionally, Luengo et al. reported the presence of a crater in the area of

* The SFA normally uses larger curved surfaces instead of micrometer-sized particles used in the AFM technique.

contact on the surface immediately after force measurements were made. The crater disappeared after some period of time and the authors suggested this was evidence of viscoelastic relaxation phenomenon [6].

Biggs and Spinks used the AFM colloidal probe technique [12] to measure the pull-off forces between a PS sphere and an atomically flat mica surface [3]. Their work showed a nearly linear increase of the pull-off force with applied loading from 5 to 15 μN. The scan rate was also shown to influence pull-off forces in this paper. Slower scan rates resulted in greater hysteresis and larger pull-off forces [3]. Their conclusion was that slower scan rates allowed nanoscale asperities on the PS particle to flatten causing an increase in the contact area [3]. The results of this study were interpreted as elasto-plastic deformation.

Along the same line of thought, Reitsma et al. studied the response of the pull-off force for a 27 μm PS sphere on a flat silica substrate using an AFM with the applied loading in the range of 5 to 100 μN [4]. These results again showed nearly a linear relationship between the pull-off force and applied load. This work also showed that the pull-off force increased with increasing loading (contact) time over the range of 0.1 to 30 seconds. The data from this study were again interpreted as the effect of elasto-plastic deformation.

More recent research has focused on the effect of viscoelastic, and to a lesser degree elasto-plastic, deformation of polymers on experimentally measured pull-off forces. Several papers have been published addressing adhesion between viscoelastic bodies and correlating the work of adhesion to pull-off force measurements using various models of viscoelasticity [13–15]. A good review of the theoretical background and experimental examples of viscoelasticity is available from Attard [16].

The problem with the studies discussing viscoelastic deformation effects on pull-off forces is the apriori assumption that viscoelastic deformation is present in the system under consideration and this may not be significant for all experimental conditions. Other studies offer experimental evidence that elastic and plastic [3,4] and purely plastic deformation [17] is observed during force measurements. To the best of the authors' knowledge, no experimental evidence has been published to prove conclusively which type of deformation controls the size of the contact area and corresponding pull-off force for a given polymer. Furthermore, detailed descriptions of all-important experimental parameters in AFM pull-off force measurements are not regularly provided in the literature.

This paper discusses the issue of polystyrene deformation during pull-off force measurements. Using the colloidal probe technique, pull-off forces between a PS particle and a compliant PS surface as well as a rigid HDT-SAM surface were explored with different AFM operating parameters. The effect of the applied load, scan rate, and the contact time between interacting bodies on the experimentally measured pull off forces were studied.

EXPERIMENTAL METHODS

Polystyrene Surface Preparation

A solution of polystyrene (PS), with an approximate molecular weight of 190,000 g/mol, (Scientific Polymer Products, Inc., Ontario, NY) in toluene (Fisher Scientific, Pittsburgh, PA) was prepared with a concentration of 2.2 wt%. A 1,500 mm diameter silicon wafer (100 orientation) was obtained from MEMS Corporation (St. Louis, MO). Pieces of silicon were cut into approximately 2 cm × 2 cm sections. Each piece was cleaned ultrasonically with acetone, methanol, and deionized water followed by exposure to ultra-violet (UV) light for 15 to 20 minutes before use.

Smooth PS surfaces were prepared by spin-coating on a cleaned silicon wafer section. Spin-coating was done on a clean brass metallography wheel at 500 rpm with the silicon wafer attached by double stick adhesive tape. Ten to twenty drops of solution were applied to the surface with an eyedropper. Toluene was allowed to evaporate as the sample spun for a period of sixty seconds. Polystyrene spin-coated samples were placed in clean, sealed containers for a period of no less than thirty minutes prior to pull-off force measurements.

HDT SAM-coated Sample Preparation

1-Hexadecanethiol ($CH_3(CH_2)_{15}SH$), henceforth referred to as HDT, was obtained from Aldrich Chemical Company (Milwaukee, WI). Pieces of silicon were cut into approximately 2 cm × 2 cm sections and cleaned in acetone, methanol, and deionized water followed by exposure to UV light and then coated with gold in a Polaron Equipment Limited SEM Coating Unit E5000. The gold target was set to a fixed distance from the samples during all gold coating sessions. The parameters of operation for the gold-coater were as follows: a vacuum of 0.05–0.07 torr was obtained, a sputtering current of 5 mA, and the coating time was 10 minutes.

Immediately after application of the gold coating, the substrates were rinsed with absolute ethanol (200 proof) and immersed in a solution of 1 to 2 mM HDT in absolute ethanol that was prepared at least 2 hours prior to placing the samples in solution. The substrates were left in the HDT solution for a period of 12 hours to allow a complete self-assembled monolayer to form. The silicon wafers were rinsed extensively with absolute ethanol upon removal from solution to remove any excess thiol components on the surface. In order to ensure that no ethanol remained on the surfaces, the samples were placed in an oven at 80°C for 20–30 minutes.

Surface Imaging with the Atomic Force Microscope

TappingMode™ etched silicon probe (TESP) cantilevers were purchased from Digital Instruments (Santa Barbara, CA). Intermittent contact mode (TappingMode™) was used to image the gold surfaces of the coated silicon substrates using a Dimension 3000 AFM (Digital Instruments). Scan sizes of 500 nm × 500 nm, 1 μm × 1 μm, and 2.5 μm × 2.5 μm were taken at a scan rate in the range of 0.75–1 Hz. The setpoint ratio, r_{sp}, was set to no lower than 0.8 to minimize sample damage caused by the AFM tip.

Intermittent contact mode (TappingMode™) was also used to image the PS surfaces. Scan sizes of 1 μm × 1 μm, 10 μm × 10 μm, and 25 μm × 25 μm were taken at a scan rate in the range of 0.75 to 0.9 Hz. The setpoint ratio, r_{sp}, was set in the range of 0.9 to 0.96.

Polystyrene Colloidal Probe Preparation

Spherical polystyrene particles were obtained from Bangs Laboratory, Inc. (Fishers, IN). The as-received particles were dispersed in solution and had a nominal diameter of 9 μm. Particles were placed in a beaker and washed with excess deionized water to remove any residual solution.

Tipless, polycrystalline silicon, V-frame cantilevers with spring constants ranging from 2.1 to 17.0 N/m were purchased from Thermomicroscopes, Inc. (Sunnyvale, CA). The PS particles were attached to the cantilevers following a gluing procedure, described by Digital Instruments [18], that was first proposed by W.A. Ducker et al. [12]. The cantilevers with attached colloidal probes were placed in clean, sealed containers and the glue was allowed to set for a minimum of four days prior to making any pull-off force measurements.

Pull-off Force Measurements with Polystyrene Colloidal Probes

All pull-off force measurements were made in air under ambient conditions. The relative humidity (RH) level during measurements ranged from less than 5 up to 15 percent. The RH level was continuously monitored during all experiments with a digital thermo/hygrometer (Cole Palmer Instrument Company, Vernon Hills, IL). Pull-off force measurements were made using a Dimension 3000 AFM equipped with Nanoscope Software version 4.43 from Digital Instruments. The pull-off force measurements were made in Force Mode with the x–y lateral motion of the piezoelectric element suspended. The deflection versus z-position curves were converted into force–separation, more commonly referred to as force–distance curves, using Scanning Probe Image Processor (SPIP) software from Image Metrology (Lyngby, Denmark).

The effect of applied load, scan rate, and contact time were examined and the AFM operating parameters for each of the experiments are described in the following subsections.

Variable Applied Load Experiments

Deflection versus z-position curves (force curves) were captured for PS particles in contact with a smooth PS surface and a HDT-SAM surface. The experiments were all done using similar operating parameters, but due to slight changes in each experiment they are described separately in the following paragraphs.

In the first set of experiments the pull-off force between a PS particle and a PS surface was measured. The measurements were carried out for three different PS particles. The PS particles were mounted on cantilevers with spring constants of 2.1 N/m, 13.0 N/m, and 17.0 N/m, respectively. The operation parameters of the AFM used for these pull-off force measurements were as follows (i) the ramp sizes (distance of piezoelectric movement in the direction normal to the surface) were 1,000 to 1,200 nm for the cantilever with spring constant of 2.1 N/m, 350 nm for the cantilever with a spring constant of 13.0 N/m, and 100 to 150 nm for the cantilever with a spring constant of 17.0 N/m; (ii) loading/unloading rate (scan rate) of 0.996 Hz; (iii) 1,024 data points per deflection versus z-position curve were recorded; and (iv) the setpoint, or zero deflection position in the non-touching regime, was set as the 0 voltage position for all curves.

One hundred force curves were collected in one location on the surface as the applied load was increased ten times. Five surface locations were sampled per cantilever for a total of five hundred force curves collected per cantilever. The force measurements in subsequent surface locations were captured immediately after the previous experiments were concluded and the cantilever was repositioned to a new location on the surface.

The second set of pull-off force measurements in the variable applied loading experiments were made between a PS particle and a PS surface. The PS particle was glued to a cantilever with a spring constant of 3.2 N/m for these experiments. The operation parameters of the AFM were as follows (i) the ramp size was 500 nm for all measurements; (ii) scan rate of 0.996 Hz; (iii) 1,024 data points per deflection versus z-position curve were recorded; and (iv) the setpoint was set as the 0 voltage position.

One surface location was used for these experiments. Ten force curves, at a predetermined applied load, were captured then the applied load was increased ten times for one "loading cycle" (set of measurements). The cantilever was then disengaged (moved high above the surface) and the next set of measurements was made exactly thirty minutes after the previous set was finished. Four sets of measurements were made following the same loading cycle routine for a total of four hundred force curves in these experiments.

Variable Scan Rate Experiments

Force curves for the contact between a PS particle attached to a cantilever with spring constant of 17.0 N/m and a smooth PS surface were captured in this experiment. The operation parameters of the AFM were as follows (i) the ramp size was 250 nm for all measurements; (ii) two loading ranges were used, 971 ± 100 nN and 1,675 ± 23 nN; (iii) scan rates were varied starting at 0.01 Hz up to 9.30 Hz; (iv) 1,024 data points per deflection versus z-position curve were recorded; and (v) the setpoint was set as the 0 voltage position.

The experiments were all made in one location on the surface and ten force curves per scan rate were captured. The lower loading experiment was done first, followed immediately by the experiment done at the higher loading range.

Variable Contact Time Experiments

Nanoscope software version 4.43r2 from Digital Instruments was installed prior to these experiments. Force curves for the contact between a PS particle attached to a cantilever with spring constant of 17.0 N/m and a HDT-SAM surface were captured in this experiment. Two new parameters are defined for these experiments, the contact time and rest time. The contact time is a user specified time the particle will remain in contact with the surface after the piezoelectric element moves toward the surface at a distance specified by the ramp size. This resulted in a specific contact time between the particle and substrate at a predetermined maximum applied load. The rest time is a pause between force curve measurements specified by the user. The operation parameters of the AFM are given as follows: (i) the ramp size was 300 nm for all measurements; (ii) the average applied load was 1,158 ± 180 nN; (iii) scan rate was 0.996 Hz; (iv) the contact time was varied beginning with 1 μsecond to 200 seconds; (v) the rest time was 10 seconds; (vi) 1,024 data points per deflection versus z-position curve were recorded; and (vii) the setpoint was set as the 0 voltage position.

The pull-off force measurements were made in one surface location. Ten force curves per contact time were captured for a total of 110 force curves. The contact time was increased by an order of magnitude beginning with 1 μsecond and ending at 200 seconds (the limit of the software). The shortest contact time experiments were done first followed immediately by the longer contact time experiments.

Scanning Electron Microscopy

A JEOL – 35C scanning electron microscope (SEM) was used to image the PS particles after making pull-off force measurements. The cantilevers with attached particles were coated with a thin gold film (10 to 40 nm) prior to imaging. The SEM was operated at 20 kV accelerating voltage at a working distance of 15 mm in secondary electron imaging mode. Images were taken from several orientations of the cantilever relative to the electron beam.

Images at various magnifications were captured digitally with a resolution of 1,024 × 1,024 pixels. Scion Image analysis software was used to determine the diameter of the particles.

RESULTS AND DISCUSSION

PS-PS System

Spin-coated PS Film Surface Roughness Analysis. The PS surface roughness was quantified using three parameters of roughness; root-mean-square roughness (RMS), average roughness (R_a), and percent surface area difference (%SAD). These parameters are defined in Equations 1–3.

$$RMS(R_q) = \sqrt{\frac{\sum (Z_i - Z_{ave})^2}{N}} \quad \textbf{(EQ 1)}$$

where Z_{ave} is the average of the Z values (Z-axis representing topographical height features) within the given area, Z_i is the current Z value and N is the number of points within the given area.

$$R_a = \frac{1}{L_x L_y} \int_0^{L_y} \int_0^{L_x} |f(x, y)| \, dx dy \quad \textbf{(EQ 2)}$$

In the equation for R_a, f(x, y) is the surface relative to the center plane, and L_x and L_y are the dimensions of the surface.

$$\%SAD = \left[\frac{\sum (\text{surface area})_i}{\sum (\text{projected surface area})_i} - 1 \right] \times 100\% \quad \textbf{(EQ 3)}$$

Table 1 presents the results of the surface roughness analysis. As the results show, the spin-coated polystyrene film prepared in this study had smooth surface with RMS roughness parameter less than 0.5 nm. Surface roughness is often a scale-dependent property [19], therefore, three scan sizes were analyzed in this work. The surface roughness was similar for all three scan sizes suggesting that the RMS roughness of the entire sample was <0.5 nm.

The thickness of the PS film on the silicon wafer is also important for pull-off force measurements. The film must be thick enough that deformation expected in the system will not exceed the thickness of the film, otherwise the analysis of PS deformation becomes a difficult task.

The coating thickness was examined by looking at edges of PS films on silicon wafers. Figure 1 shows some of the AFM images and cross-sections of the images used for thickness determination.

Five images similar to those shown in Figure 1 were examined and the film thickness was determined to be very close to 18 nm in all cases. Due to very small thickness of

TABLE 1 PS surface roughness analysis

Scan Size	RMS (nm)	R_a (nm)	%SAD
1 μm × 1 μm	0.271	0.217	0.535
10 μm × 10 μm	0.312	0.248	0.369
25 μm × 25 μm	0.416	0.341	0.388

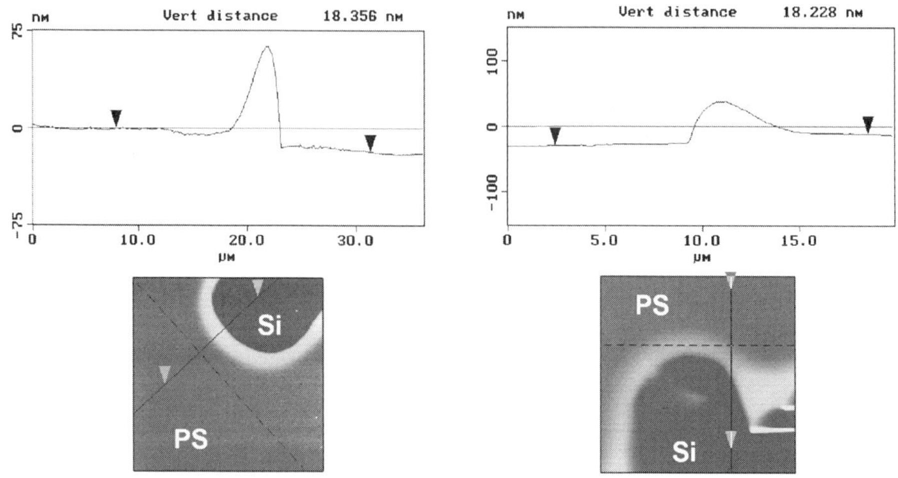

FIGURE 1 AFM images used for the PS coating thickness measurements. Arrows in the two-dimensional topography images correspond to the arrows in the cross-sectional image directly above it.

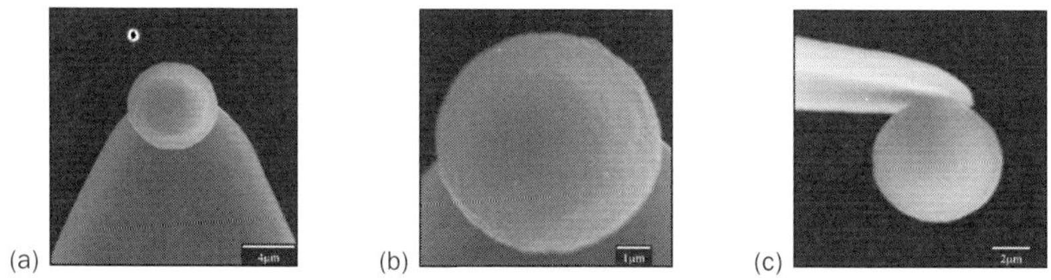

FIGURE 2 PS particle attached to the cantilever with spring constant = 2.1 N/m used to collect the pull-off force measurement results shown in Figure 6 (a). The radius of the particle was determined to be 3.39 ± 0.16 µm. The cantilever was destroyed during sample preparation after the normal (a) and (b) and left profile (c) images were captured.

the cast PS layer, it is unlikely that any excessive amount of toluene remained in the PS that could affect mechanical properties of the PS film. Nevertheless, the glass transition temperature of the PS film is expected to be slightly lowered due to the sample preparation technique used here.

Response to Various Applied Loads for the PS-PS System. PS particle-PS film pull-off force measurements were made using three PS colloidal probes. The size of the colloidal probes was determined from SEM images. The SEM images for all the PS particles used are shown in Figures 2–4.

Each cantilever had a different spring constant allowing the experiments to be done over different loading ranges. A typical force-distance curve for a PS particle on a PS surface showing the data converted from the raw AFM deflection versus z-position graph is shown below in Figure 5. The darker curve shows the deflection of the cantilever as it approaches the surface of the PS. At approximately 10 nm above the surface, the attractive surface forces cause the probe to jump into contact with the substrate surface. Following the curve at a distance of zero, the particle is pressed into the surface until it reaches its maximum applied load (approx. 825 nN). At this point, the cantilever is

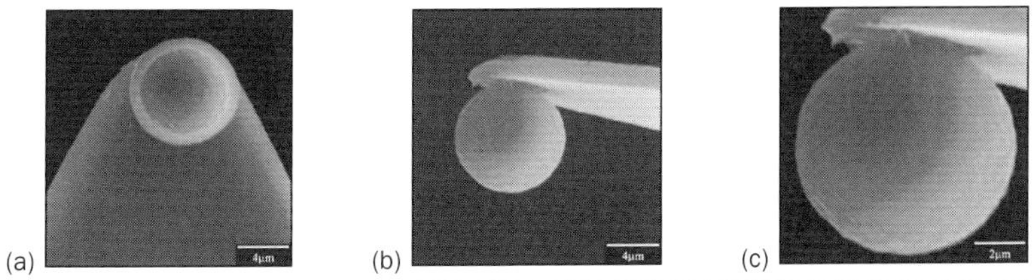

FIGURE 3 PS particle attached to the cantilever with spring constant = 13 N/m used to collect the pull-off force measurement results shown in Figure 6 (b). The radius of the particle was determined to be 4.26 ± 0.20 μm. The cantilever was destroyed during sample preparation after the normal (a) and left profile (b) and (c) images were captured.

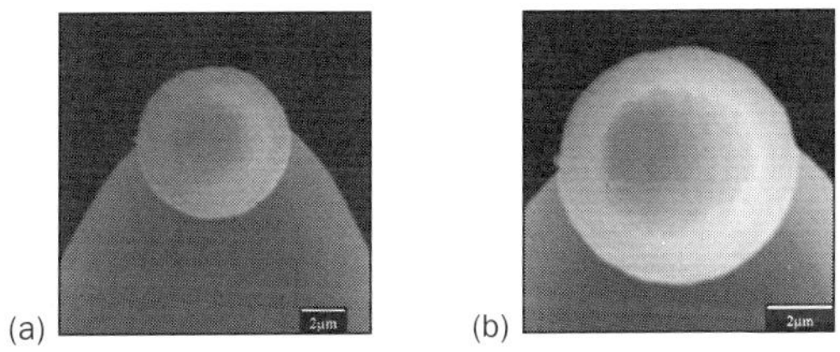

FIGURE 4 PS particle attached to the cantilever with spring constant = 17 N/m used to collect the pull-off force measurement results shown in Figure 6 (c). The radius of the particle was determined to be 3.64 ± 0.13 μm. The cantilever was destroyed during sample preparation after the normal (a) and (b) images were captured.

pulled away from the surface of the sample following the retracting curve. The pull-off force is measured as the maximum negative load to remove the particle from the surface (approx. 300 nN in Figure 5).

Pull-off forces were measured as the load was increased in one location on the surface, called a "loading cycle." The cantilever was disengaged (moved above the surface) at the end of a loading cycle and relocated to the next position on the surface where a new loading cycle started in that location (approximately 30 seconds total time elapsed between cycles while repositioning the cantilever). The results of the pull-off force measurements between PS particles and a smooth PS surface done in this manner are shown in Figure 6 (a)–(c).

The pull-off force increased as the applied load was increased during a loading cycle (Figure 6). The trend appears to be nearly linear for some of the loading cycles which is in good agreement with the results from previous studies involving PS colloidal probes [3,4]. There are, however, cases where the trend does not follow the same general pattern; for example, loading cycle #1 at low applied loads in Figure 6c and loading cycle #5 at high applied loads in Figure 6b. This is attributed to the "break-in period" where some cantilevers seem to require a significant number of loading/unloading cycles before providing reproducible pull-off force data. This phenomenon has been noted in previous studies [20]. We attribute it to the deformation of surface asperities as discussed in a later section of this paper.

The pull-off forces increased from low values at low applied loads to higher values at higher applied loads in each loading cycle. The pull-off forces at the lowest applied load of each cycle increased in each subsequent loading cycles for all three cantilevers (Figure 6). After the pull-off force measurements were completed at the highest loads during a loading cycle, the cantilever was moved several micrometers above the surface

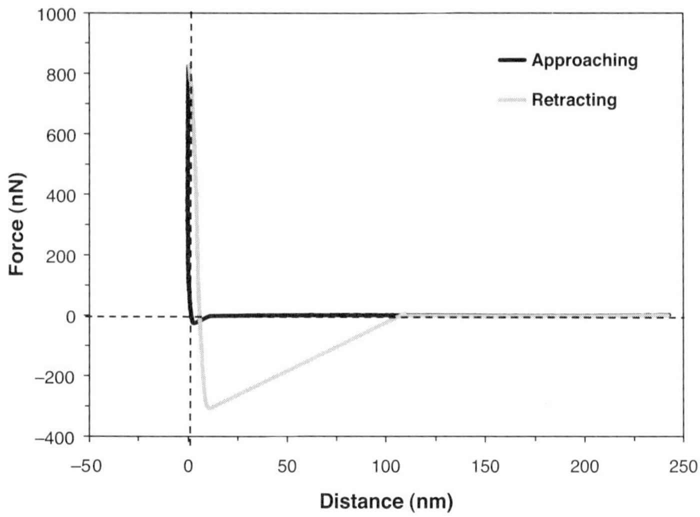

FIGURE 5 Typical force-distance curve for a PS particle on PS surface

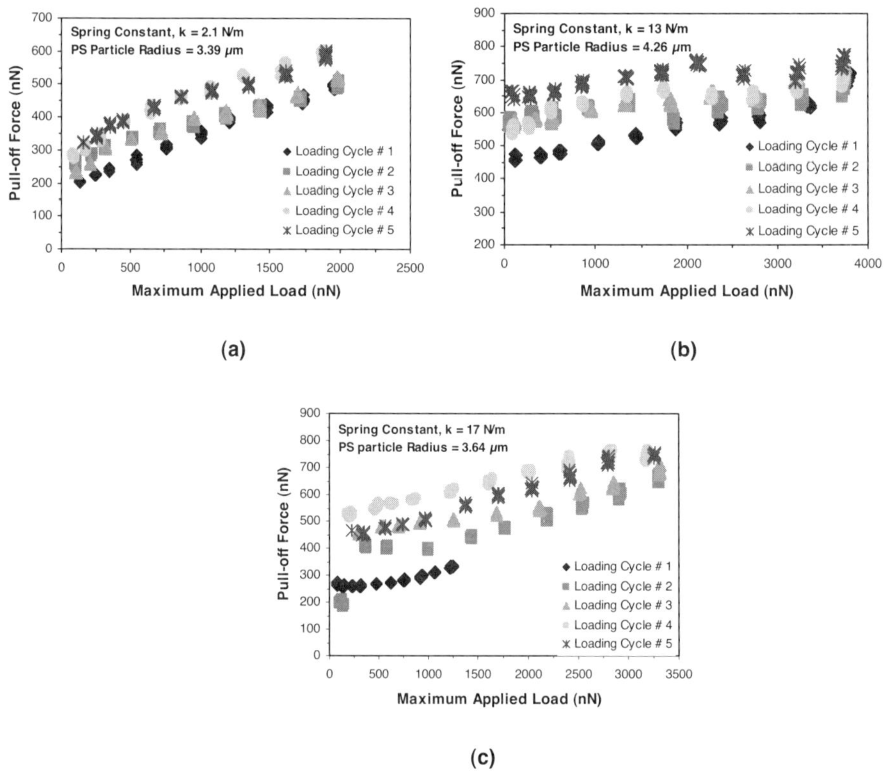

FIGURE 6 The response of the pull-off force to increasing applied load for PS particles attached to cantilevers with spring constants of 2.1 N/m (a), 13 N/m (b), and 17 N/m (c)

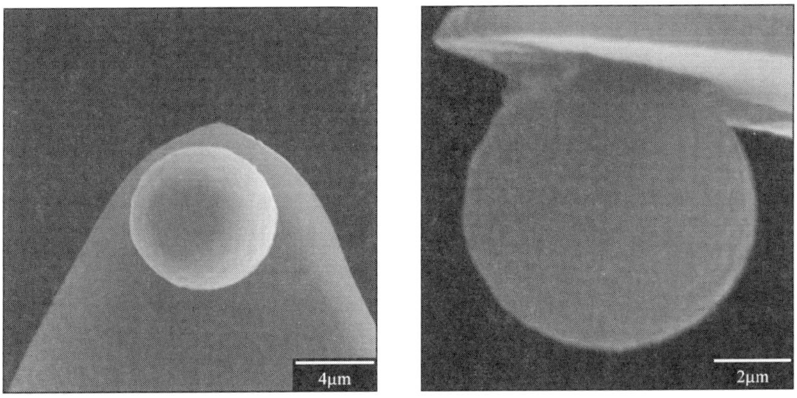

FIGURE 7 PS particle attached to the cantilever with spring constant = 17 N/m used to collect the pull-off force measurement results shown in Figure 8. The radius of the particle was determined to be 3.71 ± 0.17 µm. The cantilever was broken during sample preparation after the normal (a) and right profile (b) images were captured.

to a new location on the surface. The load was totally removed from the particle for a period of approximately 30 seconds before starting a new loading cycle. As shown in Figure 6 the pull-off forces at the lowest applied load at the beginning of a new cycle are smaller than the pull-off forces at the highest loads in the previous cycle. However, they are larger than the pull-off forces measured for comparable low applied loads of the previous cycle.

Further, as shown in Figure 6, the slope for the the pull-off force (F) versus maximum applied load (P) curve gradually decreased for data of each subsequent cycle and to lesser extent among the data collected during one loading cycle. The change in F vs. P curve slopes lead to a conclusion that irreversable processes of deformation took place. Additionally, the change in the F vs. P curves indicate that the curves for different loading cycles have a tendency to overlap in a region of high applied loads. The region of possible overlap of F vs. P curves is located at >630 nN, >750 nN, and >770 nN for the data in Figure 6a, 6b, and 6c, respectively.

Pull-off Force Response to the Loading/Unloading Rate. Measurements were performed to determine the response of the PS-PS system to various loading/unloading rates (scan rates). A PS particle attached to a cantilever with a spring constant of 17 N/m used in these measurements is shown in Figure 7. The scan rate was varied from 0.01 Hz up to 9.30 Hz over two applied loading ranges of 971.2 ± 100.3 nN and 1,675.0 ± 22.5 nN, respectively. The lower loading range measurements were made first followed immediately by the second loading range experiments. The response of the pull-off forces to various scan rates is plotted in Figure 8.

The trendlines drawn in Figure 8 are natural logarithmic fits of the data which have correlation coefficients of 0.98 for loading range 1 and 0.99 for loading range 2. The response of the pull-off force was expected to be opposite of that shown above because contact time between the interacting bodies reduces with increasing scan rate. As a consequence of reduced contact time the area of contact should be smaller at faster scan rates leading to lower pull-off force values according to previous work done by Biggs et al. [3].

Deformation of Polystyrene. The results shown in Figure 6 indicate that the PS-PS system experienced both short-term elastic and long-term plastic and/or viscoelastic deformations during the pull-off force measurements. The increase of pull-off force with

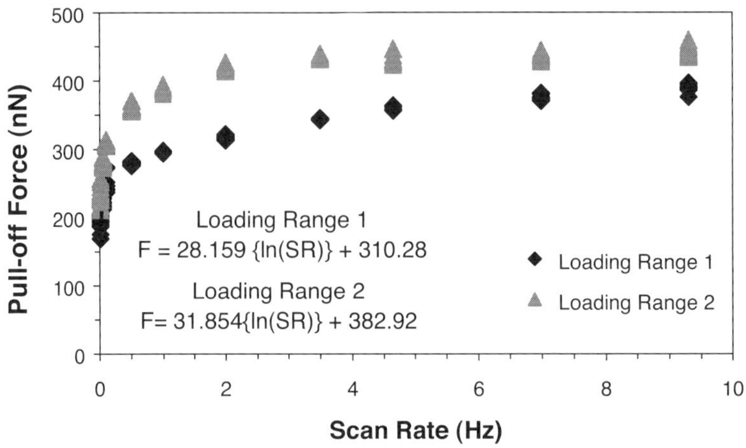

FIGURE 8 Pull-off force response to scan rate at two applied loading ranges. The term F and SR in the inset equations are the pull-off force and scan rate, respectively.

increasing applied load and then decrease of the pull-off force value after the applied load is reduced, indicate that the PS particle underwent elastic deformation. Also an elastic deformation of thin PS film is possible but to much smaller extent because it was only 18 nm thickness and cast onto a rigid silicon wafer.

Since the original pull-off force has never been restored in any experiment, results of which are shown in Figure 6, plastic and/or viscoelastic deformation of PS particle and PS film is obvious. When the load is applied between a particle and a substrate, the particle will deform. After a loading cycle is complete and the cantilever is disengaged and relocated to the next position on the PS film, some of the deformation introduced during the previous measurements can relax in the materials due to its viscoelastic properties and the particle may need time to return to original shape. The time between loading cycles in our experimentation was only approximately 30 seconds, which may not be sufficient time for complete viscoelastic relaxation of the PS particle to occur.

Pull-off forces between a PS particle and PS film were measured with a 30 minute time period allowed between loading cycles to allow viscoelastic relaxation of the components. If the viscoelastic effect is dominant, the pull-off force should return to the original value at the beginning of each new loading cycle. The results of pull-off force measurements with a 30 minute time period between loading cycles are shown in Figure 9.

The pull-off force values do not return to the original values when a 30 minute time period between loading cycles was allowed. This suggests that viscoelastic effects are not primarily responsible for the variation of pull-off forces that were measured in subsequent cycles. This also indicates that permanent plastic deformation of particle took place and requires closer consideration for the PS-PS system.

One possible explanation for the continual increase in the pull-off force for PS particle–PS film measurements is that the surface irregularities on the PS particle and film are flattened during loading. This would result in an increased contact area as shown schematically in Figure 10.

Roughness has been shown to decrease adhesion by a factor of ten in previous research [21, 22]. Flattening of surface roughness on the PS particle and PS film might cause the increase in the pull-off forces for successive loading cycles seen in the PS-PS measurements. Visible surface irregularity of the PS particles is shown in the SEM

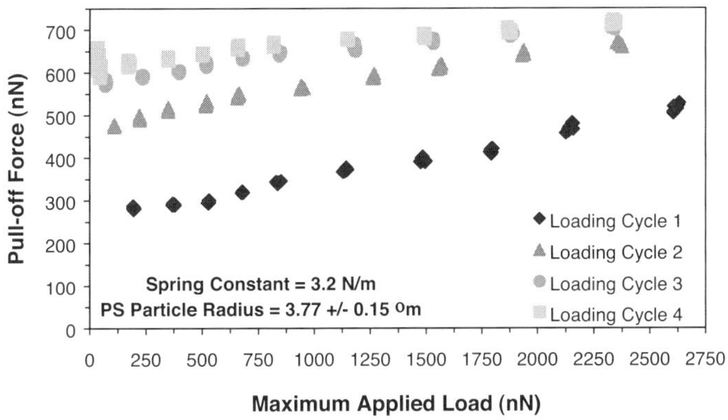

FIGURE 9 Pull-off forces between a PS particle and PS film with 30 minute time period between loading cycles allowed

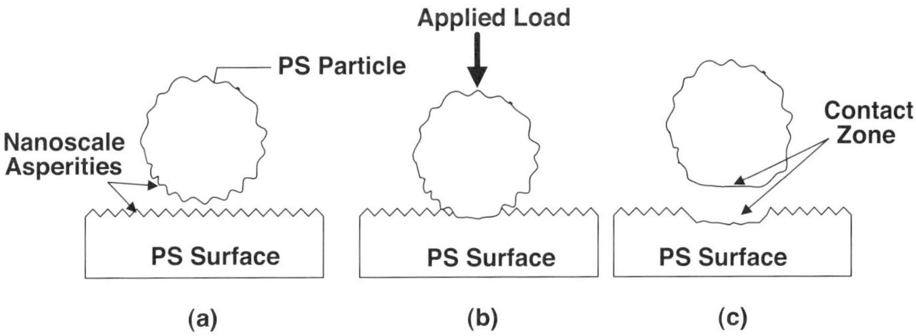

FIGURE 10 Drawing showing one possible mechanism of deformation in PS-PS system. The particle and surface are shown with nanoscale roughness (not drawn to scale) before contact (a). After the particle has been pressed into the surface with some applied load (b), the nanoscale asperities in the contact area are flattened (c).

images of these particles (Figures 2–4). The PS thin film was smooth with the RMS values less than 0.5 nm (Table 1).

No significant permanent deformation of the colloidal probe was observed in SEM images. This observation suggests that indeed the deformations, if present, should occur at submicroscopic level and mainly affect asperities of rough particle rather than the bulk particle.

If the asperities on the surface of the PS particle were flattened, an increase in the pull-off forces from one loading cycle to the next would be expected. This effect was observed during the experimentation, results of which are shown in Figure 6. Also results shown in Figure 8 can be explained based on this effect. Subsequent flattening of the irregularities on the surface of colloidal probe caused the increase of pull-off force with increasing scan rate; a trend that was completely opposite of the effect expected.

In addition to simply flattening the irregularities of the surface, it could be expected that the PS colloidal probe may indent some small distance into the PS film. However, this does not explain the results in Figure 6. The particle was moved to a new location before measuring the pull-off forces in each loading cycle and every time the pull-off

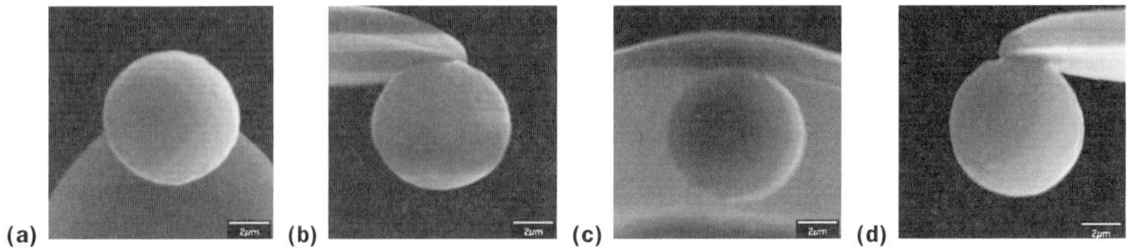

FIGURE 11 PS particle attached to the cantilever with spring constant = 17 N/m used to collect the pull-off force measurements at various contact time results shown in Figure 14. The radius of the particle was determined to be 3.39 ± 0.09 μm.

force measured at lowest applied load for a given loading cycle increased as compared to the pull-off force measured at previous location using similar applied load.

As already well accepted in the literature, the PS-PS system (with deformable particle) can be treated by the Johnson-Kendall-Roberts (JKR) contact mechanical model [23]. For example, using the measured pull-off forces (F), the radius of particle (R) in contact with a smooth surface can be calculated based on the JKR model [23]:

$$F = 3BR\gamma \quad \textbf{(EQ 4)}$$

where γ is surface energy of polystyrene; $\gamma = 40.7\ mJ/m^2$ [24].

The pull-off force versus applied load graphs were extrapolated to the zero load intercept on the pull-off force axis during the first loading cycle and this (F_o) value was used in calculations of R value. The calculated R value varies from R = 0.5 to 1.2 μm depending on the F_o value taken from Figures 6 and 9.

As can be seen from SEM micrographs in Figures 2–4 and 7, the polystyrene particles used in this study were spherical lacking any sharp asperities. However, these particles had wave-shaped submicroscopic irregularities on their surface with radius of curvature of the irregularities of about 0.1–1 μm which is similar magnitude of the R value calculated above.

The R value increases to 1.6 to 2.0 μm if instead of pull-off force values at zero load, the pull-off force values obtained at highest applied loads (Figures 6 and 9) are used in calculations. Although these R values are less by about 50% than the radii of the PS particles determined from the SEM micrographs, they indicate that the effective contact area between particle and substrate increases with increasing applied load.

PS Particle on HDT-SAM Surface: Variable Contact Times

An additional experiment was performed in which the contact time between a PS particle (Figure 11) and HDT-SAM chemically adsorbed on a gold-coated silicon wafer surface were held for various times at an applied load in the range of 1,157.8 ± 180.3 nN. The scan rate approaching the surface and retracting from the surface was 1 Hz before and after the predetermined contact time with the surface. The results of this experiment are shown in Figure 12.

The results of this experiment show there is virtually no effect of the contact time on pull-off forces in the range of 1μsecond to 1 second (Figure 12). A slight increase in the pull-off force was observed at 10 seconds. Contact times of 100 and 200 seconds caused a substantial increase in the pull-off force. The average value of the pull-off force was 487.7 ± 16.4 nN for contact times ranging from 1 μsecond to 1 second. This value nearly

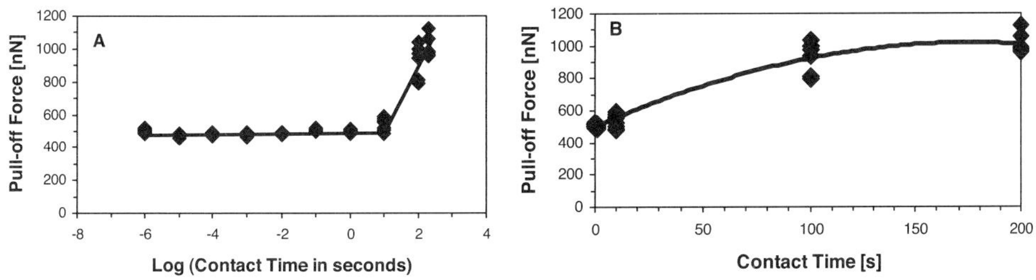

FIGURE 12 Pull-off force response to various contact times between a PS particle and HDT-SAM substrate. Spring constant = 17 N/m and PS particle radius is 3.39 ± 0.09 μm.

doubled at 100 seconds with an average value of 927.2 ± 93.0 nN. Then the average value of the pull-off force increased only slightly to 1,008 ± 61.7 nN at 200 seconds suggesting that some kind of leveling of pull-off force takes place at 100–200 s. The trend for the pull-off force to increase with increasing contact time agrees with results from previous research with polymeric components (particles or surfaces) [4–7].

It is possible to estimate the radius of particle curvature being in contact with self-assembled monolayer of hexadecanethiol using Equation 4. Because the PS-HDT system is asymmetrical and HDT layer is apolar, Equation 4 takes the following form:

$$F = 3BR(\gamma_{HDT}\gamma_{PS}^{LW})^{1/2} \quad \textbf{(EQ 5)}$$

where γ_{HDT} is the surface energy of hexadecanethiol on a gold film, γ_{HDT} = 25 mJ/m^2 [2], and γ_{PS}^{LW} is the Lifshitz-van der Waals component of the polystyrene surface energy, γ_{PS}^{LW} = 40 mJ/m^2 (calculated based on the contact angle data from [24] and using the Lewis acid-base interfacial free energy theory [2]).

The data in Figure 12 and Equation 5 suggest that R = 1.6 μm at contact times smaller than 10 seconds. This R value is slightly larger than any R value calculated from the previous set of experiments (R = 1.6 μm versus R = 0.5 – 1.2 μm) presented in the earlier section. There could be two reasons for this discrepancy. First, the particle used in this experiment had smoother surface in the area of contact with the substrate. Second, the silicon wafer coated with a gold film and modified with the self assembled mono layer of hexadecanethiol is substrate with smoother surface than spin-coated PS film. Indeed, the RMS value for the gold film before modification with hexadecanethiol was found to be very consistent at all scan sizes and varied from 0.33 to 0.35 nm [2] as compared to 0.27–0.42 nm for the PS film (Table 1).

When the R value was calculated for the pull-off force values recorded at 200 seconds contact time (Figure 12), it was found to be approximately R = 3.38 μm. Surprisingly, this R value matches perfectly the value for the radius of PS colloidal probe used in the experiments and determined through scanning electron microscopy examination (R = 3.39 μm, Figure 11). This result suggests that in order to establish a well-defined contact area, without interference coming from surface irregularities, the polystyrene particle must remain under high applied load for 100–200 seconds. In other words, appropriate time and parameters of AFM operation during pull-off force measurements can satisfy the conditions posed by the JKR model. This opens the prospect of using the AFM for studying surface energies of polymers at microscopic and submicroscopic scales.

CONCLUSIONS

Pull-off forces were determined for polystyrene (PS) particles on smooth polystyrene and hexadecanethiol (HDT) self-assembled monolayer surfaces. The pull-off forces were shown to be dependent on the applied load, loading/unloading rate, and the contact time. The dependence on each operating parameters in the PS-PS system is not fully understood at this point, although general conclusions and new hypotheses are proposed from this research.

The pull-off forces increased with increasing applied loading, increasing loading/unloading rate, and increasing contact time for the PS particle on PS film and on HDT monolayer. It is hypothesized that the particle deforms both elastically and plastically during loading. The effect of particle-substrate contact time on pull-off forces suggests that viscoelastic effects are important in PS particle–PS surface pull-off forces as well. It is believed that flattening of surface irregularities is the primary reason for an increase in the particle-substrate contact area.

Results from examining the effect of increasing applied load and increasing contact time on pull-off force suggest that there is a minimum load and/or time required in the AFM pull-off force measurements in order to establish well-defined particle-substrate contact area, if deformable particles with submicroscopic surface irregularities are used. The results of pull-off force measurements carried out for a PS particle in contact with the HDT monolayer for 100–200 seconds satisfied the predictions of the JKR theory. This result, although based on one set of experiments, should reactivate research in exploring atomic force microscopy for studying surface energies of deformable microparticles. This research direction was not explored over the last few years due to absence of perfectly spherical particles on market and, thus, difficulties associated with the ill-defined and often irreproducible contact area between a particle and a substrate during pull-off force measurements. The results from this research indicate that the effect of surface irregularities can be eliminated or reduced if long contact time and high loads, relative to the mechanical properties of the material, are used during the pull-off force measurements.

ACKNOWLEDGMENT

Acknowledgment is made to the donors of the American Chemical Society Petroleum Research Fund and Dura Pharmaceuticals Inc. (currently Elan) for partial support of this research.

REFERENCES

1. D. Maugis, in Modern Tribology Handbook, Vol. 1, B. Bhushan (Ed.), CRC Press, Boca Raton, 2001, pp. 163–203.
2. E.R. Beach, G.W. Tormoen, and J. Drelich, *J. Adhesion Sci. Technol.* 16, 845 (2002).
3. S. Biggs and G. Spinks, *J. Adhesion Sci. Technol.* 12, 461 (1998).
4. M. Reitsma, V. Craig, and S. Biggs, *Intern. J. Adhesion Adhesives* 20, 445 (2000).
5. F.-J. Schmitt, T. Ederth, P. Weidenhammer, P. Claesson, and H.-J. Jacobasch, *J. Adhesion Sci. Technol.* 13, 79 (1999).
6. G. Luengo, J. Pan, M. Heuberger, and J.N. Israelachivili, *Langmuir* 14, 3873 (1998).
7. V. Mangipudi, M. Tirrell, and A.V. Pocius, *J. Adhesion Sci. Technol.* 8, 1251 (1994).
8. J.P. Amié, Z. Elkaakour, C. Odin, T. Bouhacina, D. Michel, J. Curely, and A. Dautant, *J. Appl. Physics* 76, 754 (1994).
9. J.L. Parker, H.K. Christenson, and B.W. Ninham, *Rev. Sci. Instruments* 60, 3135 (1989).
10. J.N. Israelachivili and G.E. Adams, J. Chem. Soc., *Faraday Trans. I* 74, 975 (1978).

11. K. Feldman, T. Tervoort, Paul Smith, and N.D. Spencer, *Langmuir* 14, 372 (1998).
12. W.A. Ducker, T.J. Senden, and R.M. Pashley, *Nature* 353, 239 (1991).
13. A. Falsfi, P. Deprex, F.S. Bates, and M. Tirrell, *J. Rheology* 41, 1349 (1997).
14. J.M. Barney, C.-Y. Hui, and C. Cohen, *Langmuir* 17, 681 (2001).
15. J.A. Hammerschmidt, W.L. Gladfleter, and G. Haugstad, *Macromolecules* 32, 3360 (1999).
16. P. Attard, *J. Adhesion Sci. Technol.* 16, 753 (2002).
17. T. Aoike, H. Uehara, T. Yamanobe, and T. Komoto, *Langmuir* 17, 2153 (2001).
18. *Attaching Particles to AFM Cantilevers.* 1995. Support Note 226, Rev. A, Digital Instruments Inc., Santa Barbara, CA.
19. A. Majurndar and B. Bhushan, in Characterization and Modeling of Surface Roughness and Contact Mechanics, Vol 3, B. Bhushan (Ed.), Boca Raton, CRC Press. CRC Series Mechanics and Material Science. 1995, pp. 109–165.
20. Y.I. Rabinovich, J.J. Adler, A. Ata, R.K. Singh, and B.M. Moudgil, *J. Colloid Interface Sci.* 232, 17 (2000).
21. Y.I. Rabinovich, J.J. Adler, A. Ata, R.K. Singh, and B.M. Moudgil, *J. Colloid Interface Sci.* 232, 10 (2000).
22. E.R. Beach, G.W. Tormoen, J. Drelich, and R. Han, *J. Colloid Interface Sci.* 247, 84 (2002).
23. K.L. Johnson, K. Kendall, and A.D. Roberts, Proc. R. Soc. London, A 324, 301 (1971).
24. S. Wu. Polymer Interface & Adhesion, Marcel Dekker, Inc., New York, 1982.

Silanes and Other Silane-like Surface Treatments for Inorganic Fillers

J.J. Kellar*, C.D. Struckman*, D.A. Boyles†, H. Du‡, and W.M. Cross*

Silane coupling agents have been used in the filler industry for years. One particularly unique feature is the ability of the coupling agents to enhance the adhesion between filler and polymer matrices. This paper will review the use of coupling agents for polymer matrix composite applications and discuss a new approach to coupling agent chemistry, namely, direct formation of a silicon-carbon non-hydrolyzable linkage between glass and coupling agent. The grafted silane monolayers were evaluated by infrared spectroscopy to ascertain their surface coverage and relative stability. Finally, composite specimens were fabricated and their mechanical properties were compared with traditional silane coupling agent treatment.

BACKGROUND OF POLYMER MATRIX COMPOSITES

Polymer matrix composites (PMCs) are physical mixtures of a polymer matrix and a discontinuous phase such as fillers and fibers that serve to alter some aspect, such as the mechanical properties of the matrix. Fillers and fibers may be inorganic (e.g., glass) or organic (e.g., Kevlar fibers). Typical matrices for high performance PMCs are thermosets such as epoxy or (unsaturated) polyesters. Polymer composites have been widely used because of their excellent weight-to-strength ratio. Principal applications for composites are in construction and transportation. [1]

Fillers are added materials that serve to alter some of the properties of PMCs. Fillers serve a variety of roles: [1, 2]

---

* Dept. of Materials and Metallurgical Engineering, South Dakota School of Mines and Technology, Rapid City, S.D.

† Dept. of Chemistry and Chemical Engineering, South Dakota School of Mines and Technology, Rapid City, S.D.

‡ Dept. of Metallurgical Engineering, University of Utah, Salt Lake City, Utah

- Cost reduction (e.g., glass beads)
- Improved processability in exothermic thermosetting reactions (e.g., mica)
- Dissipation of heat in exothermic thermosetting reactions (e.g., sands)
- Modification of the polymer's electrical- and heat-insulating properties (e.g., mica)
- Reduced mold shrinkage (e.g., graphite)
- Minimize electrostatic charging (e.g., carbon black)
- Improvement in specific mechanical property or properties like strength and abrasion resistance (e.g., silica)

In addition, new fillers based on advanced technologies (e.g., carbon nanotubes) are being studied and produced for special applications.

Mechanical properties of PMCs are primarily dependent upon three factors: (1) strength and modulus of the fiber/filler, (2) strength and chemical stability of the resin, and (3) effectiveness of the bond between resin and fiber/filler in transferring stress across the interface. [3] Given the importance of item (3) it follows that the interfacial region is also considered to be of great importance in PMCs as it is in the interfacial region where stress concentration develops because of differences between the reinforcement and the matrix-phase thermal expansion coefficients, where loads applied to the structure act, and where cure shrinkage and crystallization can occur. [4] Complicating the understanding of PMC interfaces is the possible formation of a so-called interphase region adjacent to the reinforcement which has properties different than that of the bulk polymer matrix. [5] In addition, the interface/interphase can also serve as a nucleation site, a preferential adsorption site, and a locus for chemical reaction. [6] For all of these reasons the interface/interphase is often considered as a major contributor to the mechanical and various physical properties of PMCs. [4] More specifically, the interface is also very critical to the adhesion between mineral fillers/fibers and organic resins. It has been shown conclusively that improved adhesion can serve to enhance mechanical strength and chemical resistance (hygrothermal stability) of the PMC. [7] In order to improve the mechanical properties and hydrothermal stability of the interface, silane coupling agents are often utilized.

SILANE COUPLING AGENTS

When two materials are incompatible compatibility can be brought about by introducing a third material that has properties intermediate between those of the other two. Beginning in earnest in the early 1960s organosilicon compounds were chosen as potential coupling agents for glass-reinforced polymers because the inorganic portion of the molecules are similar to glass, and organic functionality allows compatibility with organic polymer. [8] Since their initial use, organo-functional silanes have been found to be useful as adhesion promoters between organic polymers and inorganic substrates such as fibers/fillers under a variety of circumstances.

There have been many theories or mechanisms proposed concerning the methods by which silanes impart their influence on PMCs including: [9]

- Creation of weak boundary layers—coupling agents in some way may eliminate weak boundary layers
- Wettability—coupling agents may improve wetting (and hence the opportunity for improved adhesion) between polymers and the reinforcement
- Acid-base effects—coupling agents may alter the acidity/basicity of a surface and thereby alter the reactivity with the polymer

- The button-down effect—coupling agents may agglomerate/aggregate into islands on the surface, causing roughening and possibly altering adhesion tendencies
- Deformable layer theory—coupling agents may produce a tough, flexible layer between the reinforcement and the polymer
- Restrained layer theory—coupling agents may develop a highly cross-linked interphase region with a modulus intermediate between that of the reinforcement and the polymer
- Chemical bonding theory—the coupling agent may form covalent bonds with both the reinforcement and the polymer, coupling the two together

The chemical bonding theory is the oldest and is still the most often recognized of the theories or silane attributes listed above. More specifically, within this theory it is maintained that the coupling agent contains chemical functional groups that should react with silanol groups on glass and other oxide-based reinforcements. Under this scenario attachment should thus be made by strong covalent bonds. In addition, most coupling agents contains at least one organo-functional group which should co-react with the resin during cure. Assuming the above has occurred, the coupling agent acts as a vehicle to bond the reinforcement to the resin with a chain of primary bonds (50–100 kcal/mol). [3] Recent infrared spectroscopic analysis however have demonstrated that this theory is more complex than first presented. [10] Specifically, rarely do silanes adsorb as well-organized mono-layers (more likely they adsorb in patches or in very disorganized multi-layers) and only a fraction of the available organo-functional groups react with the matrix polymer. Researchers have been able to deposit silanes as well-defined mono-layers, but only under very controlled conditions. [11–18]

Regardless of the specific coupling mechanism, it is generally accepted that coupling agents do form, under optimal conditions such as judicious choice of coupling agent, reinforcement and polymer, covalent chemical bonds with both the reinforcement and the resin. Schrader and coworkers [19–22] found physisorbed and chemisorbed silane layers as well as a last smooth layer having a thickness of about one monolayer on the surface of the glass. The physisorbed silane consisted of small oligomers with almost no linkage between them. In the next chemisorbed inner layers there are some bonds connecting the oligomers. In the last silane layer on the surface of the interfacial siloxane linkage to the glass as well as those between the silanes themselves helped develop a stable network.

Schrader et al. also found that the physisorbed layers were easily washed away with water. The second chemisorbed layers were more resistant to hydrothermal attack and removed only by extraction with boiling water. The highly cross-linked layers were more resistant to hydrothermal attack. But, most importantly, it was found that failure of the interface resulted from hydrolysis of siloxane bonds. Thus, PMCs made with silane coupling agents have drawbacks including the hydrolytic instability due to the lability of the Si–O linkage. This instability ultimately results in debonding between the matrix and fiber and premature composite failure. To mitigate the severe limitation associated with silane coupling agents researchers have attempted to alter the chemistry so as to minimize failure due to hydrolysis of adsorbed silane layers.

STRATEGIES TO LIMIT HYDROLYSIS OF SILANES

In recent years there have been attempts to improve the hydrolytic stability of silane coupling agents. For example, Stratmann et al. [11–13] utilized long chain alkylsilanes deposited via the Langmuir-Blodgett method to protect metal surfaces in corrosive environments. Presumably, the alkylsilane headgroup (siloxane) became chemisorbed to the hydroxylated metal surfaces. This bonding, along with van der Waals bonding between adjacent silane molecules, helped prevent corrosion of the metal substrate. Another noteworthy study where traditional silane coupling agent chemistry was modified to improve the hydrolytic stability of the system is the work of DiBenedetto et al. [23] Here, these researchers improved the hydrolytic stability (as gauged by measurement of interfacial fracture energy changes) of glass fibers in a dimethacrylate resin by pretreating the fibers with silicon tetrachloride prior to traditional adsorption of silane coupling agents. They attributed the improved hydrolytic stability to replacement of hydroxyl groups on the glass by covalent O–Si–O bonds.

Sukenik [17] has attempted to pursue the self-assembly of long-chain alkanes as a strategy by which to overcome the problems of oligomerization and inhomogeneity in the coupling agent layer. The fundamental strategy was the deposition of eleven-carbon non-functionalized silylamine coupling agent on the glass surface. Conceivably such a strategy could provide a more consistent interface resembling self-assembly, owing to van der Waals interactions among such long chain lengths. Although such a strategy might overcome certain problems, it itself introduces new problems. Furthermore, although it affords a well-organized layer on the glass surface, it fails to address the fundamental problem of the destabilizing and hydrolytically susceptible siloxane network which has plagued all silane coupling agent strategies to date.

Additionally, the literature has seen a rise of interest in alkylated layers on heterogeneous materials. While many of these applications have utilized self-assembled monolayer (SAM) strategies to give uniform coatings, the newer references have emphasized grafting organic molecules to solid substrates. In many cases the substrates have been silicon with an oxide layer and the subsequent grafting of carbon chains via the oxygen, as in the case of polysiloxanes deposited on silicon wafers [24], in the fabrication of metal insulator-semiconductor devices, [25] in grafted planar siloxanes for silicon microsensors, [26] in the fabrication of ion-sensitive field effect transistor monolayer membranes, [27] for their tribological properties, [28, 29] and in traditional siloxane-bound silica for chromatography. [30–33]

In all the above cases in which attempted modifications to the traditional use of the grafted organic have been made, the coupling agent is nonetheless bonded to the hydroxyl glass surface by an O-Si-O-C bond, providing a hydrolytically labile couple between reinforcement and coupling agent.

Interestingly, although other linkages—such as the O–Si–C linkage reported in the chromatographic literature—have been made a decade ago, this strategy has yet to be applied to PMCs. It is this approach that is reported, with the overall objective of being improved PMC performance.

EXPERIMENTATION

The purpose of this research is to explore novel coupling agents that can potentially enhance the hydrolytic stability of the polymer composites. These new coupling agents are expected to substitute the Si–O–Si bonds formed by the traditional silane coupling on fillers' surfaces with more stable Si–C bonds as shown in Figure 1.

Toward this end, glass beads were treated by the Speier reaction and Grignard reaction and characterized using Fourier transform infrared (FT-IR) spectroscopy. The mechanical properties of the resulting PMCs were measured under various environmental conditions and the results were compared to evaluate the hydrolytic stability of these novel coupling agents.

EPON resin 828 (Resins-Versatics) cured by methylene dianiline (MDA) was chosen as the polymer matrix. The recommended curing procedure was followed, namely, an

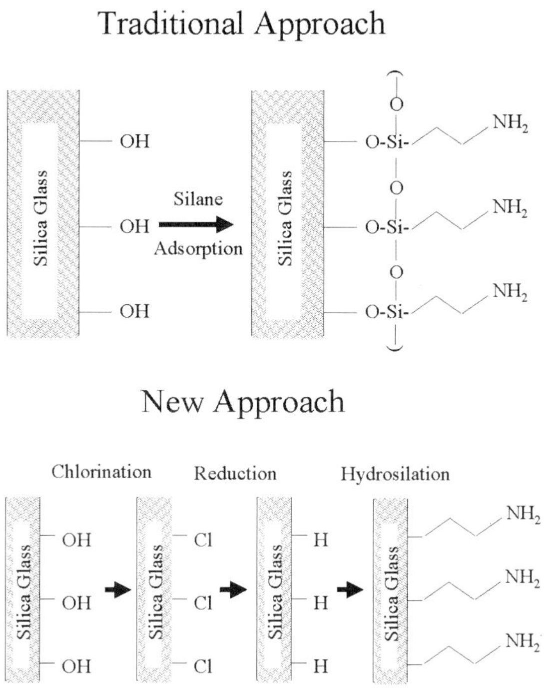

FIGURE 1 Schematics showing traditional adsorption of silane coupling agents (top) and a new approach for grafting of silane coupling agents (bottom)

TABLE 1 Properties of E-glass beads

Sales name	Spheriglass® glass spheres E-glass
Type/component	Borosilicate glass
Appearance and odor	Odorless white powder
Diameter	4 μm
Specific gravity	2.50–2.55 g/cc

initial cure at 80°C for two hours followed by post cure at 150°C for another two hours. Spheriglass 10004E purchased from Potter Industries Inc. was chosen as a model filler. Some typical properties of 10004E are listed in Table 1. [34]

Speier and Grignard reactions were performed on the E-glass beads to graft the expected novel coupling agents. As a comparison, the beads were also treated by the traditional silane coupling agent γ-aminopropyltrimethoxysilane (γ-APS).

The Speier route is tedious work composed of a series of chemical reactions including:

1. Chlorination reaction by refluxing the E-glass beads in thionyl chloride/toluene solution for 96 hours
2. Hydridization reaction by contacting the chlorinated E-glass beads with lithium aluminum hydride in 1M ether solution
3. Hydrosilation (Speier) reaction by refluxing the E-glass beads with hydrochloroplatinic acid (catalytic) phthalimidoalkenes in toluene solution
4. Phthalimide deprotection reaction by refluxing the E-glass beads with H_2NNH_2 in ethanol solution

Because phthalimidoalkenes were not commercially available, laboratory synthesis was performed by refluxing undecenylbromide with phthalimide and DMF.

Finally, undecenylamine and allylamine were grafted on the E-glass surfaces following the Speier route.

The Grignard route has advantages over the Speier route as the E-glass bead surfaces do not need to be hydridized after chlorination. Therefore, this route can be accomplished in two steps:

1. Chlorination reaction by refluxing the E-glass beads in thionyl chloride/toluene solution for 96 hours
2. Grignard reaction by reacting the chlorinated E-glass beads with alkenylmagnesium halides

Vinylbenyl, undecenyl and allyl were grafted on the E-glass surfaces following the Grignard route. E-glass modified by the Grignard reaction can be reacted with monomers such as vinyl esters and others containing reactive double bonds.

γ-aminopropyltrimethoxysilane (γ-APS) was purchased from Sigma Chemical company and a 1% γ-APS solution at natural pH was used to adsorb the coupling agent on the E-glass beads. After adsorption the beads were dried in an oven at 110°C for 2 hours. More details on the experimental procedure can be found elsewhere. [35]

Diffuse reflectance Fourier transform infrared (DRIFT) spectroscopy was used to characterize the treated E-glass beads. Because the E-glass beads (4 μm in diameter) were fine powders, no grinding was needed before the FT-IR characterization. FT-IR spectra were taken before and after each step of chemical treatments allowing a comprehensive study of the surface chemistry.

The hydrolytic stability of the grafted material on E-glass beads was tested by exposing the grafted beads to a 0.75 mmol solution of trifluoroacteic acid (TFAA) for 1 day. The FT-IR spectra of the grafted material were compared before and after exposure to TFAA. The amount of the grafted material remaining after 1 day in TFAA was determined by measuring the height of the symmetric and asymmetric CH_2 stretching bands at about 2,855 cm^{-1} and 2,925 cm^{-1}, before and after TFAA exposure. The results were compared to the same beads with adsorbed γ-APS that had been treated in the same manner.

PMCs filled with E-glass beads with various surface treatments were mixed, cast and degassed prior to curing. [35] The mechanical properties of the composites under different environmental condition were tested and compared to evaluate the hydrolytic stability properties.

A miniature materials tester was used to perform the tensile testing of the composites filled with E-glass beads. The test speed was set to 0.2 mm/min according to ASTM standard [36] and the load beam for the test was 1,000 N. Five types of samples including pure polymer, polymer filled with untreated E-glass beads, polymer filled with APS treated E-glass beads, polymer filled with allylamine treated E-glass beads and polymer filled with undecylamine treated E-glass beads were tested. Five samples of each type were tested. The thickness, ultimate tensile strength (at break), ultimate stain (strain at break) and Young's modulus were measured and recorded for each sample. The standard deviations of the data were calculated.

One sample of each type was exposed to boiling distilled water for 24 hours prior to the tensile testing. Testing results were recorded and compared with those without boiling to study the hydrolytic stability of various surface treatments. Also, the change of weight and size before and after boiling was recorded.

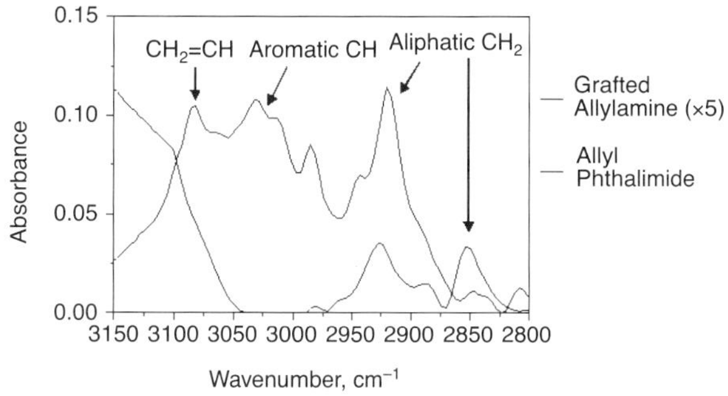

FIGURE 2 Comparison of allyl phthalimide (top spectrum) with grafted allylamine (bottom spectrum)

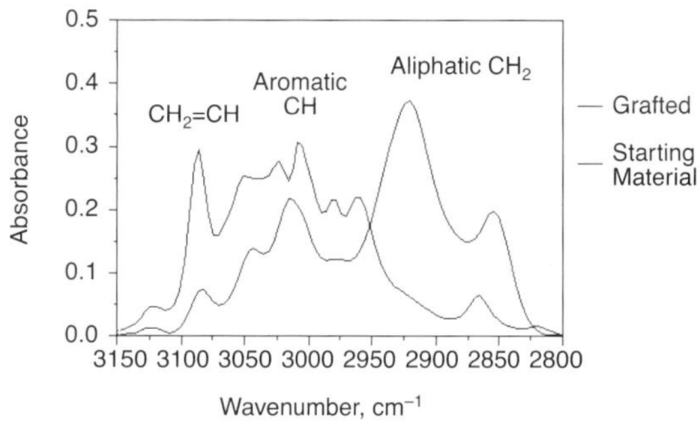

FIGURE 3 Comparison of chlorovinylbenzene (bottom spectrum) with grafted vinylbenyl (top spectrum)

RESULTS AND DISCUSSION

FT-IR spectroscopy was used to determine whether the grafting reactions were successful. For example, Figure 2 shows the comparisons of the allyl phthalimide before grafting with the grafted materials after deprotection of phthalimide to amino groups. For both the allyl material, the loss of the bands labeled CH_2=CH indicate the success of direct grafting by reacting the CH_2=CH group of the coupling agents with the SiH group on the E-glass beads surfaces. The loss of the aromatic CH groups indicates that the phthalimides have been removed by the deprotection reaction. The retention of the bands labeled aliphatic CH_2 indicates that the grafted material remains on the surface. Similar IR spectra were recorded for the undecenyl phthalimide and undecenyl grafted material.

For the Grignard route, the FT-IR spectra of the initial components were also compared with those of the grafted compounds to determine whether grafting had occurred. Figure 3 shows the spectra for vinylbenzyl Grignard grafting. In Figure 3, successful grafting can be seen by the retention of the bands labeled CH_2=CH, aromatic CH and aliphatic CH_2. Similar IR spectra were recorded for the grafted allyl and undecenyl grafted materials.

IR spectra of E-glass beads with and without γ-APS treatment are compared in Figure 4. The appearance of CH_2 bands at 2,855 cm^{-1} and 2,925 cm^{-1} indicate the success of γ-APS treatments. The band labeled CH_3 at 2,950 cm^{-1} is caused by unhydrolized γ-APS.

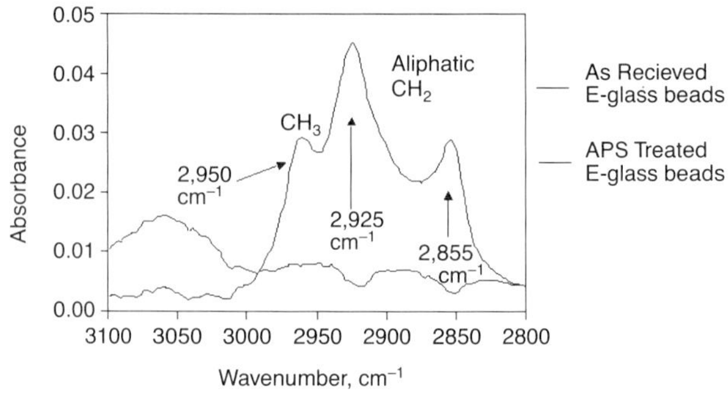

FIGURE 4 Comparison of as received beads (top-left spectrum) with γ-APS treated beads (bottom-left spectrum)

TABLE 2 Summary of FT-IR hydrolytic stability data

Surface Treatment	Percent Remaining After 1 Day in TFAA, %
Grafted vinylbenzyl (Grignard route)	66–75
Grafted undecenyl (Grignard route)	50–66
Grafted undecenylamine (Speier route)	50–66
Grafted allyl (Grignard route)	50–66
Grafted allylamine (Speier route)	75–80
Adsorbed γ-APS (1 wt% solution)	35–50

Spectra such as those shown in Figures 2–4 were evaluated for loss of adsorbed/grafted material as a function of exposure time to TFAA. The strong aliphatic IR stretching bands (for example, the asymmetric CH_2 band at 2,929 cm^{-1}) were used to determine loss of material. Table 2 contains a summary of the average amount of band height decrease for the grafted E-glass and for γ-APS treated E-glass. Table 2 shows that the grafted material was clearly superior to γ-APS in terms of the amount of material removed from the surface by TFAA treatment. For instance, about 75–80% of grafted allylamine remained at the E-glass surface while only 35–50% of the γ-APS remained adsorbed to the E-glass. Thus, grafting was approximately twice as successful at keeping material on the surface compared to γ-APS adsorbed from solution. Among the directly grafted coupling, allylamine formed by the Speier route showed the best hydrolytic stability after TFAA treatments.

Next, the E-glass beads with different surface treatments were compounded into epoxy polymers. Mechanical properties under various environmental conditions were tested to study the hydrolytic stability.

Tensile tests were performed on polymer composites filled with E-glass beads having different surface treatments. In general, it was found that addition of E-glass beads increased the Young's modulus (initial slope of the stress-strain curve), while decreasing the tensile (maximum) strength. Table 3 shows these trends more clearly. Specifically, the Young's modulus has increased by about a factor of 2 from about 1.7 GPa to about 3.5 GPa. The different surface treatments do not exhibit much difference in Young's modulus. Although the grafted fillers composites do exhibit statistically different Young's moduli compared to the γ-APS treated filler composite, none of the treated composites can be differentiated from the composite fabricated with untreated E-glass

TABLE 3 Summary of mechanical testing data

Material	Young's Modulus, GPa	Tensile Strength, GPa
Epon 828 with MDA	1.68 ± 0.70	81.5 ± 3.0
Epoxy with undecenylamine grafted E-glass beads	3.40 ± 0.40	54.3 ± 20
Epoxy with allylamine grafted E-glass beads	3.71 ± 0.27	62.6 ± 1.6
Epoxy with untreated E-glass beads	3.44 ± 0.39	73.7 ± 5.3
Epoxy with APS treated E-glass beads	3.16 ± 0.13	80.5 ± 7.5

TABLE 4 Young's modulus of composites after hydrolytic exposure

	Young's Modulus, GPa		
Filler Treatment	No Exposure	Hydrolytic Exposure	% Change
None	3.4 ± 0.4	2.2	–35.3
Adsorbed γ-APS	3.1 ± 0.1	2.5	–19.4
Grafted allylamine	3.7 ± 0.3	2.3	–37.8
Grafted undecenylamine	3.4 ± 0.4	2.8	–17.6

beads. For the tensile strength, little difference is seen between the pure epoxy, the untreated E-glass beads in epoxy and the γ-APS treated E-glass beads in epoxy. The grafted E-glass bead composite exhibited slightly lower tensile strengths, which is probably related to the observation that, after curing, the tensile test specimens were still tacky. Thus, curing was not completed for the grafted material following the same cuing procedure for the pure epoxy and the non-grafting treatments. This indicates that the curing cycle was not optimized for the grafted E-glass bead composites, leading to a decrease in mechanical properties.

After successful compounding of epoxy matrix composite materials, these samples were tested to determine the grafted materials effect on the hydrolytic stability of the composite materials. Hydrolytic stability testing was performed by boiling tensile test specimen for 8 hours in distilled water. The size and weight change of the samples were measured, but were found to be less than 1% gain for all samples tested. Thus, the material does not incorporate much water during this exposure. Following hydrolytic exposure, the samples were subject to tensile testing. As before, the Young's modulus and the tensile strength were determined from the stress-strain curves. Table 4 shows the results for Young's modulus obtained from tensile testing of hydrolytically exposed samples, while Table 5 shows the resulting tensile strengths. With respect to the Young's modulus, all treatments displayed significant loss of stiffness. The undecenylamine grafted E-glass and the adsorbed γ-APS treated E-glass filled composites exhibited the least loss of stiffness and the change from the non-hydrolytically treated composite was about the same in both cases. The loss in stiffness could not be attributable to changes in the matrix as the pure matrix properties exhibited essentially the same Young's modulus whether or not it had been hydrolytically exposed. The tensile strength data exhibited a different trend. As shown in Table 5, the degree of change was essentially constant for no filler treatment, γ-APS adsorption and allylamine grafting. These three data are somewhat greater than the change observed for undecenylamine grafted E-glass bead filler. Similar to the Young's modulus data, the tensile strength of the pure epoxy did not decrease significantly upon hydrolytic exposure. Thus, when both data sets are considered simultaneously, the undecenylamine grafting treatment appears to give the best resistance to the

TABLE 5 Tensile samples of composites after hydrolytic exposure

	Tensile Strength, MPa		
Filler Treatment	**No Exposure**	**Hydrolytic Exposure**	**% Change**
None	73.7 ± 5.3	36.8	–50.1
Adsorbed APS	80.5 ± 7.5	45	–44.1
Grafted allylamine	62.6 ± 1.6	34	–45.6
Grafted undecenylamine	54.3 ± 20	32	–31.5

type of hydrolytic exposure performed. The superiority of grafting with respect to no surface treatment and γ-APS adsorption was expected due to the stability of the Si-C bond formed during grafting. The difference between the allyl- and undecenyl-amine grafting was unexpected, but may be due to the difference in chain length. The difference may be of particular importance on E-glass surfaces as full surface coverage is not necessarily expected and thus the longer chain of the undecenylamine graft may cover more of the surface, in the areas in which no grafting occurs, than the relatively short-chained allylamine graft. Thus, more water is kept away from the surface by the undecenylamine graft and the hydrolytic stability of the surface is improved.

CONCLUSIONS

Traditional silane coupling agent treatments on fillers for PMCs has proven to be a robust method to modify the properties of PMCs. However, one limitation associated with traditional silane coupling agents is their lack of hydrolytic stability. Alternatives to traditional silane coupling agents are emerging. In this regard, research presented herein showed that Speier and Grignard reactions are applicable in forming directly grafted coupling agents on glass surfaces and that the Grignard route is superior to the Speier route in that it is easier to perform and the products are easier to characterize using FT-IR spectroscopy. It was also demonstrated that the hydrolytic stability of various surface treatments on E-glass beads can be indirectly studied by performing mechanical testing on polymer composites filled with those beads. It was also found that directly grafted coupling agents had better hydrolytic stability than traditional silane coupling agents like γ-APS presumably due to the Si-C bonds formed on the glass surfaces. Of the grafted coupling agents, those with longer aliphatic chains show better hydrolytic stability, possibly due to higher surface coverage.

ACKNOWLEDGMENT

This work was funded through the Air Force Office of Sponsored Research, Grant No. F49620-00-1-0282.

REFERENCES

1. J.R. Fried, *Polymer Science and Technology*, Prentice Hall, New York, 1995, pp. 252–257.
2. E.P. Plueddemann, *Composite Materials*, Academic Press, New York and London, 1974, Vol. 6, pp. 132–133.
3. E.P. Plueddemann, *Composite Materials*, Academic Press, New York and London, 1974, Vol. 6, pp. 2–18.
4. C.D. Arvanitopoulos, and J.L. Koenig, Applied Spectroscopy, Vol. 50, Number 1, 1996.
5. W.M. Cross, F. Johnson, J. Mathison, C. Griswold, L. Kjerengtroen, and J.J. Kellar, Journal of Adhesion, 78, 2002, pg. 1–21.

6. J.L. Kardos, *Molecular Characterization of Composite Interfaces*, H. Ishida and G. Kumar, Eds. Plenum Press, New York, 1985, pp. 891–895.
7. E.P. Plueddemann, *Composite Materials*, Academic Press, New York and London, 1974, Vol. 6, pp. 183–185.
8. K.L. Mittal, *Silane and Other Coupling Agents*, VSP BV, 1992, pp. 4–7.
9. E.P. Plueddemann, *Silane Coupling Agents*, 2nd edition, Plenum Press, Inc., 1982, p. 4.
10. F.J. Johnson, W.M. Cross, D.A. Boyles, and J.J. Kellar, Composites–Part A, 2000, 31(9), pp. 959–68.
11. M. Wolpers, M. Stratmann, H. Viefhaus, and J. Fresenius, Anal. Chem, 1991, 341, pp. 337.
12. M. Wolpers, H. Viefhaus, and M. Stratmann, Applied Surface Science, 1991, pp. 47.
13. M. Wolpers, M. Stratmann, H. Viefhaus, and H. Streckel, Thin Solid Films, 1992, 210/211, pp. 592.
14. F. Effenberger, and S. Heid, Synthesis, September 1995, pp. 1126–1130.
15. H. Watson, P.J. Mikkola, J.G. Matisons, J.B. Rosenholm, Colloids and Surfaces–Part A., 161, 2000, pp. 183–192.
16. A.N. Parikh, J. Wood, R. Sharma, and D.L. Allara, "Prepolymerized Langmuir–Blodgett Films of n-Octadecylsiloxane Monolayers," in *Surfactant Adsorption and Surface Solubilization*, ACS Symposium Series 615, R. Sharma Ed., 1995, pp. 355–363.
17. N. Balachander, and C. Sukenik, Langmuir, 1990, 6(11), 1621.
18. Y.W. Lee, J. Reed-Mundell, C.N. Sukenik, and J.E. Zull, Langmuir, 1993, 9, pp. 3009–3014.
19. M.E. Schrader, I. Lerner and F.J. D'oria, Med. Plast, 45, 1967, pp. 195–197, 268, 272, 276, 280 and 282.
20. M.E. Schrader, J. Adhesion, 2, 1970, pp. 202–212.
21. M.E. Schrader, and A. Block, J. Polym. Sci., Part C, 34, 1971, pp. 281–291.
22. M.E. Schrader, J. Polym. Sci., C34, 1973, p. 281.
23. W. Wang, and A. DiBenedetto, Journal of Adhesion, 1998, 68, p. 183.
24. G. Mao, F. Sun, and D. Grainger, "Synthesis and Characterization of Alkylsilane-branched Polysiloxanes and Their Self-Assembling Monolayers on Silicon Wafers," American Chemical Society, Division of Polymer Chemistry, Papers Presented at Meeting; 1993, 34(1), p. 134.
25. D. Vuillaume, and R. Rondelez, Mat. Res. Soc. Symp. Proc., 1994, 328, p. 203.
26. Y. Duvault-Herra, N. Jaffrezic-Renault, D. Morel, J. Serpinet, J-L. Duvalut, and G. Hollinger, Colloids and Surfaces, 1991, 53, p. 169.
27. H. Perrot, N. Jaffreqic-Renault, A. Gagnaire, J. Dubault, G. Hollinger, N. De Rooij, G. Racine, S. Jeanneret, A. Maaref, Thin Solid Films, 1992, 214, p. 110.
28. H. Zarrad, P. Clechet, M. Belin, C. Martelet, N. Raffrezic-Renault, Tribology International, 1995, 28(4), p. 241.
29. B. Bhushan, V. Ashok, V. Kulkarni, V. Vilas, N. Koinkar, M. Boehm, L. Odoni, C. Martelet, M. Belin, Langmuir, 1995, 1, p. 3189.
30. J. Pesek, J. Sandoval, and S. Minggong, Journal of Chromatography, 1993, 630, p. 95.
31. J. Sandoval, and J. Pesek, Anal. Chem., 1991, 63, p. 2634.
32. J. Sandoval, and J. Pesek, 1991, US Pat. 5,017,540.
33. J. Sandoval, and J. Pesek, Anal. Chem., 1989, 61, p. 2067.
34. Potters Industries Inc, "Materials Safety Data Sheet," *www.pottersbeads.com/MSDS/SpheriglassE.pdf*, 07/30/2002.
35. Hao Du, M.S. Thesis, "Novel Grafting Strategies for Silane Coupling Agents," South Dakota School of Mines and Technology, 2002.
36. ASTM, "Standard Test Method for Tensile Properties of Plastics (Metric), "Annual Book of ASTM Standards, volume, 08.01, D 638, p 61.

Adhesion of Talc and Ink Particles in Paper De-inking Flotation

J. Liu*, **J. Zinterer***, **J. Masliyah***, **J. Yordan†**, **and Z. Xu***

Effective removal of ink particles by froth flotation is of up-most importance in paper recycling. In this study, the potential use of talc particles as carriers of difficult-to-float fine ink particles is investigated. In order to accomplish ink particles removal, the interactions between talc and oily-ink particles in aqueous solutions containing flotation chemicals was studied by zeta potential distribution measurements and with the impinging jet apparatus. The chemical conditions under which talc-ink adhesion occurs are identified and reported.

INTRODUCTION

In the paper recycling industry, ink removal is a critical step to improve the quality of the recycled fibers. Froth flotation has been successfully used commercially for the removal of ink particles from the re-pulped paper slurry [1, 2]. However, the removal efficiency of very fine ink particles is typically low owing to the low collision-attachment efficiency of small ink particles being floated with flotation bubbles. In this regard, it has been recently shown that either agglomerating ink particles with oil [3] or using coarse talc particles to carry (piggy-back) ink particles [4] are promising avenues for improving the removal efficiency of fine ink particles by flotation.

OBJECTIVE

This study is concerned with the use of talc particles as carriers of fine ink particles in flotation. The aim of the investigation is to identify the chemical conditions required to promote the adhesion of talc and ink particles in carrier (piggy-back) flotation. It is believed that this information is vital in making the talc-assisted de-inking flotation concept a successful commercial technology. In order to accomplish this, the chemical conditions leading to

* Dept. of Chemical and Materials Engineering, University of Alberta, Edmonton, Atlanta, Canada.

† Luzenac America, Centennial, Colo.

adhesion or lack thereof between talc and ink particles were identified from zeta potential distribution measurements made with individual ink and talc particles as well as with their binary mixtures. The measurements were conducted in chemical environments pertinent to de-inking flotation. The adhesion of talc and ink particles was then visually confirmed by deposition tests conducted with an impinging jet apparatus.

PRINCIPLES OF THE TECHNIQUE

Theoretically, the classical DLVO (Deryaguin-Landau-Verwey-Overbeek) theory considering summation of van der Waals and electro double layer interactions was often used to predict colloidal particle interactions. Calculation of electrostatic double layer and van der Waals interactions is relatively straightforward for well defined systems with known system parameters such as surface potential of interacting colloidal particles, Hamaker constant or the parameters needed to calculate the resultant Hamaker constant of the systems and electrolyte composition and concentration. However, in many cases, the classical DLVO could not accurately predict the colloidal interactions and an extended DLVO theory has to be used, which may include repulsive hydration force for hydrophilic surfaces, attractive hydrophobic force for hydrophobic surfaces, repulsive steric force and attractive bridge force for polymer bearing surfaces, and acid-base (AB) interaction. The theory for describing these additional non-DLVO forces is less well developed. Alternatively, these forces were often inferred from the deviation of the experimentally measured colloidal forces from those predicted by the classical DLVO theory. In this regard, many experimental methods have been developed to study the colloidal interactions, most of them for well-defined systems. For instance, visual observation such as settling test can directly observe coagulation behavior, but is limited to the particle density. Direct force measurement with atomic force microcopy (AFM) and surface force apparatus (SFA) are the most direct method, but rather limited to the systems that can be studied by the requirement of regular geometry or optical transparency. In this paper, we started by recognizing that different particles would possess different surface properties, and therefore tracing the change of surface properties could provide us with an important avenue to study colloidal particle interactions. Zeta potential is one the most important surface properties in a colloidal system. Based on this idea, zeta potential distribution measurement was developed in this study into a technique to investigate talc-ink particle interactions. This surface properties-based technique does not suffer from the limitations of particle density and geometry.

Previous researchers [1–4], using conventional electrophoretic zeta potential meters, have investigated electrokinetic properties of talc and ink particles extensively. In these early measurements, only the average electrophoretic mobility or zeta potential for a single component system was reported, even though binary systems were studied. In the case of a binary mixture system, the measured average zeta potentials of the two components often either over or under-estimate the electrical behavior of the system, thereby leading to misguided information regarding the interpretation of interactions between the two components. With the capability of measuring electrophoretic mobility or zeta potential distributions, it is possible to identify the deposition conditions of one component on the other in a binary suspension system. The concept for this application is described as follows.

When zeta potential distributions are measured separately using talc and oily ink particles, each has its unique zeta potential distribution centered at ζ_T and ζ_I, respectively. For illustrative purposes, these two distributions are overlaid schematically in Figure 1a. When talc and oily ink particles are mixed together under the same physicochemical condition, the measured zeta potential distribution of the resultant suspension can be interpreted in

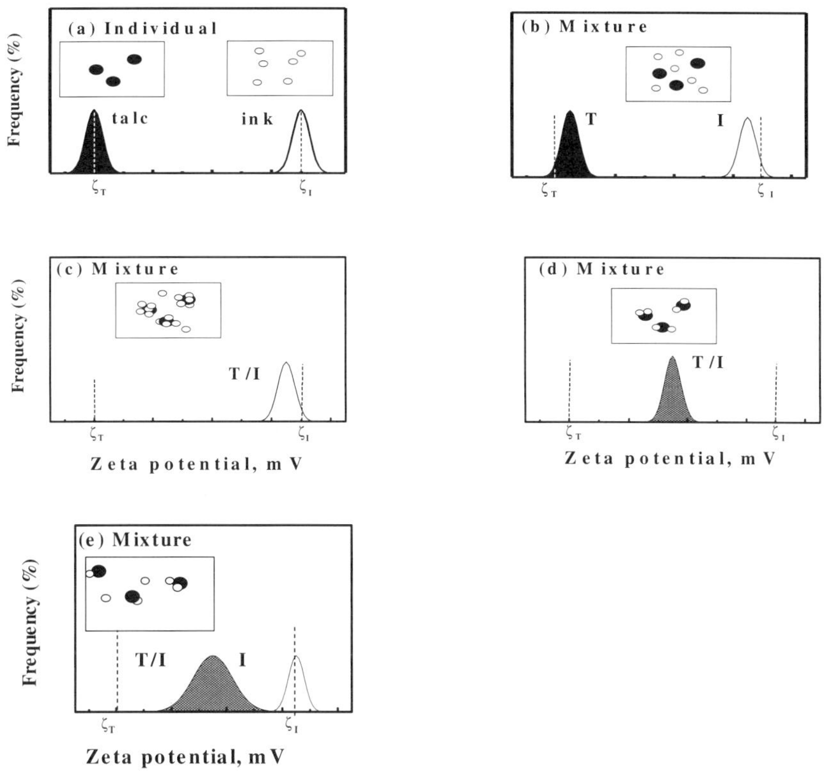

FIGURE 1 Schematic zeta potential distributions for a binary particulate component system that can be interpreted for particle interactions. In this figure, the black and white circles represent talc particles and ink particles, respectively. (a) Zeta potential distribution of the two components measured separately; (b) binary mixture without attraction; (c) strong attraction (talc particles fully covered and excessive ink particles present); (d) strong attraction (talc particles partially covered with insufficient amount of ink particles); (e) weak attraction (talc particles partially covered with a substantial amount of individual ink particles unattached).

terms of the interactions between the two components as described below. If the talc and the ink particles do not interact with each other, a bimodal zeta potential distribution with two peaks centered at ζ_T and ζ_I as shown in Figure 1b is anticipated. Due to hydrodynamic interaction of moving particles at different electrophoretic mobilities, a slight shift of ζ_T and ζ_I towards each other may be observed. This phenomenon can be considered as electro-kinetic retardation (reduction of $|\zeta_T|$ of fast moving talc particles in the presence of slow moving oily ink particles) and enhancement (increase of $|\zeta_I|$ of slow moving ink particles in the presence of fast moving talc particles).

In the case where the talc particles strongly attract the ink particles, the measured zeta potential distribution would show only one distribution peak with its position depending on the talc-to-ink ratio. For illustration purposes, we will take ink coating on talc particles as an example for the analysis. At a high ink-to-talc ratio, the strong attraction would ensure a complete surface coverage of talc particles by ink particles as such that only one distribution peak centered around ζ_I would be observed as shown in Figure 1c. At a lower ink to talc ratio, talc would be only partially covered by the ink although all the ink particles are depleted from the suspension due to the strong attraction for the ink droplets. As a result, both distribution peaks for the original talc particles and oily ink particles disappear, accompanied by a new peak corresponding to partially ink-covered talc particles as shown in Figure 1d. In this case, the position of the zeta potential distribution peak depends on the coverage of talc surfaces by oily ink particles, but falls in between the ζ_T and ζ_I.

For a system of weak attraction between talc particles and ink particles, a bimodal zeta potential distribution is anticipated as shown in Figure 1e. In this case, the most left-hand peak representing the original talc particles disappears due to partial coverage by ink particles, accompanied by a new distribution peak. Since the weak attraction would result in a less efficient deposition of ink particles, some of the ink particles would remain as individual particles, contributing to the presence of the remnant (right) ink distribution peak in the measured overall zeta potential distribution. The position of the new peak depends on both the strength of attraction between talc particles and ink particles and their ratio. At a given talc-to-ink particles ratio, a stronger interaction would lead to a greater shift in the peak position towards the distribution peak of ink particles, with a higher area ratio of the new peak to the remnant peak for ink particles. A similar shift is anticipated when increasing the ink-to-talc ratio. A similar analysis can be equally applied to the case of talc coating on ink surfaces. It should be noted that the term strong or weak used in this context is rather loosely defined and highly qualitative. Although the theory of zeta potential is well developed for individual colloidal particles [14, 15], the quantitative description of zeta potential distribution in a coagulated colloidal system is rather complex if not impossible to comprehend from the first principle and is beyond the scope of this communication. Nevertheless, the measured zeta potential distribution pattern of a complex colloidal system does provide an important indication with regard to slime coating status as shown in the inserts of Figure 1.

Based on the previously discussed concept, it is clear that the zeta potential distribution measurement can be a simple yet powerful tool to study interactions in a binary colloidal system, as illustrated in the following example of talc particles interacting with oily ink particles.

EXPERIMENTAL PROCEDURES

Materials

Luzenac America, USA, provided the talc sample used in this study. Talc A is a commercially available naturally occurring, finely ground talc product. Rieger Printing Inks (Edmonton, Alberta, Canada) provided the oily ink particles used in the study. Reagent grade HCl and NaOH (Fisher Scientific) were used as pH modifiers. Ultrahigh purity KCl (>99.999%, Aldrich) was used as supporting electrolytes while reagent grade $CaCl_2$ (99.9965%, Fisher) was used as the source of calcium ions. Sodium oleate (NaOl, 98%) was purchased from Aldrich Chemicals. Reagent grade toluene (Fisher) was used as the dilution solvent. The de-ionized water with a resistivity of 18.2MΩ cm, prepared with an Elix 5 followed by a Millipore-UV plus ultra water purification system was used throughout this study.

Talc Slurry Preparation

The dry talc particles were first wetted with potassium chloride solution to form a paste. The obtained paste was slurried in a potassium chloride containing aqueous solution to approximately 10 wt% solids under high shear mixing for several minutes to ensure proper talc particle dispersion. This concentrated suspension was further diluted as needed for the zeta potential and the impinging jet measurements.

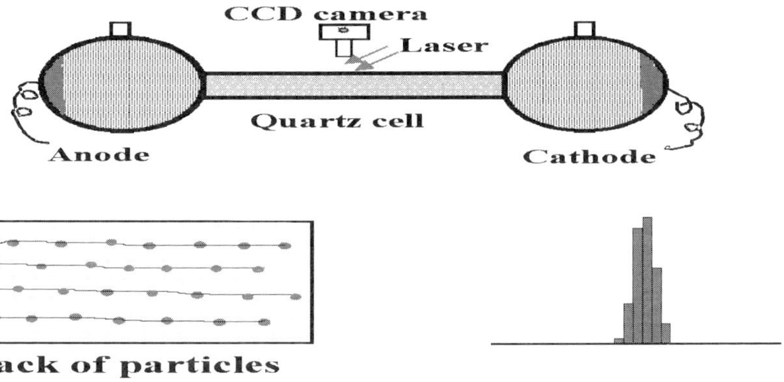

FIGURE 2 Illustration of the zeta potential distribution measurements

Ink Suspension Preparation

The oily ink was dispersed in a potassium chloride solution to contain about 2 wt% ink using an ultrasonic bath. The prepared suspension was diluted as required for the zeta potential distribution measurements.

Oily Ink Collecting Surface Preparation

For preparation of the collecting surface used in the impinging jet measurements, the ink sample was dissolved in toluene at ca. 10 mg ink per ml toluene dilution ratio. About 0.5 ml ink-in-toluene solution was placed drop-wise onto a 5 × 5 cm transparent, cleaned glass slide spinning at 5,000 rpm for 1 minute. The ink-coated surface was dried for about one hour in a particle-free horizontal laminar hood under ambient conditions. The ink-coated glass slide had a smooth, reddish appearance, yet sufficiently transparent to be seen through. The uniform coating of ink was confirmed by a water contact angle value of 73° measured with a drop shape analyzer (Kruss, USA). The surface was used as the collecting surface for talc particles in the impinging jet experiment.

Zeta Potential Distribution Measurements

The zeta potential distributions of the talc slurries and ink suspension were measured with a Zetaphoremeter (SEPHY/CAD). The operating system in this instrument allows tracking 40~100 particles simultaneously and provides a zeta potential distribution for these particles [13]. If these particles are homogeneous, the distribution peak is narrow as illustrated in Figure 2. Otherwise, the peak would be broad. A 40 ml suspension containing 0.1% solids was conditioned for 5 minutes and then used to fill the electrophoresis cell. The suspensions of ink and talc were measured first individually, then as a mixture with a weight ratio of ink to talc being 1:1. The average of four independent measurements was reported.

Impinging Jet Measurements

The impinging jet measurement was carried out to visually observe the deposition of talc particles on the ink surface with an impinging jet apparatus [10–11]. It consists of a flow cycle system with a flow rate control device, an impinging jet cell with a capillary tube, a collector (glass slide coated with probing sample) and a view system (optic microscope with a CCD camera and video recorder). The suspension flows from the capillary tube in the impinging jet cell at a constant flow rate and impinges on the collecting surface. In this manner, the particles in suspension may deposit on the collecting surface, should an attractive force between the collecting surface and impinging particles exist. The particle deposition can be observed through the recorded impinges. About 1.5 L talc suspension at 0.01 wt% solids was used for each test with a pressure driven flow at a Reynolds number of 180. Impinging process at the first 3 minutes was recorded for the analysis. The deposition over the stagnation region (near the center of impinging jet) was used for coagulation analysis.

ANALYSIS AND DISCUSSIONS OF RESULTS

Zeta Potential Distributions

The interaction in KCL-containing water between the as-received talc and ink particles was studied by zeta potential distribution. The results obtained are presented in Figure 3. As noted from Figure 3a (top), both the talc A particles and ink particles are negatively charged with the zeta potential distributions peaking at −35 mV and −10 mV, respectively. Also included in Figure 3a (bottom) is the zeta potential distribution of their binary mixture which exhibits two distinguishable peaks at −30 mV and −10 mV, respectively. The characteristic of the zeta potential distributions for the binary mixture is indicative of a non-coagulating talc-ink system as schematically illustrated in the insert of the figure. In other words, under the chemical environment investigated, there is no attraction leading to adhesion between the talc and ink particles involved. This was expected as both the talc and the ink particles carry sufficiently high negative surface charges to make the talc and ink particles repel each other.

Also in Figure 3b are the zeta potential distributions in the presence of 0.1 mM sodium oleate for both talc and ink particles and their binary mixture. Sodium oleate is a commonly used fatty acid collector in paper de-inking practice. Note that the addition of sodium oleate shifted the zeta potential distributions of talc and ink particles (Figure 3b top) slightly to the more negative direction. As a result, when the talc and ink particles were mixed together in the presence of the flotation collector, a bimodal zeta potential distribution was again observed as is shown in Figure 3b (bottom). This indicates that presence of sodium oleic acid alone could not induce the hetero-coagulation of talc particles with ink particles. This lack of interaction was expected between talc and ink particles due to the electrostatic repulsion existing between the negatively charged talc and ink particles.

Calcium chloride is also a common additive used in paper de-inking flotation. Therefore, it would be important to study its effect on talc-ink particles interaction. As shown in Figure 4a, the addition of calcium ions to the talc-ink system results in a noticeable shift in zeta potential distributions of both talc A and ink particles rendering the particles less negatively charged. In fact, adding 1 mM calcium ions caused zeta potential reversal of ink particles to +5 mV. In this case, only a single zeta potential distribution peak was observed for the mixture as shown in Figure 4a (bottom). Clearly, calcium addition induced hetero-coagulation of talc A with ink particles, as anticipated for a binary mixture system of particles carrying opposite electrical charges.

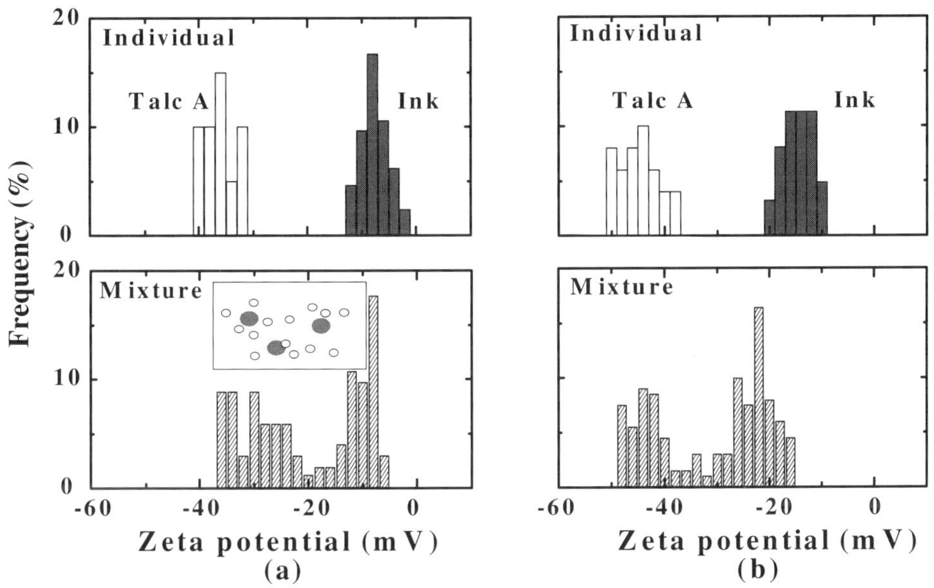

FIGURE 3 Measurements of zeta potential distribution for talc and oily ink particles in (a) 1mM KCl solution and (b) 1mM KCl and 0.1 mM sodium oleic acid solution

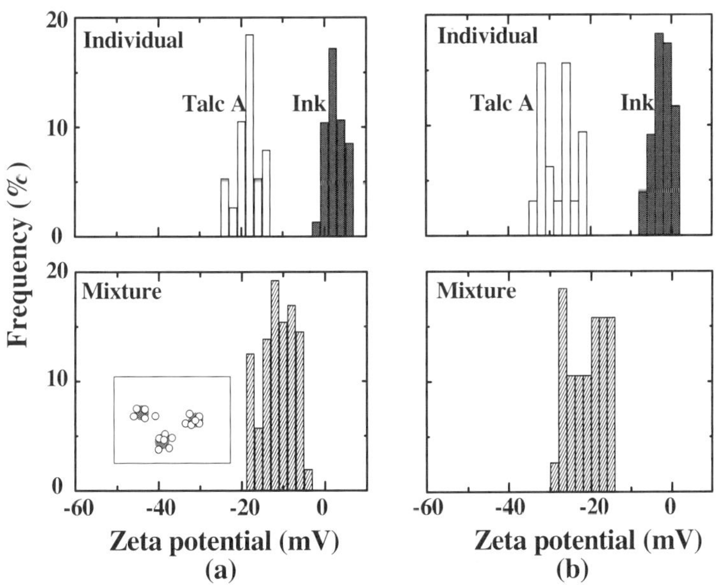

FIGURE 4 Measurements of zeta potential distribution of talc and ink particles in (a) 0.5 mM calcium chloride solution and (b) 0.5 mM calcium chloride and 0.1 mM sodium oleic acid solution

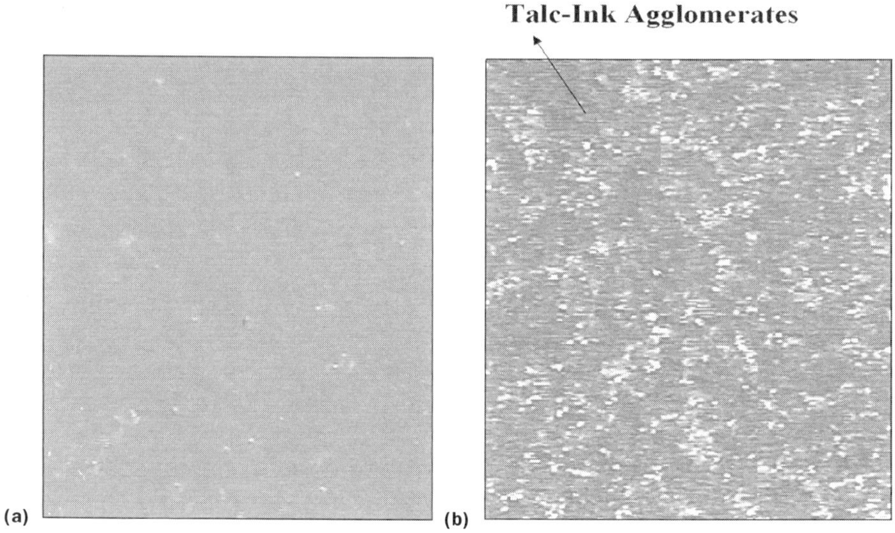

FIGURE 5 The visual observations of talc A particles on the ink-coated collector surface in (a) 1 mM KCl solution and (b) 0.5 mM of calcium chloride and 0.1 mM of sodium oleate in a 1 mM KCl solution

A more common practice in de-inking flotation is the co-addition of sodium oleate and calcium chloride as flotation collectors. This increases ink hydrophobicity resulting in an improved ink removal efficiency in the flotation cells. For this reason, it was important to study the effect of this practice on the talc-ink interactions. The results of the zeta potential distributions obtained with such a system are also included in Figure 4b. Note that as expected the co-addition of calcium and sodium oleic acid produced the hetero-coagulation of talc and ink particles as shown in Figure 4b. Surprisingly, talc-ink hetero-coagulation occurred in the latter case despite the ink particles carrying a marginal negative surface charge. It is important to point out that in both cases the talc particles coated the ink particles with close to a monolayer coverage as schematically illustrated in the insert of Figure 4a.

The previously presented zeta potential distribution data demonstrates the feasibility of enhancing flotation of ink particles with talc as a carrier. One way of adhering talc and ink particles is through the addition of chemical additives such as calcium and calcium/oleic acid mixtures.

Impinging Jet Measurements

The zeta potential distribution results presented in the foregoing section suggested that a very weak interaction exists between the talc and ink particles in water due to strong repulsive electrostatic forces. In order to confirm this observation, impinging jet measurements were conducted in a similar environment. The result of such a measurement is presented in Figure 5. Note that the deposition of talc A particles on the ink collecting surface is very small. This seems to experimentally confirm the inference made from the zeta potential distribution measurement.

Similarly, the zeta potential distribution measurements presented in the foregoing section suggested that it would be feasible to enhance the deposition of negatively charged talc particles on the ink particles by the co-addition of sodium oleate and calcium cations.

In order to confirm that indeed a hetero-coagulation of talc and ink particles can occur, impinging jet measurements were conducted in the presence of previously described chemistry. As shown in Figure 5b, a significant increase in talc deposition was observed with the co-addition of 0.5 mM calcium and 0.1 mM oleic acid. In this case, the talc is deposited on the ink-collecting surface as large aggregates. It is evident that at this condition, calcium induced the adsorption of oleic acid (probably as calcium oleate precipitate) on talc, rendering the talc surface more hydrophobic. The hydrophobic force between talc and ink surface appears to be responsible for the improved talc-ink adhesion.

It should be noted that the analysis of impinging jet test results is rather qualitative in this study. Detailed quantification can be done with imaging analysis procedures. The discussion of interaction is also limited to qualitative nature in this paper. The calculation can be performed using the classical DLVO theory to quantify the contributions of hydrophobic interactions in de-inking system using carrier flotation. With the quantification of deposition, a phenomenological relationship can be derived for practical purposes.

CONCLUSIONS

This study showed that zeta potential distribution measurement is a simple, yet effective method to study particle interactions in a colloidal system of complex nature, such as the talc-assisted de-inking flotation.

The zeta potential distribution measurement showed two avenues to accomplish the adhesion of talc and ink particles: (i) inducing electrostatic attraction by modifying talc surfaces to be oppositely (positively) charged from ink particles by adding divalent calcium ions, and (ii) inducing hydrophobicity by means of calcium and oleic acid co-addition, with the later being more effective.

Impinging jet tests confirmed the deposition of talc particles on ink collecting surfaces under the favorable conditions inferred from the zeta potential distributions measurements conducted with individual components and their binary mixtures.

The attractive electrostatic double layer and hydrophobic forces are identified as the major driving forces controlling the talc deposition on ink surfaces.

REFERENCES

1. Borchard, J.K., "Paper De-inking Technology," *Chem. Ind.,* Vol. 19, 273–276, 1993.
2. Fuerstenau, D.W., "Fine Particle Flotation" in fine particle processing, P. Somasundaran, Ed., SME, New York, NY, USA, 669–705, 1980.
3. Snyder, B.A. and Berg, J.C., "Liquid Bridge Agglomeration: A Fundamental Approach to Toner Deinking," *Tappi J.,* Vol. 77, 79–84, 1994.
4. Maiolo, J. and Pelton, R., "Aerosol-Enhanced Flotation–A Possible Approach to Improve Flotation Deinking," *J. Pulp Paper Sci.,* Vol. 24, 324–328, 1998.
5. Sparks, B.D. and Puddington, I.E., "Deinking of Waste News by Adsorption of Contaminants on a Hydrophobic, Particulate Solid," *Tappi J.,* Vol. 59, 117–119, 1976.
6. Markham, L.D., Ala, M. and Srivatsa, N.R., "Methods for Removing Ink from Printed Paper Using Agglomeration Agent, Followed by Addition of Talc," US patent 5,340,439(1994).
7. Aston, D.E. and Berg, J.C., "Fluid Interfacial Separations for Secondary Fiber Recovery as Probed with Atomic Force Microscopy," *J. Pulp Paper Sci.,* Vol. 24, 121–125, 1988.
8. Derlich, J., Nalaskowski, J., Gosiewska, A., Beach, E. and Miller, J.D., "Long-range Attractive Forces and Energy Barriers in De-inking Flotation: AFM Studies of Interactions between Polyethylene and Toner," *J. Adhesion Sci. Technol.,* Vol. 14, 1929–1843, 2000.

9. Chi, R., Xu, Z., Difeo, T., Finch, J.A. and Yordan, J.L., "Measurement of Interaction Forces between Talc and Toner Particles," *J. Pulp Paper Sci.*, Vol. 27, 152–157, 2001.
10. Sanders, R.S., Chow, R.S. and Masliyah, J.H., "Deposition of Bitumen and Asphaltene-stabilized Emulsions in an Impinging Jet Cell," *J. Colloid Interface Sci.*, Vol. 174, 230–345, 1995.
11. Yang C., Ph.D. thesis: Attachment of Fine Gas Bubbles onto a Solid Surface and Electrokinetics of Gas Bubbles, University of Alberta, 2000.
12. Liu, J., Zhou, Z. and Xu, Z., "Electrokinetic Study of Hexane Droplets in Surfactant Solutions and Process Water of Bitumen Extraction Systems," *Ind. Eng. Che. Res.*, Vol. 41, 52–57, 2002.
13. Liu, J., Zhou, Z., Xu, Z. and Masliyah, J., "Interactions between Clays and Bitumen in Aqueous Media from the Measurement of Zeta Potential Distributions," *J. Colloid Interface Sci.,* Vol. 252, 409–418, 2002.
14. Hunter, R.J., Zeta Potential in Colloid Science: Principle and Applications, Academic Press, London, 1981.
15. Sposito, G. and Grasso, D. in Interfacial Forces and Fields: Theory and Application, ed. J.P. Hsu, Marcel Dekker, New York, 1999.

Effect of Surface Treatment on the Acid-base Properties of Fillers and Pigments

Ismail Yildirim*, **Jinhong Zhang***, **and Roe-Hoan Yoon***

Inorganic fillers and pigments such as kaolin, diatomaceous earth silica, and talc were treated with various surfactants, and the changes in their surface properties monitored. When they were treated with a cationic surfactant, they became hydrophobic and the overall surface free energies were decreased substantially. The decrease in surface free energy was mostly due to the decrease in the acid-base component (γ_S^{AB}) of the surface free energy, indicating that hydrophobizing an inorganic pigment entails a reduction in γ_S^{AB}. It was found that the surface characterization technique employed in the present work was useful for studying the effects of using various other surfactants for surface treatment.

INTRODUCTION

Minerals are widely used as the filler and reinforcements in many different industrial sectors, including plastics, rubber, polymer manufacturing, composite materials, etc. The purpose of using minerals as fillers and reinforcements is to improve the mechanical properties of polymer/filler composites such as tensile modulus and impact strength and extend the utility of polymers. In general, stronger filler-polymer matrix interactions result in improved processability, impact strength, and surface quality, while too weak interactions lead to decreased strength and increased deformability of the composite (1).

When the mineral fillers are used as the filler or reinforcements in the polymer manufacturing, proper control of the adhesion between filler and polymer matrix is important in controlling the property of the composite material. Most of the mineral fillers, however, are naturally hydrophilic and there is a lack of affinity between hydrophilic mineral and the hydrophobic polymer surface. Because of the poor affinity between mineral and polymer, it is difficult to achieve a homogenous mixture. The poor interaction leads to aggregate formation of mineral particles within the polymer matrix.

* Center for Advanced Separation Technologies, Virginia Polytechnic Institute and State University, Blacksburg, Va.

As a means of controlling the filler-matrix interactions and achieving a good compatibility between the hydrophilic mineral particles and polymers, minerals are treated with appropriate surfactants (2, 3). Organic and inorganic cations, silanes, and nonionic surfactants are the most widely used regents for the mineral surface treatment. As a result of surface treatment, the surface free energy of mineral will be reduced and the wetting properties with the polymer matrix will be improved. It has been shown that acid-base interactions play a crucial role in controlling these interactions (4). The knowledge of surface free energy components (acidic, basic, and nonpolar components) of mineral filler and polymers, originating from different kinds of intermolecular forces is, therefore, very useful for understanding the surface interaction mentioned previously.

The objective of the present work was to study the effect of surface treatment on the hydrophobicity and surface free energies of the mineral fillers and pigments. Talc, kaolin clay, and diatomaceous earth silica were chosen as the minerals for this study and are widely used as filler and pigment in a variety of industrial applications. The mineral fillers were treated with varying amounts of cationic amine reagent, and the changes in their surface properties monitored. The contact angle values of untreated and surface treated mineral powders were determined using the thin layer wicking technique. The contact angle values were then used to calculate the surface free energy parameters of untreated and surface treated mineral powders using van Oss-Chaudhury-Good (OCG) acid-base theory.

It was found that after surface treatment, the mineral fillers became increasingly hydrophobic and the overall surface free energies were decreased substantially. The decrease in surface free energy was mostly due to the decrease in the acid-base component (γ_S^{AB}) of the surface free energy, indicating that hydrophobizing an inorganic pigment entails a reduction in γ_S^{AB}. It is hoped that such information regarding the surface free energy properties of mineral fillers will not only help to understand the surface hydrophobicity and surface free energies, but will also be useful for understanding the adhesion properties between two interacting surfaces, i.e., mineral filler and polymer matrix. The surface characterization technique employed in the present work was found to be useful for studying the effects of using various other surfactants for surface treatment.

THEORETICAL BACKGROUND

Van Oss–Chaudhury–Good (OCG) Equation

Fowkes (5–7) proposed that the surface tension of a material i can be subdivided into separate components:

$$\gamma_i = \sum_j \gamma_i^j \quad \textbf{(EQ 1)}$$

where j refers to types of surface tension component, e.g., dispersion, H-bonding, and metallic interactions. Thus,

$$\gamma_i = \gamma_i^{LW} + \gamma_i^{AB} \quad \textbf{(EQ 2)}$$

where γ_i^{LW} and γ_i^{AB} refer to the non-polar and acid-base (polar) surface free energy components, respectively. The former can be represented by the Lifshitz-van der Waals (or LW) interactions that include the dispersion (London), induction (Debye), and orientation

(Keesom) components. The latter represents interactions between Lewis acids (electron acceptor) and bases (electron donor) on the surface.

van Oss et al. (8–10) assumed that the geometric mean combining rule can be applied to express the γ_i^{AB} as follows:

$$\gamma_i^{AB} \equiv 2\sqrt{\gamma_i^+ \gamma_i^-} \quad \text{(EQ 3)}$$

where γ_i^+ is the acidic component of the surface tension and γ_i^- is the basic component. Therefore, Equation 2 becomes

$$\gamma_i = \gamma_i^{LW} + 2\sqrt{\gamma_i^+ \gamma_i^-} \quad \text{(EQ 4)}$$

It can be readily shown also that the free energy change (ΔG_{SL}) associated with the adhesion of a liquid (e.g., polymer melt) on a solid (mineral filler) in the air medium is given by:

$$\Delta G_{SL} = -2\left(\sqrt{\gamma_S^{LW} \gamma_L^{LW}} + \sqrt{\gamma_S^+ \gamma_L^-} + \sqrt{\gamma_S^- \gamma_L^+}\right) \quad \text{(EQ 5)}$$

where γ^{LW} is the Lifshitz-van der Waals component of the surface energy, γ^+ the acidic component, and γ^- is the basic component of the interacting solid (represented by subscript S) and liquid (represented by subscript L) surfaces.

Combining this with the Dupre equation (11):

$$\Delta G_{SL} = \gamma_{SL} - \gamma_S - \gamma_L \quad \text{(EQ 6)}$$

one obtains:

$$\gamma_{SL} = \gamma_S - \gamma_L - 2\left(\sqrt{\gamma_S^{LW} \gamma_L^{LW}} + \sqrt{\gamma_S^+ \gamma_L^-} + \sqrt{\gamma_S^- \gamma_L^+}\right) \quad \text{(EQ 7)}$$

which can be used to obtain interfacial tensions.

Combining Equation 7 with Young's Equation 11,

$$\gamma_L \cos\theta = \gamma_S - \gamma_{SL} \quad \text{(EQ 8)}$$

where θ is contact angle, one obtains the OCG equation (8–11):

$$(1 + \cos\theta)\gamma_L = 2\left(\sqrt{\gamma_S^{LW} \gamma_L^{LW}} + \sqrt{\gamma_S^+ \gamma_L^-} + \sqrt{\gamma_S^- \gamma_L^+}\right) \quad \text{(EQ 9)}$$

which is very useful for characterizing a solid surface in terms of its surface free energy components, i.e., γ_S^{LW}, γ_S^+, and γ_S^-. To determine these values, it is necessary to determine contact angles of three different liquids of known properties (in terms of γ_L^+, γ_L^-, γ_L^{LW}) on the surface of the solid of interest. One can then set up three equations with three unknowns, which can be solved to obtain the values of γ_S^{LW}, γ_S^+, and γ_S^-. Table 1 gives a list of liquids that can be used for the contact angle measurements, along with the values of γ_L, γ_L^{LW}, γ_L^+, and γ_L^-.

TABLE 1 Values of the surface tension components (in mJ/m^2) of the liquids

Liquid	γ_L	γ_L^{LW}	γ_L^{AB}	γ_L^+	γ_L^-
Bromonaphthalene	44.4	44.4	0	0	0
Methylene iodide	50.8	50.8	0	0	0
Water	72.8	21.8	51.0	25.5	25.5
Formamide	58.0	39.0	19.0	2.28	39.6
Ethylene glycol	48.0	29.0	19.0	1.92	47.0

Contact Angles

Equation 9 suggests that the contact angles of three different liquids must be obtained, if one wishes to characterize the surface of a solid in terms of its surface free energy components. The thin layer wicking technique was shown to be a very useful method of determining contact angles on the powdered mineral surfaces (12, 13). In this work, the contact angles of various liquids on untreated and surface treated minerals (talc, silica, and kaolin clay) were measured using thin layer wicking technique.

In this technique, a small quantity of mineral in water suspension is deposited on a microscopic glass slide to form a thin layer of powdered mineral on the slide. After oven drying the glass slide at 105–110 °C temperature, one end of the slide is immersed vertically in a liquid of known surface tension (γ_L). The liquid will creep up the mineral-coated slide driven by capillary forces. The distance (l) traveled by the liquid during a given time (t) is measured and then converted to contact angle, using Equation 10 given below (12, 13):

$$\cos\theta = \left(\frac{2\eta l^2}{\gamma_L r^* t}\right) \qquad \textbf{(EQ 10)}$$

where, η is the liquid viscosity and r^* the mean pore radius of the powdered mineral layer on the glass slide. The value of r^* is determined by performing the thin layer wicking tests with non-polar liquids such as saturated hydrocarbons that will wet mineral surface completely. In the case of complete wetting, the contact angle (θ) is zero and hence, $\cos\theta = 1$. Therefore, a plot of $2\eta l^2/t$ vs. γ_L would yield a straight line for which r^* is the slope.

EXPERIMENTAL

Mineral Samples

A finely ground talc sample was received from Luzenac America, Englewood, CO. The talc sample had a mean particle size of 3.4 μm. A finely ground diatomaceous earth silica was also supplied by Luzenac America. Talc and diatomaceous earth silica samples were used in the surface free energy characterization studies as received.

A crude east Georgia kaolin clay was obtained from Thiele Kaolin Company located in Sandersville, Georgia. Before the surface free energy characterization studies, the crude kaolin was subjected to selective flocculation to remove discoloring impurities (i.e., anatase and iron oxides). The details of removing the discoloring impurities from kaolin clay using selective flocculation technique are given elsewhere (14). Before the characterization studies, the selective flocculation product (beneficiated kaolin clay) was subjected to bleaching using 5 lb/ton aluminum sulfate ($Al_2(SO_4)_3 \times 18H_2O$) and

9 lb/ton of sodium hydrosulfite ($Na_2S_2O_4$) in a dilute sulfuric acid solution at pH 2.5–3.0. 90% of the beneficiated kaolin clay was minus 2 μm in size. Talc, kaolin clay, and diatomaceous earth silica are widely used as the fillers in paint, plastics, polymer, and rubber manufacturing.

Wicking Liquids

The thin layer wicking measurements were conducted using the following non-polar and polar liquids: heptane, octane, decane, dodecane, tetradecane, methylene iodide and 1-bromonaphthalene, water, formamide, and ethylene glycol. The organic liquids used in the present work were HPLC grade (>99.5% purity) and dried overnight over 3 to 12 mesh Davidson 3-A molecular sieves before use. Heptane, octane, decane, dodecane, tetradecane, 1-bromonaphthalene, and methylene iodide were obtained from Aldrich Chemical Company while formamide and ethylene glycol were received from Fisher Scientific. All experiments were conducted using Nanopure water produced with a Barnstead Nanopure II water purification system.

Reagents for Surface Treatment

In the present work, a cationic reagent was used for the surface treatment of powdered mineral surfaces. The cationic reagent investigated, Armeen C a commercial brand amine, was obtained from Akzo Nobel. It is a cocoalkyl amine (RNH_2, R = cocoalkyl) and had purity greater than 97% (97% primary amines and 3% secondary amines), and was used as received.

Surface Treatment

In the present work, only diatomaceous earth silica and kaolin clay were subjected to surface treatment. In these series of tests, an aqueous suspension of diatomaceous earth silica or kaolin clay (5% weight by volume) was prepared by dispersing a known amount of powdered mineral in nanopure water. After adjustment of pH to 8.5–9.0 using sodium hydroxide, Armeen C was added in the amount of 0.5%, 1.0%, and 2.0% (based on dry weight of mineral), and the suspension was agitated for an hour with a magnetic stirrer at pH = 9.0–9.5.

Contact Angle Measurements

An aqueous suspension of untreated talc, silica, or kaolin clay (5% weight by volume) was prepared by dispersing a known amount of powdered mineral in nanopure water. Approximately 3–3.5 ml of the suspension was withdrawn with a pipette and sprayed over a glass slide (1" × 3" in dimension). Similarly, approximately 3–3.5 ml of the surface treated mineral suspension as described previously was withdrawn with a pipette and sprayed over the glass slide for the contact angle measurements.

Upon evaporating the water at room temperature, the samples were further dried in an oven at 105–110 °C to remove any residual pore water. As a result, a uniform thin layer of powdered mineral firmly adhering to the glass was obtained. The mineral-coated glass slide was stored in a desiccator until further use.

The wicking experiments were performed by immersing the mineral-coated glass slides in the wicking liquid in the vertical position to a depth of about 5 mm. The time required for the wicking liquid to travel a pre-set distance (e.g., between 1 to 5 cm) on the powdered mineral layer was recorded. With each liquid, the tests were repeated at

TABLE 2 Contact angles (θ, degree) of various liquids on untreated and treated mineral powders measured using thin layer wicking technique

	Contact Angle (θ)				
Mineral	**1-Br**	**DIM**	**W**	**FO**	**EG**
Talc untreated	74.6 ± 1.0	78.6 ± 1.0	89.4 ± 0.4	77.2 ± 0.9	74.3 ± 0.8
D. E. silica					
Untreated	45.8 ± 1.2	60.0 ± 1.2	41.3 ± 1.3	0 (spreads)	35.1 ± 1.4
0.5% Armeen C	49.9 ± 1.0	62.4 ± 1.4	64.7 ± 1.7	55.7 ± 1.8	—
1.0% Armeen C	59.3 ± 1.3	67.9 ± 1.5	74.6 ± 1.6	65.4 ± 1.5	—
2.0% Armeen C	68.8 ± 1.1	77.4 ± 1.1	81.6 ± 1.0	74.3 ± 1.2	—
Kaolin clay					—
Untreated	23.7 ± 1.3	43.3 ± 1.5	47.4 ± 1.3	40.9 ± 1.8	—
0.5% Armeen C	52.4 ± 1.7	61.3 ± 1.2	72.9 ± 1.2	63.1 ± 1.3	—
0.5% Armeen C	65.4 ± 1.4	75.5 ± 1.3	84.6 ± 1.1	74.7 ± 1.4	—
2.0% Armeen C	78.6 ± 1.2	84.3 ± 1.3	87.0 ± 0.8	84.3 ± 1.5	—

1-Br = 1-Bromonaphthalene, DIM = Methylene Iodide, W = Water, FO = Formamide, EG = Ethylene glycol.

least three times. The value of r^* in the Equation 10 was determined using low surface tension non-polar liquids (heptane, octane, decane, dodecane, and tetradecane), which completely spread over the mineral surface (i.e., $\cos\theta = 1$). For the mineral powders used in this work, the $2\eta l^2/t$ vs. γ_L yielded a straight line. Once the value of r^* was obtained, it was then possible to calculate the values of contact angles using Equation 10.

RESULTS AND DISCUSSION

Contact Angles

Table 2 shows the contact angles of various non-polar (1-bromonaphthalene and methylene iodide) and polar (water, formamide, and ethylene glycol) liquids on the untreated and surface treated mineral powders as obtained from thin layer wicking measurements. As shown, the water contact angle on talc is 89.4°, while it is only 41.3° on the untreated diatomaceous earth silica and 47.4° on the untreated kaolin clay. The high contact angle value of water on talc confirms that talc is a very hydrophobic mineral, while the low values of water contact angles on kaolin and silica suggest these minerals are hydrophilic.

Also shown in Table 2 are the contact angle values of kaolin and silica treated with varying amounts of amine (0.5–2.0% Armeen C, based on dry weight of mineral). The results given in this table show that as the amount of amine used for the surface treatment increased from 0.5% to 2.0%, the values of water contact angles were gradually increased both on the kaolin and silica surfaces, and so does the hydrophobicity. A similar contact angle trend was observed for the other test liquids, e.g., 1-bromonaphthalene, methylene iododide, and formamide. In the case of kaolin clay, it seems that 1% of amine is enough to obtain a very hydrophobic surface, after which the effectiveness of reagent is limited.

The data given in Table 2 suggest that as low as 2% by weight of amine is required to render the surface of diatomaceous earth silica surface hydrophobic (i.e., to increase water contact angle from 41.3° on untreated surface to 81.6° on the surface treated silica). The water contact angle data clearly indicate that silica requires higher

TABLE 3 Surface free energy (SFE) and its components for untreated and surface treated mineral powders (mJ/m^2)

	SFE Parameter				
Mineral	γ_S^{LW}	γ_S^-	γ_S^+	γ_S^{AB}	γ_S
Talc untreated	18.2	6.5	0.6	4.0	22.2
D. E. silica					
Untreated	32.0	47.0	0.30	7.5	39.5
0.5% Armeen C	30.0	20.7	0.46	6.2	36.2
1.0% Armeen C	25.3	14.9	0.37	4.7	30.0
2.0% Armeen C	20.6	12.7	0.25	3.5	24.1
Kaolin clay					
Untreated	37.9	33.7	0.35	6.8	44.7
0.5% Armeen C	27.8	13.5	0.60	5.7	33.5
1.0% Armeen C	19.9	9.7	0.43	4.1	23.9
2.0% Armeen C	15.3	13.4	0.09	2.2	17.5

amine dosages than kaolin for surface hydrophobization. It is interesting to note also that the value of water contact angles on both kaolin and silica become similar to those of talc, only after these minerals were treated with substantially higher reagent additions.

Surface Tension and Its Components on Mineral Powders

Table 3 gives the surface tension and its components on the untreated and surface treated minerals calculated from the contact angle data using the methodology described in the foregoing section.

The data given in Table 3 indicate that the values of γ_S^{AB} on the untreated kaolin and silica surfaces are higher than the value of γ_S^{AB} on the talc surface. Also shown in Table 3, the values of γ_S^{AB} components on both kaolin clay and diatomaceous earth silica decrease substantially as the amount of amine increased for the surface treatment. Most likely, the polar (acid-base, γ_S^{AB}) component of the surface free energies of silica and kaolin decrease with increasing degree of hydrophobicity due to the substitution of surface polar groups by non-polar groups that carry low or no permanent dipole.

Similarly, there is a significant reduction in the values of γ_S^{LW} components on both kaolin clay and diatomaceous earth silica with increasing amine concentration used for the surface treatment. The observed reduction in the value of γ_S^{LW} with increased surface hydrophobicity may be due to the replacement of easily polarizable high surface energy groups by non-polarizable low surface energy groups. Similar reductions in the values both of γ_S^{LW} and γ_S^{AB} were obtained by Yoon and Yordan (15), who treated the fused silica plates with varying amounts of trimethyl chlorosilane (TMCS) or dodecylamine hydrochloride (DAHCl).

Also shown in Table 3 are the values of γ_S^- and γ_S^+ on the untreated mineral fillers. As shown, all of the mineral fillers studied in the present work contain both basic (γ_S^-) and acidic (γ_S^+) surface free energy components. However, the values of γ_S^- are much higher than the value of γ_S^+, indicating that the mineral fillers are essentially more basic than acidic. It should be also noted that the magnitude of γ_S^- both on silica (47 mJ/m^2) and kaolin clay surfaces (33.7 mJ/m^2) is considerably higher than talc surface (6.5 mJ/m^2). The surface basicity of diatomaceous earth silica, kaolin clay, and talc should most likely be attributed to the basic oxygen molecule that is exposed as a result of adsorbed layer of water molecules on the surface of the minerals (16–18).

The presence of such a large number of especially basic component indicate that the interaction between untreated silica or kaolin clay and various polymers (such as polyvinylchloride and polymethylmethacrylate etc) is most likely driven by acid/base interactions. Fowkes (4) has experimentally proven that the acid-base interactions are the main driving force between basic mineral fillers (i.e., calcium carbonate) and acidic polymers (i.e., polyvinyl chloride), and vice versa.

As shown in Table 3, after the surface treatment, the values of γ_S^- component decrease on kaolin clay and silica as the amount of amine increased from 0.5% to 2.0% by wt used for the surface treatment, while the value of γ_S^+ remains constant. The values of γ_S^-, however, are much higher than the values of γ_S^+ in the entire range of amine concentration studied.

Table 3 also gives the values of the surface free energies (γ_S). As shown, the surface free energies of untreated diatomaceous earth silica and kaolin clay are substantially higher than talc. Low surface free energy (only 22.2 mJ/m^2) of talc is a good indication for its natural hydrophobicity. The data presented in Table 3 shows that talc has a non-polar component (γ_S^{LW}) of the surface free energy that is significantly higher than the polar acid-base components (γ_S^{AB}). This suggests that when using talc as filler, the interaction between talc and various polymers (that are also hydrophobic) is most likely driven by non-polar (LW) interactions and not by acid/base interactions.

The values of γ_S both on the kaolin clay and silica surface decreased, however, as the amount of amine increased from 0.5% to 2.0% by wt for the surface treatment. For example, γ_S of kaolin changes from 33.5 mJ/m^2 at 0.5% Armeen C to 17.5 mJ/m^2 at 2.0% addition. It has to be noted that the reduction in the value of γ_S is achieved due to the decreases in the values of both the γ_S^{LW} and γ_S^{AB} components. This finding suggests that the value of γ_S becomes smaller as the surface hydrophobicity of solid increases, which is in line with the literature (15, 19).

As shown, a hydrophilic mineral such as kaolin clay and diatomaceous earth silica can be made hydrophobic after conditioning with hydrophobic reagents (e.g., surfactant). As a result of the reagent adsorption, the surface free energy of the solid (γ_S) is reduced. Such a decrease is necessary for a better adhesion between mineral filler and polymer matrix. Therefore, it is necessary to determine the changes in the surface free energies of solids (i.e., mineral filler) taking place upon treatment with various reagents.

CONCLUSIONS

1. The surface free energy (γ_S) and its non-polar (γ_S^{LW}) and polar (γ_S^- and γ_S^+) components on various mineral fillers and pigments were determined using OCG acid-base theory. As expected, talc was found to be a low surface free energy mineral (hydrophobic) with the non-polar component of surface energy being significantly larger than its polar (acid/base) component. On the other hand, untreated kaolin clay and diatomaceous earth silica were found to be high surface free energy (hydrophilic) minerals.
2. It was shown that naturally hydrophilic minerals such as kaolin clay and diatomaceous earth silica can be made hydrophobic after treating the mineral with hydrophobic reagents (e.g., amine). It was also found that the adsorption of surfactant molecules at the mineral surface reduces both the Lifshitz-van der Waals (γ_S^{LW}) and Lewis acid-base (γ_S^{AB}) components of the surface free energy (γ_S) of the mineral filler.

3. This study showed that both Lifshitz-van der Waals (non-polar) and acid-base (polar) interactions may be the driving forces between two interacting surfaces (i.e., polymer and mineral filler). Therefore, it is critically important to determine the surface free energy parameters of both mineral fillers and polymers for better understanding of the adhesion properties of filler/polymer composites.

REFERENCES

1. Pukansky, B., in: *Polypropylene: Structure, Blends and Composites*, Ed: J. Karger-Kocsis, Vol. 3, pp. 1, Chapman and Hill, London, 1995.
2. Nakatsuka, T., *Polymer Sci. Technol.*, 27, 51, 1985.
3. Ishida, H., Campbell, S., and Blackwell, J., *Chem. Mater.*, 12, 1260–1267, 2000.
4. Fowkes, F.M., in: *Physicochemical Aspects of Polymer Surfaces*, Ed: K.L. Mittal, Vol. 2, p. 583, Plenum Press, New York, 1983.
5. Fowkes, F.M., *J. Phys. Chem.*, 66, 382, 1962.
6. Fowkes, F.M., *J. Phys. Chem.*, 67, 2538–2541, 1963.
7. Fowkes, F.M., *Ind. and Eng. Chem.*, 56/12, 40, 1967.
8. Van Oss, C.J., Good, R.J., and Chaudhury, M.K., *J. of Colloid and Interface Science*, 111, pp. 378, 1986.
9. Van Oss, C.J., Chaudhury, M.K., and Good, R.J., *J. Separation Sci. and Technology*, 22, 1, 1987.
10. Van Oss, C.J., Good, R.J., and Chaudhury, M.K., *Langmuir*, 4, 884, 1988.
11. Van Oss, C.J., *Interfacial Forces in Aqueous Media*, Marcel Decker Inc., New York, 1994.
12. Costanzo, P.M., Giese, R.F., and van Oss, C.J., *Int. Symp. on Colloid and Surface Engineering*, Eds: Williams, R.A., and de Jeager N.C., Butterworth Heinemann, London, 223–232, 1991.
13. Van Oss, J.C., Giese, R.F., Li, Z., Murphy, K., Norris, J., Chaudhury, M.K., and Good, R.J., *J. Adhesion Sci. Technol.*, Vol. 6, No. 4, pp. 413–428, 1992.
14. Yildirim, I., Ph.D. Thesis, *Virginia Tech*, April 2001.
15. Yoon, R.-H., and Yordan, J.L., *J. of Colloid and Interface Science*, Vol. 146, No. 2, p. 101–108, 1991.
16. Pazhianur, R., Ph.D. Thesis, *Virginia Tech*, 1999.
17. Holysz, L., *Colloids and Surfaces*, 134, 321, 1998.
18. Rayss, J., Gorgol, A., Podkoscienly, W., Widomski, J., and Cholyk, M., *J. Adhesion Sci. Technol.*, 12, 293, 1998.
19. Chibowsi, E., and Holysz, L., *J. of Colloid and Interface Science*, Vol. 112, p. 15–23, 1986.

. . . .

Particle Attachment and Detachment in the Presence of Linear Polyphosphates

Adam Feiler* **and John Ralston***

Optical reflectometry was used to study the attachment and subsequent detachment of silica particles (diameter 25 nm) from the surface of titanium dioxide wafers under well-defined hydrodynamic conditions.

Atomic force microscopy was used to measure the interaction between silica spheres (diameter 7 μm) and titanium dioxide wafers under the same solution conditions. In addition to repulsive electrical double layer forces, adsorbed polyphosphates provide a short-ranged steric layer that reduces the lateral interaction between the surfaces. The use of these two complementary techniques has given valuable insight into the process responsible for fine particle detachment and has particular application to surface cleaning.

INTRODUCTION

Surface contamination due to submicron particulate matter is of concern in many areas including silicon wafer fabrication, mineral processing, water purification, and detergency. The permanent removal of these particles from surfaces is a critical factor in these processes. In solution, the combined effects of Van der Waals attractive forces and electrical double layer forces govern the interaction between particles and a surface. By varying the solution conditions it is possible to alter the surface chemistry of interacting materials and change their interaction from attractive to repulsive. In this work the attachment and detachment of nanosized silica particles onto titanium dioxide surfaces has been studied as a function of pH and addition of solutions of linear polyphosphates. The latter has the general formula $[P_nO_{3n+1}]^{(n+2)-}$ where n is the number of phosphate

* Ian Wark Research Institute, the ARC Special Research Centre for Particle and Material Interfaces, University of South Australia, Mawson Lakes, South Australia.

atoms in the molecule. Previous in-situ infrared studies[1] have shown that linear polyphosphates selectively adsorb onto titanium dioxide forming a strong chemical bond. Streaming potential measurements[2] have shown that for a fixed polyphosphate concentration, the titanium dioxide becomes more negative with increasing n, which was attributed to an increased charge density with n. In addition, direct force measurements[2–4] have shown that adsorbed polyphosphate introduces a steric layer which leads to a short ranged repulsive interaction.

Reflectometry combined with a stagnation point flow cell was used to measure silica particle attachment and detachment. The stagnation point flow is ideally suited for the study of colloidal particle attachment and detachment processes in which their transport is governed only by diffusion.[5]

EXPERIMENTAL

Materials

Titanium Dioxide Wafers. A titanium dioxide layer was deposited on the surface of silicon wafers by magnetran sputtering (Philips Research, The Netherlands).[2] Ellipsometry measurements showed the titanium dioxide layer to be 30 nm thick and XPS analysis showed the composition of the deposited layer to be pure titanium dioxide. X-ray diffraction showed the deposited TiO_2 to be amorphous. Imaging by AFM showed a rms roughness of 0.3 nm over an area of 1 μm^2 with a maximum peak height of 2 nm. The wafers were cleaned by detergent washing followed by rinsing with high-purity water, ethanol, heptane and copious amounts of more high-purity water. Finally, the wafers were blown dry in a stream of nitrogen and plasma cleaned (Harrick Plasma Cleaner) for 1 minute immediately prior to use. The isoelectric point of the titanium dioxide covered wafers was determined to be pH 4.2 using streaming potential measurements.[2]

Silica. Solutions of silica particles used for reflectometry experiments were prepared from Ludox™ AS40 (Dupont). The particles were dialyzed for 2 days in MQ water and then made up in solutions of KNO_3 (10^{-3} M) with a particle concentration of 100 mg/l. The mean particle radius was $R = 12 \pm 2$ nm, determined by TEM (CESMA, Adelaide University). The silica spheres used for AFM measurements were obtained from Allied Signal (Chicago, Illinois). XPS analysis (CSIRO Division of Molecular Science) showed the composition of the sample to be pure silica. The typical diameter was found to be 7 μm. AFM imaging of the spheres over an area of 500 nm^2 showed the rms roughness to be 0.8 nm.

Linear Polyphosphates. The linear polyphosphates were provided by Albright and Wilson (Australia) as dry sodium salts and were made up into solutions with a polyphosphate concentration of 10^{-5} M with background electrolyte concentration of KNO_3 (10^{-3} M). The linear polyphosphates referred to here as P_1, P_2, P_3 denotes the monodisperse species (Na_3PO_4), ($Na_4P_2O_7$) and ($Na_5P_3O_{10}$); $P_{<10>}$ refers to a polydisperse sample with n ranging from 1–19, with an average of 10.

Other Reagents. Analytical grade KNO_3, HNO_3 and KOH were obtained from BDH Chemicals (Australia) and were used as supplied. High purity water (surface tension of 72.8 mN/m and resistivity 18 MΩ at 20°C) was obtained from an Elga UHQ system. The solution pH was adjusted with drop wise addition of 0.01 M HNO_3 and KOH via micropipette.

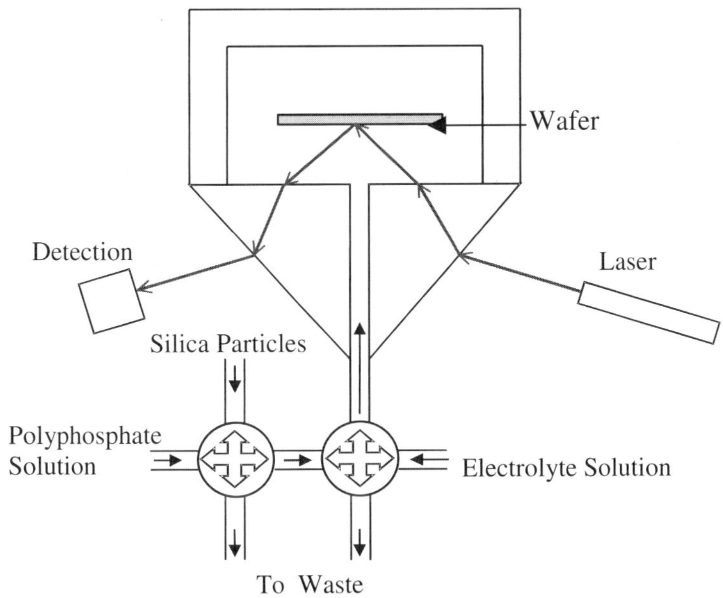

FIGURE 1 Schematic diagram of the combined reflectometry and stagnation point flow cell

METHODS

Reflectometry

The experimental setup for the combined reflectometry and stagnation point fluid cell is shown schematically in Figure 1. A detailed description of the experimental procedure and the theory behind the technique is given elsewhere.[6, 7] Here a brief description of the essential features is given. The reflected light from a plane polarized He/Ne laser striking a substrate surface is measured. The substrate surfaces were titanium dioxide covered silicon wafers. The reflected light is split into its parallel (p) and perpendicular (s) components and the measured signal, S, is the ratio of these intensities:

$$S = \frac{I_p}{I_s} \quad \text{(EQ 1)}$$

This ratio depends on the refractive index profile close to the surface of the substrate. Material adsorbed at the interface, in this case silica nanoparticles, will change the refractive index profile and hence result in a change in S. Quantitative measurements of the attached amount (Γ) can be obtained from the change in signal via:

$$\Gamma = \frac{1}{A_s}\frac{\Delta S}{S_0} \quad \text{(EQ 2)}$$

where S_0 is the intensity ratio prior to adsorption and ΔS is the change in intensity ratio. The sensitivity factor, A_s, takes into account the explicit refractive index contributions from the surface, the adsorbed material and the aqueous medium.

Stagnant Point Flow

In the stagnant point fluid cell, the collector surface is positioned at a critical distance from the inlet tube such that a stagnant point flow is generated at the point where the fluid impinges the surface. Under these conditions the hydrodynamics can be very well defined.[8] The particles arrive at the surface under the influence of surface forces and Brownian diffusion only. Solutions were gravity fed into the reflectometry cell from high-density polyethylene (HDPE) containers (250 ml) mounted on adjustable lab jacks. The height of the liquid above the inlet port to the fluid cell determined the flow rate. The flow rate was maintained at 1.5 cm^3/min with a height of the liquid 13 cm above the inlet port. A valve was used to switch between solutions entering the cell. The cell volume was 30 cm^3. A vacuum pump was used to suck excess solution from the cell.

AFM

A silica sphere attached to the end of an AFM cantilever comprises a colloid probe. A sphere was attached to a cantilever using a heat softening resin (Epikote 1004, Shell) using a micromanipulation arm attached to a metallurgical microscope (Olympus BH2). The cantilevers were silicon nitride, tipless, 200 μm long, wide legged from Nanoprobe (Park Scientific, USA). Spring constants were determined to be 0.1 ± 0.05 N/m by measuring the resonance frequency of the cantilevers with added known masses.[9] Prior to measurement the colloid probe was rinsed with ethanol, dried in a stream of nitrogen and plasma cleaned (Harrick Plasma Cleaner/Steriliser PDC-32) for 1 minute. A Nanoscope III controller and Atomic Force Microscope (Digital Instruments, Santa Barbara, USA) equipped with a fluid cell was used to measure the forces of interaction. Electrolyte solutions at the required pH and containing the desired polyphosphate solutions were introduced into the fluid cell via Teflon tubing. The solution was allowed to equilibrate to at least 15 minutes prior to measurement. The experiments were conducted employing standard measurement procedures comprehensively described by other authors.[10–12] Measurements of the cantilever deflection against scanner (z-piezo) displacement were taken. The z-piezo was calibrated via an optical interference technique.[13] The cantilever deflection data were subsequently converted to force (F) as a function of apparent surface-surface separation (h), simply called separation hereafter. The force of interaction was normalised by the radius of the sphere, i.e., F/R employing the Derjaguin approximation for sphere-flat interactions.[14]

RESULTS AND DISCUSSION

An example of typical reflectometry data is shown in Figure 2. Electrolyte solution at pH 4 is flowed into the reflectometry cell for 20 minutes before the start of the measurement. This ensures that the solution in the cell has reached thermal equilibrium with the solution in the reservoirs. The baseline signal, S_0, was monitored during this time to ensure that the signal was stable prior to measurement. The background electrolyte solution is permitted to flow into the cell for a further 100 seconds after the measurement begins. After this time (point (a) in Figure 2), the valve is switched to allow silica particles into the cell (the electrolyte concentration remained unchanged). The reflectometer signal increases due to the attachment of the silica particles. Initially the signal increases linearly with time. The rate of attachment decreases markedly close to saturation coverage. At point (b), the valve is switched to introduce a new solution, containing either a particle-free electrolyte solution at high pH or a solution of linear polyphosphate. The decrease in

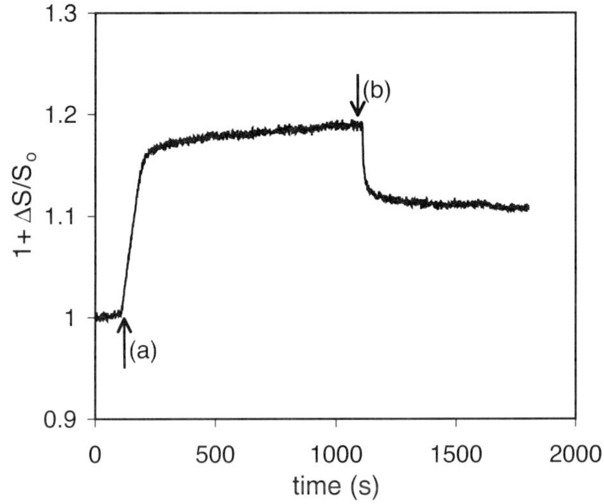

FIGURE 2 Typical reflectometry raw data showing change in signal with time upon introduction of particles at (a) followed by the introduction of a displacing solution at (b)

the signal at point (b) is due to the detachment of the silica particles. Initially, the decrease in signal is very rapid but the rate of detachment slows as the maximum detached amount is reached. The signal reaches a "plateau detached amount" before complete detachment of the particles has been obtained.

The attachment of silica particles to a titanium dioxide wafer at pH 4, followed by their detachment upon introduction of electrolyte solution at higher pH, is shown in Figure 3. The rate of attachment and the maximum attached amount, Γ_{max}, was the same for each experiment. The linear attachment regime is indicative of a rate limited only by the mass transport of the particles from solution to the surface.[15, 16] A similar observation was found in other reflectometry studies of nanoparticles.[7, 17, 18] No detachment was measured when electrolyte solutions below pH 6 were introduced into the cell. Upon switching to electrolyte solutions at pH 6 and higher, detachment of silica particles was detected. Both the rate of detachment and the maximum detached amount increased with increasing pH. At pH 6 only a small quantity (10%) of silica particles were detached and the detachment rate was slow compared to initial rate of attachment at pH 4. At pH 9, half of the pre-attached particles were detached and the initial rate of detachment was faster than the initial rate of attachment.

In Figure 4 the detachment force measured by AFM between a silica sphere and titanium dioxide substrate as a function of pH under the same solution conditions as in Figure 3 is presented. A pH dependent adhesion force is evident. Previous studies have shown[2, 19] that the interaction force between silica and titanium dioxide surfaces is well described by the DLVO theory of colloidal stability. The adhesion results presented here may be rationalised in terms of the combined effects of an attractive van Waals force and pH-dependent electrical double layer interactions. The isoelectric point (iep) of the silica particles and the titanium dioxide wafers have been measured to be at pH~2 and pH~4.5, respectively.[2, 19] At pH values below the iep of the titanium dioxide, in addition to the attractive van der Waals force, there will be an attractive electrical double layer interaction between the oppositely charged surfaces. At pH values above the iep of titanium dioxide both surfaces will be negatively charged and the electrical double layer

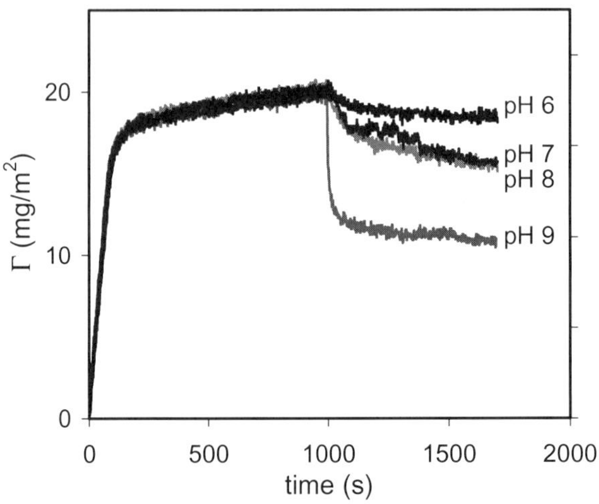

FIGURE 3 The attachment of silica particles onto titanium dioxide substrates at pH 4 followed by their detachment upon introduction of solutions at higher pH in 10^{-3} M KNO_3

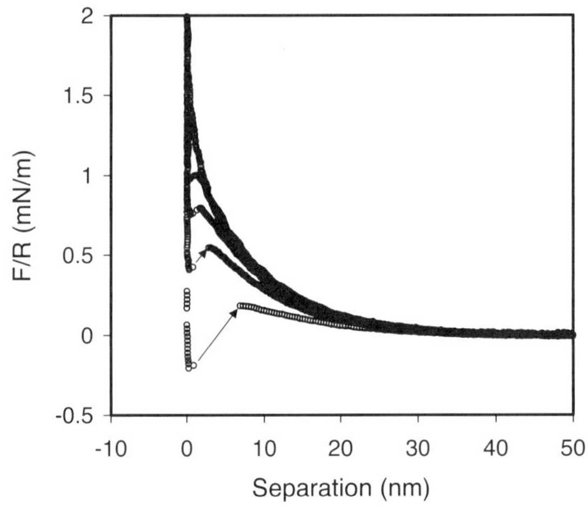

FIGURE 4 Force distance curves for the retraction of a titanium dioxide wafer from a silica sphere (*R* = 3.5 μm) as a function of pH in 10^{-3} M KNO_3. The curves correspond to, from top to bottom, data at pH 9, 8, 7, 6, and 5.6.

interaction will be repulsive. A tensile force is needed to separate the surfaces from intimate contact at pH 5.6. This is indicated by the negative value of the normalised force at the point at which the surfaces jump out of contact. At this pH, the electrical double layer interaction will be weakly repulsive and the adhesion is due to the attractive van der Waals force. At higher pH values the electrical double layer interactions become increasingly repulsive and the surfaces are seen to separate from contact even in the presence of a positive applied force. The separation force curves correlate well with the detachment data seen in Figure 3 and explain why no detachment of silica particles was detected below pH 6 and also why the detached amount increased with pH.

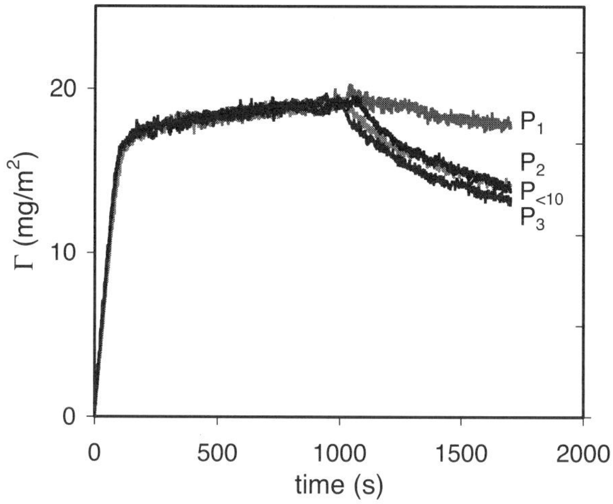

FIGURE 5 The attachment of silica particles onto titanium dioxide substrates at pH 4 followed by their detachment after introduction of solutions of linear polyphosphates (10^{-5} M) of varying *n* at pH 4 in 10^{-3} M KNO_3

The attachment of silica particles at pH 4 onto titanium dioxide substrates followed by their detachment upon switching to solutions of linear polyphosphates (10^{-5} M) is presented in Figure 5. As discussed in the introduction, adsorption of polyphosphate onto titanium dioxide modifies the surface, rendering it negatively charged. Note that whereas changes in pH affected both the silica and titanium dioxide surface potentials, the specific adsorption of polyphosphate to titanium dioxide leaves the silica unmodified. Thus, any detachment of silica particles can be attributed solely to chemical changes at the titanium dioxide surface. For n = 2, 3 and <10> the rate of detachment and the detached amounts of particles are similar to each other and larger than P_1 on the time scale of the experiments. The discrepancy seen in the detachment profile in the presence of P_1 can be understood in terms of the adhesion measurements.

Figure 6 shows the interaction force curves during separation between a silica colloid probe and titanium dioxide at pH 4 in the presence of solutions of polyphosphate (10^{-5} M) of varying n. For comparison, the interaction force between silica and titanium dioxide at pH 4 in the absence of polyphosphate is also shown. At pH 4 in the absence of polyphosphate there is a large adhesion due to combined attractive electrical double layer and van der Waals forces. The presence of polyphosphate clearly modifies the interaction force, dramatically reducing the adhesion. In the presence of P_1 the detachment force is negative indicating a significant adhesion force. For $n > 1$ the detachment force is positive and the force curves show that the separation is dominated by a repulsive interaction. The magnitude of the repulsive interaction increases with n. The trends seen in the interaction force curves are in accord with previous AFM studies and streaming potential measurements[2] and can be directly related to chemical modification of the titanium dioxide surface due to adsorbed polyphosphate. The increase in the repulsive interaction with n is due to an increased negative surface potential on the titanium dioxide and also the presence of a steric layer, whose thickness, δ, increases with n ($\delta \approx 0.4$–0.6 nm for n = 1–3 and $\delta \approx 1.6$ nm for n = <10>).[2]

A good correlation between the detached amounts of silica particles measured by reflectometry and the normalised detachment force measured by AFM is seen in Figure 7.

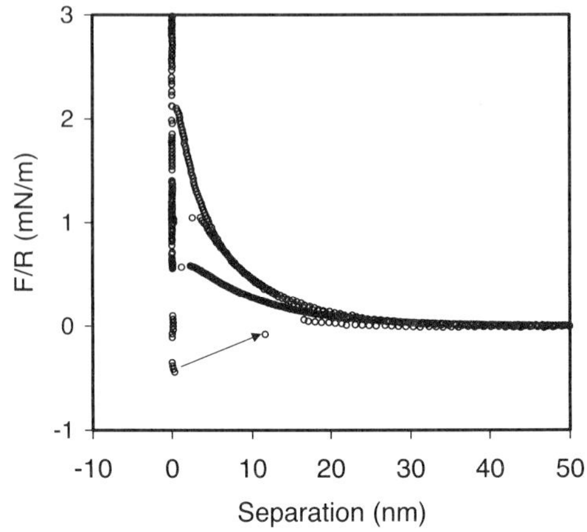

FIGURE 6 Force distance curves for the retraction of a titanium dioxide wafer from a silica sphere (R = 3.5 μm) at pH 4 in 10^{-3} M KNO_3 in the presence of linear polyphosphate solutions (10^{-5}M) from top downwards $P_{<10>}$, P_3, P_2 and P_1. The bottom force curve is in the absence of polyphosphate.

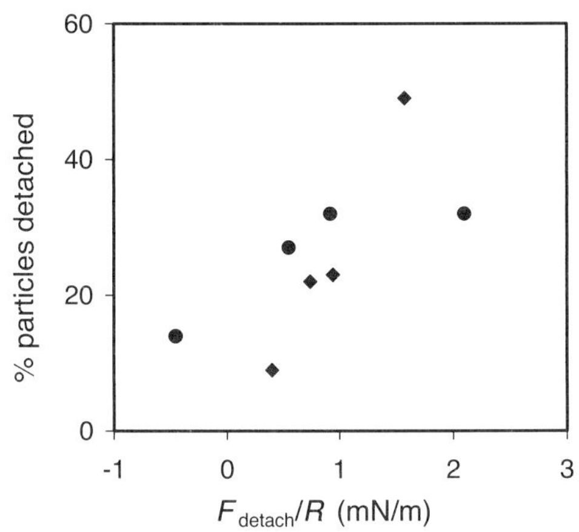

FIGURE 7 Detached amount of particles measured by reflectometry against the normalised detachment force measured by AFM under similar solution conditions. The symbols refer to the data measured as a function of pH (♦) and in the presence of polyphosphate (•).

The detached amount of particles is plotted as a percentage of the total attached amount of particles prior to switching to the displacing solution. The detached amount of particles increases as the normalised detachment force becomes more positive (more repulsive). The detachment of the silica particles is clearly sensitive to the variation in the electrical double layer interactions brought about by changes in pH and due to adsorbed polyphosphate. The fact that detachment of silica particles is detected at all in the presence of P_1 despite the force curves showing an adhesive interaction is evidence that the steric property of adsorbed polyphosphate is important in the detachment process. The slight discrepancy from a linear trend between the detached amount and detachment force seen for the $P_{<10>}$ data point can be understood in terms of the polydisperse nature of the $P_{<10>}$

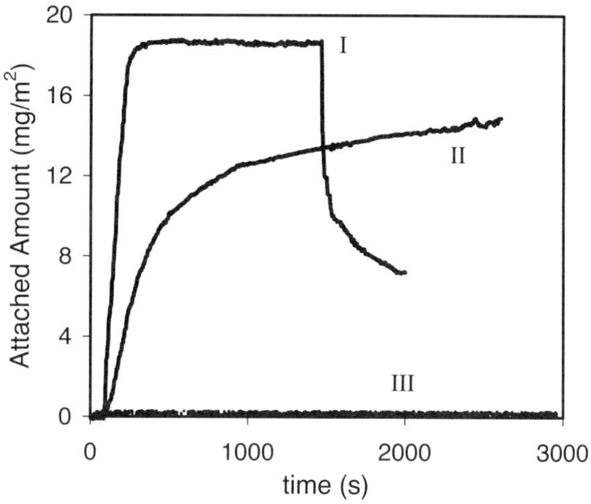

FIGURE 8 The effect of the sequence of addition of silica particles and P_3 solution. Curve I is for the introduction of silica particles followed by the introduction of P_3 (10^{-5} M) at t = 1500. Curve II is for the introduction of silica particles after pretreatment by flowing P_3 solution into the cell for 1000 seconds and rinsing with electrolyte solution for 100 seconds. Curve III is for the introduction of a solution containing a mixture of silica particles and P_3 (10^{-5} M). All experiments were conducted in 10^{-3} M KNO_3 at pH 4.

sample which contains a range of linear polyphosphates from n = 1–19. In the reflectometry measurements the diffusion rate of the polyphosphate species becomes a critical factor. The diffusion rate of the polyphosphate species will decrease with n. Thus, although the larger n species will impart a more repulsive force on the silica-titanium dioxide system, the smaller n species will diffuse to the surface more quickly. The consequence is a competition between the rate of polyphosphate adsorption and the modification of the resultant polyphosphate adsorption.

Finally, it is of interest from an application viewpoint as to the most efficient use of a dispersing agent such as polyphosphate in preventing particle attachment. In Figure 8 the order of addition of silica particles and P_3 solution (10^{-5} M) to the cell is investigated. Curve I shows the attachment of particles in the absence of polyphosphate followed by their detachment upon introduction of a P_3 solution (10^{-5} M). This sequence of addition is identical to that shown in previous figures. Curve II shows the attachment of silica particles on a titanium dioxide wafer that was conditioned with polyphosphate. For this experiment, a solution of P_3 (10^{-5} M) was introduced into the cell for 1,000 seconds followed by rinsing with electrolyte solution for 100 seconds immediately prior to introducing the silica particles. Particle attachment was detected; however, the initial rate of attachment was less than that seen in Curve I onto the bare surface. The maximum attached amount is also reduced. The most striking effect is seen in Curve III, in which silica particles and P_3 solution was introduced simultaneously into the cell. No particle attachment was detected, even after 50 minutes. The results indicate that equilibrium considerations are relevant in determining the amount of attached particles and adsorbed polyphosphate that takes place at the surface. Preconditioning the titanium dioxide surface (Curve II) reduces particle attachment by reducing the available surface sites at which the particles can attach. In Curve III the bulk solution contains both excess silica particles and polyphosphate. As a much smaller species, the polyphosphate possesses a much higher diffusion coefficient than the particles. Thus it is expected that the polyphosphate would arrive and adsorb at the titanium dioxide surface before the particles. The excess concentration of polyphosphate ensures that even if polyphosphate desorbs from one surface site, another polyphosphate molecule will quickly adsorb in its place and so prevent the attachment of particles over long periods of time.

CONCLUSIONS

The attachment of nanosized silica particles onto titanium dioxide surfaces and their subsequent detachment due to changes of solution pH or in the presence of linear polyphosphates was studied using reflectometry supported by atomic force colloid probe microscopy. The rate of detachment and maximum detached amount increased with both pH and n. It was seen that increasing the pH led to repulsive electrical double layer interactions, which was responsible for the detachment of the particles. In addition to electrical double layer interactions, adsorption of linear polyphosphate onto titanium dioxide provided a steric component, which facilitated the particle detachment. A good correlation was seen between the AFM measured detachment force and the amount of detached particles. Furthermore, it was shown that a solution containing excess linear polyphosphate could totally prevent particle attachment.

REFERENCES

1. Michelmore, A.P., Gong, W., Jenkins, P. and Ralston, J. *Phys. Chem. Chem. Phys,* 2, 2985, 2000.
2. Feiler, A., Jenkins, P. and Ralston, J. *Phys. Chem. Chem. Phys.,* 2, 5678, 2000.
3. Feiler, A., Larson, I., Jenkins, P. and Attard, P., *Langmuir,* 16, 10269–10277, 2000.
4. Leong, Y.K., Scales, P.J., Healy, T.W., Boger, D. and Buscall, R. *J. Chem. Soc. Faraday Trans.,* 89, 2473, 1993.
5. Adamczyk, Z., Siwek, B., Zembala, M. and Belouschek, P. *Adv. Colloid Interface Sci.,* 48, 151, 1994.
6. Dijt, J.C. PhD thesis, Wageningen University, The Netherlands, 1993.
7. Bohmer, M.R.J. *Colloid Interface Sci.,* 197, 251, 1998.
8. Dabros, T. and van de Ven, T.G.M. *Colloid Polym. Sci.,* 261, 694, 1983.
9. Cleveland, J.P., Manne, S., Bocek, D. and Hansma, P.K. *Rev. Sci. Instrum.,* 64, 403, 1993.
10. Ducker, W.A., Senden, T.J. and Pashley, R.M. *Nature,* 535, 239, 1991.
11. Larson, I., Drummond, C.J., Chan, D.Y.C. and Grieser, F. *J. Am. Chem. Soc.,* 115, 11885, 1993.
12. Hartley, P.G., Larson, I. and Scales, P.J. *Langmuir,* 13, 2207, 1997.
13. Jaschke, M. and Butt, H.-J. *Rev. Sci. Instrum.,* 66, 1258, 1995.
14. Israelachvili, J.N., "Intermolecular & Surface Forces," 2nd ed., Academic Press, London, 1992.
15. Adamczyk, Z., Szyk, L. and Warszynski, P.J. *Colloid Interface Sci.,* 209, 350, 1999.
16. Kallay, N., Tomic, M., Biskup, B., Kunjasic, I. and Matijevic, E. *Colloids and Surfaces,* 28, 185, 1987.
17. Bohmer, M.R., van der Zeeuw, E.A. and Koper, G.J.M. *J. Colloid Interface Sci.,* 197, 242, 1998.
18. Hayes, R.A., Bohmer, M.R. and Fokkink, L.G. *Langmuir,* 15, 2865, 1999.
19. Larson, I., Drummond, C.J., Chan, D.Y.C. and Grieser, F. *J. Phys. Chem.,* 99, 2114, 1995.

SECTION 5

Effects of Fillers on Mechanical, Electrical, Thermal Stability, and Other Properties

Solution Phase Surface Modification of Filler Particles

Richard Partch*, Steven Brown†, Richard Stephenson‡, and Hongyu Wang§

Filler technology has enjoyed a revolution in attention and achievement in the past half century. The genesis of the advancements is greater fundamental understanding of the synthetic chemistries for and physical properties of colloidal matter and instrumentation capable of characterizing particle size, shape and surface features. Several advanced chemical methods for processing powders are now employed in ceramic, electronic, imaging, medicinal and plastic technology. This presentation will highlight solution phase synthetic methods for in situ coating of fillers such as SiC, Al_2O_3, AlN and SiO_2 and some properties of the product solids.

INTRODUCTION

Colloid and surface chemistries play pivotal roles in whether a particle-matrix composite functions optimally. Both the core and surface compositions of such filler particles must be selected to have the desired physico-chemical properties and interfacial compatibility with the matrix. The latter property is generally achieved by surface modification, otherwise referred to as coating or encapsulation, or by covalent attachment of gradient molecular functionalities to their surfaces. Physical deposition methods such as incipient impregnation, mechanofusion, spray drying and physical vapor transfer are common, as well as are chemical methods where the coating material is synthesized in the presence of the cores.

Techniques for improving the commercial value of several types of powders have been successfully developed in the speaker's laboratory over the past two decades. The

* Dept. of Chemistry, Center for Advances Materials Processing, Clarkson Unviersity, Potsdam, N.Y.

† Cabot Corp., Billerica, Mass.

‡ PGP Industries, Inc., Santa Fe Springs, Calif.

§ Rodel, Inc., Newark, Del.

processes include gas, aerosol droplet and solution phase reactions, and the substrates coated or modified are particles used for ceramics, electronics, imaging, printing, medicine and optics (Mayville 1987; Partch 1991; Oyama 1993; Groat 1993; Huang 1995; Stephenson 1996; Avella 1997; Dilsiz 1997; Partch 1998; Partch 1997; Partch 1997; Wang in preparation; Partch 2000). Two examples are in situ synthesis of polydivinylbenzene (PDVB) for continuous coating and moisture protection of particles of thermally conducting aluminum nitride (Partch 1997), or of cesium bromide phosphor (Bryan 1997). Another is tethering of 18-crown-6 or of C_{60} fullerene molecular functionalities to anatase particles to mediate their photosemiconductor activity (Randive in preparation). Presented in this paper are data on, Part 1, coating silicon carbide nanoparticles with polycarbonate; Part 2, passivation of aluminum nitride particles with boron nitride; and Part 3, encapsulating alumina spheres in zirconium carbide. All product powders have "core-shell" morphology. Characterizations include use of scanning electron microscopy (SEM), x-ray diffraction, thermal gravimetry, mobility, spectroscopy and light scattering.

PROCEDURES AND DISCUSSION

Part 1: Coating Silicon Carbide Nanoparticles with Polycarbonate

Silicon carbide, SiC, is one of several advanced materials having promise as filler in ceramic, metal and polymer composites. It is hard and resistant to extensive corrosion under normal conditions. In whisker or fiber form it serves to inhibit crack propagation in stressed materials and is therefore useful for strengthening (Avella 1997; Hsueh, 1988).

Characteristic of all silicon containing materials, SiC has a propensity to acquire a surface layer of silica (SiO_2) when exposed to oxygen. Generally this layer is only a few molecules thick and passivates the core from further oxidation. Depending on the environment in which SiC filler nanoparticles are to be placed, it may be necessary to coat the particles with an interfacial material. An example is coating SiC whiskers with a thin continuous layer of aluminum nitride or boron nitride so the high aspect ratio cores can be used to strengthen aluminum metal (Brown 1998). The coatings mentioned protect SiC from reacting with molten aluminum and improve dispersibility of the whiskers. In the work described, SiC nanoparticles, considered desirable for evaluation as filler in polycarbonate body armor composites, were dispersed in a reactive medium in which polycarbonate is sequentially synthesized and coated onto the polyhedra.

Discussion of Coating Procedures and Results. Two experimental variables had to be maximized for success in this project. One was to disperse the as-received SiC polyhedra into their primary <50 nm state or the smallest aggregates as possible. The second was to uniformly coat each nanoparticle or smallest aggregate with a controlled amount of polymer.

Nanocrystalline SiC was obtained from Nanophase Technologies Corp. and used as received. The powder had a specific surface area of 67.5 m^2/g and thermogravimetric analysis (TGA) showed it to contain about 3 weight percent volatiles over the range of 100–800°C. Transmission electron microscopy revealed that the primary particle sizes were quite polydispersed, 10–50 nm in diameter.

The dispersibility and aggregate sizes of the SiC particles were evaluated in several liquid media using a Coulter Quantisizer at 0.01% (w/v) solids using simple shaking agitation as well as sonication with the ACE 600W horn (Didenko 2000). Solvents included water, dimethylsulfoxide, ethylene glycol, methanol, ethanol, 2-propanol, 2-butanone, methylene chloride, toluene, hexane and mineral spirits. In all of these the SiC aggregates were 400–600 nm and the dispersions were less stable over short time in the more nonpolar solvents. Water gave the best results.

Four types of surfactants were dissolved in water, dimethylsulfoxide and ethylene glycol to determine their ability to further reduce aggregate particle size (Sato 1980; Goodwin 1981; Ottewill 1990). The surfactants were poly(vinyl alcohol), poly(vinylpyrrolidone), Daxad 19 (Hampshire Chemical Co.) and Lomar P62 (Henkel Chemical Co.). The latter two are polymeric condensates of formaldehyde with naphthalene sulfonic acid. Results were that only with Daxad and Lomar in water and using horn sonication could SiC aggregate sizes be reduced to a new minimum of 150–160 nm. These experiments employed 0.1% (w/v) solids and 1%, 5% and 10% (w/v) of either surfactant. Dispersions using Lomar were stable for longer time (days) than those using Daxad.

The next phase of the study involved coating the minimum-sized SiC species with polymer. Three exemplary coating procedures and their outcome are described.

Encapsulation with Commercial Polycarbonate by Non-solvent Addition. To a solution of 10.0 g of poly(Bisphenol A carbonate) (Aldrich 18,162–5) dissolved in 200 g methylene chloride was added 20.0 g SiC powder. After stirring for 30 min, 400 mL of petroleum ether was added dropwise. The dispersion was stirred for an additional 12h after which time the solids were collected and analyzed as described above. As expected, TGA analysis showed the product contained 30 wt% polymer. SEM photographs revealed that the morphology was similar to that of the SiC encapsulated in globular poly(divinylbenzene).

Encapsulation Using in situ Synthesis of Polycarbonate in Diethyl Ether. A mixture of 2.8 g bisphenol A and 2.4 g of pyridine in 300 mL of diethyl ether was stirred at room temperature until the bisphenol A dissolved. To this solution was added 20.0 g SiC particles and the dispersion was stirred for 1 h. A solution of 0.48 g diphosgene in 20 mL diethyl ether was added dropwise, followed by stirring for 24 h. The particles were then filtered, washed with dilute HCl and water, and dried at 50°C in vacuum. TGA gave 7–8 wt% polymer and SEM showed that the aggregates of coated SiC were in the range of 300–400 nm.

Encapsulation Using in situ Synthesis of Polycarbonate in Aqueous Sodium Hydroxide. A dispersion of 1.0 g of SiC particles in 100mL of water containing 0.01 g Lomar P62 surfactant was ultrasonicated with the horn apparatus, then stirred under nitrogen purge while 0.5 g (2 mmole) bisphenol A was added, followed by 1.75 g (0.04 mole) sodium hydroxide. The mixture was heated to 80°C and three sequential additions of 0.40 g (2 mmole) each of diphosgene were added dropwise over 6 h. After cooling to room temperature the solid product was collected by filtration, washed with dilute sodium hydroxide and water and dried at 60° under vacuum. Sampling of aliquot portions of the reaction mixture after each addition of diphosgene, followed by product isolation and TGA, revealed that about 8 wt% polymer coating formed as coating at each stage. Particle size analysis gave average diameters at 153 nm with 100% of the particles below 231 nm. Fourier transform infrared (FTIR) data on the coated powder had the required aliphatic and aromatic C-H, carbonyl and C-O absorptions. SEM photos showed that the polycarbonate deposited as a continuous and relatively uniform coating.

Discussion. Experimental conditions by which SiC particles can be both dispersed into relatively small 150 nm aggregates of the primary species and encapsulated with a polymer have been determined. The in situ synthesis of bisphenol A polycarbonate in water containing a surfactant and excess sodium hydroxide is the method of choice. Subsequent discussion elaborates on that process in more detail.

Figure 1 shows the general synthetic scheme and reactants used to prepare polycarbonates. The extremely useful polymers are produced commercially by a solvent-free

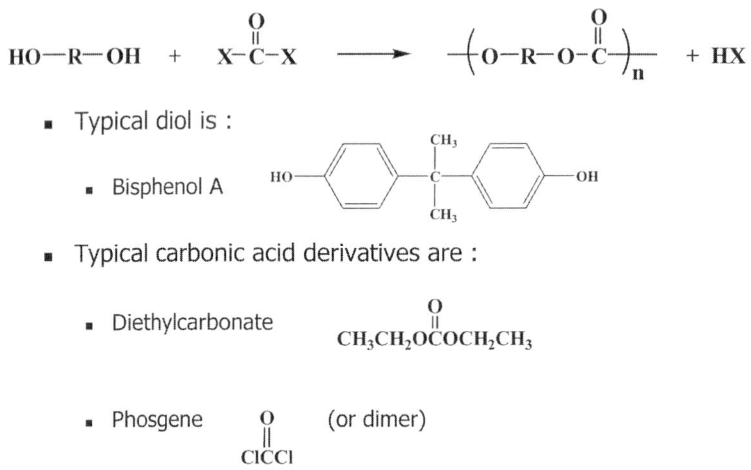

FIGURE 1 Generic synthesis of polycarbonates

process at melt temperature. Such a procedure is not conducive to coating particles added to the melt, so a solvent-based procedure was investigated. Non-solvent ethanol addition to a dispersion of as-received SiC particles containing commercial bisphenol A polycarbonate dissolved in methylene chloride yielded aggregates of coated SiC ranging in size 400–500 nm. Thermogravimetric analysis was employed to verify that the observed weight percent loss of polymer matched the calculated value if all polycarbonate was precipitated. By this procedure the amount of polymer isolated with the SiC could be controlled at 1–15% depending on ratios of reagents mixed together. Solids having less than 2–3 wt% polymer generally contained smaller aggregates of 300–400 nm. Light scattering analysis of the dispersions before ethanol was added showed that the primary SiC particles were unaggregated. Similar results were obtained using methylene chloride as solvent and petroleum ether as non-solvent.

An attempt was made to reduce the SiC aggregate size by an in situ synthetic procedure. SiC nano-particles were dispersed in a solution of bisphenol A dissolved in methylene chloride. Some aggregation occurred. Comonomer diphosgene, a dimeric liquid form of phosgene, was added to the stirred dispersion. An exotherm and HCl gas evolution was observed. The same phenomena occurred when the addition sequence was reversed, e.g., when the phenol reagent was added to a dispersion of particles in a solution of diphosgene in methylene chloride. After the mixture cooled to ambient, non-solvent ethanol was added to nucleate the synthesized polycarbonate and deposit it on the particles. Again, 200–250 nm SiC-polymer aggregates were formed.

Scanning and transmission electron microscopy was employed to determine the disposition of the precipitated polycarbonate in association with the SiC particles from both the above procedures. In both cases the polycarbonate appeared as globular and was discontinuous over the surface of individual particles and their aggregates.

Comment is in order regarding the different surface chemistries expected between the two addition sequences for polycarbonate synthesis mentioned above. The molecular amount of native SiO_2 on the surface of the SiC nanoparticles exposes silanol (Si-OH) groups to the reaction medium. Bisphenol A is not reactive with the silanol groups. When the comonomer diphosgene is added it can be assumed it reacts faster with the dissolved phenol than with the heterogeneous Si-OH groups. As a result, the oligomers-polymers of polycarbonate being generated and finally precipitated by addition of ethanol should not be covalently associated with the particle surfaces.

The opposite situation was predicted to take place when the SiC particles are dispersed in methylene chloride containing dissolved diphosgene. The latter reagent, like thionyl chloride and organic acid chlorides, is capable of reacting with Si-OH groups. If the diphosgene is in excess, the surface of the SiC particles should be reacted to produce some Si-OCOCl groups, which could serve as a covalent linker between the SiC and the polycarbonate. However, even when diphosgene addition preceded that of bisphenol A, no polymer was observed to form on the dispersed particles until ethanol non-solvent was added. That is, the expected covalent linkage failed. We plan to investigate the interaction of diphosgene with Si-OH groups in more detail to clarify what chemistries are taking place in the current situation.

A third attempt to produce primary SiC nanoparticles or minimum-sized aggregates uniformly coated with polycarbonate involved use of water as solvent. Light scattering results showed that use of an anionic surfactant like sodium naphthalene sulfonate, but not a variety of nonionic or cationic surfactants, gave 150 nm aggregates of SiC. Water reacts with diphosgene so it was necessary to use an excess of the reagent for competitive reaction with bisphenol A to be successful. To enhance the phenol reactivity, as well as to dissolve the reagent in water, two equivalents of sodium hydroxide were added before diphosgene. This reaction procedure yielded the desired SiC in the smallest 150 nm aggregates obtainable by this work. Importantly, the polycarbonate coating was uniform and continuous. Their application as nanoscale fillers in various composites is under investigation.

Part 2: Passivation of Aluminum Nitride Particles with Boron Nitride

Aluminum nitride (AlN) is well known as a refractory ceramic with good strength and very high thermal conductivity for a non-metal. Because its coefficient of thermal expansion is similar to that of silicon, it has been employed as a heat transfer body and sink in microelectronics (Mroz 1994; Wirth 1991). Surface hydrolysis and oxidation diminish thermal conduction across grain boundaries in compacted and sintered devices (Bowen 1990; Jackson 1997), hence industrial environment (Takada 1992; Li 1996; Howard 1996). These are generally not robust enough and opportunities exist for improving the protecting shell.

The overall method employed in the present research involves carbothermal reduction of mixtures of boron oxide and polymer coated AlN powders in ammonia at elevated temperature. The polymer coating reactions were carried out as previously published (Partch 1997). The weight percent polymer can be controlled by adjusting the ratio of solids to monomer and initiator, and the reaction time. In this case 10 g AlN powder (10–50 μm) dispersed in mineral spirits and 5 g divinylbenzene reacting at 100°C for 12 hours yielded about 25 wt% polymer coating. The PDVB coated powder (10 g) was then mixed with 5 g B_2O_3 and heated in ammonia at 1300°C for 4–6 h. At 500–600°C the polymer pyrolyzed to carbon coating on the AlN, which at higher temperature served to assist in conversion of the B_2O_3 into BN. This process produced an easily granulated powder having excellent mechanical durability, moisture stability and very high thermal conductivity when used as filler in polysiloxane rubber (Figures 2–3).

Rubber samples filled with either pure AlN or AlN coated with boron nitride (BN) have about the same relative thermal conductivity, 24.4 versus 21.2 W/m°K, even

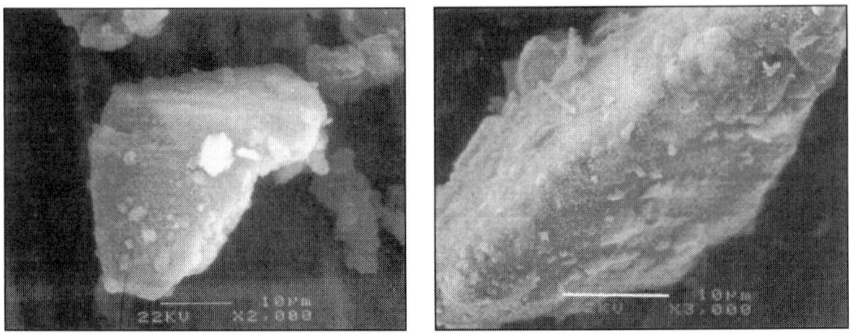

FIGURE 2 SEM micrographs of AlN coated with BN from reaction of 1.0 g B_2O_3 and 1.40 g carbon coated AlN in ammonia at 1300°C

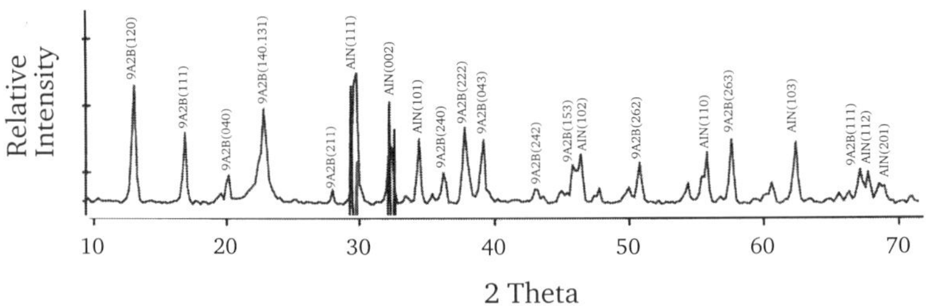

FIGURE 3 XRD spectrum of AlN coated with BN formed as in Figure 2. Note the absence of alumina.

though the latter sample has only 50% the amount of AlN in composition. The BN coating serves as a moisture barrier and conducting interface between particles in contact in the composite. When the coating on the AlN is aluminum borate, formed when boron oxide (B_2O_3) and pure AlN powders are mixed and heated in air, the filler (50% AlN) has only 9.2 W/m°K conductivity.

Part 3: Encapsulating Alumina Spheres in Zirconium Carbide

Aluminum oxide (Al_2O_3) and its hydrous derivatives such as aluminum hydrous oxide [$AlO(OH)_2$] have long been fillers in many composites either alone or in complex form (e.g., aluminosilicates). Being amphoteric, it reacts with and dissolves in both acidic and basic media. In some applications this may be an undersireable property which can be reduced or eliminated by encapsulating the core material in shell capable of inhibiting contact of the core with the environment. Oxides of other elements in column IIIA of the Periodic Table also require similar protection. Fuels for nuclear power generation are typically uranium oxides packed into containment rods and may require encapsulation (Morley 1994; Ogawa 1992). In this work relatively large 500 μm diameter spheres of high density alumina, encapsulated with zirconium carbide in a 2-step process, are inert when placed in hot nitric acid.

Several experimental procedures were employed in order to determine which would produce the optimum coating of ZrC on the Al_2O_3 cores. All of the procedures involved deposition of zirconium oxide (ZrO_2) followed by carbiding the oxide coating in methane at 1,700°C. Since the morphology, packing density, uniformity and thickness of the final ZrC layer depends on the quality and amount of precursor ZrO_2 deposited, much attention

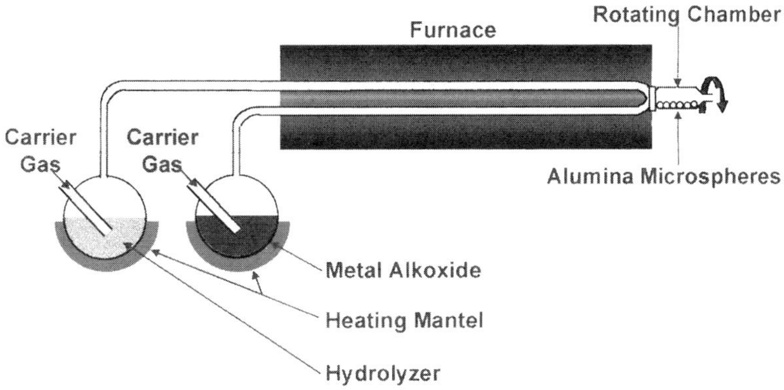

FIGURE 4 Apparatus for co-vapor-hydrolysis method of coating Al_2O_3 spheres with ZrO_2

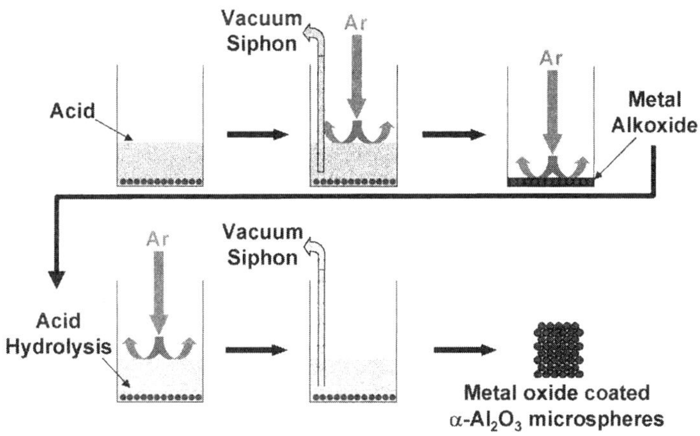

FIGURE 5 Optimum procedure for encapsulating Al_2O_3 spheres with ZrO_2

was given to the method of hydrolyzing zirconium ispropoxide [$Zr(OPr)_4$]. The variations included sol-gel hydrolysis of the propoxide in several alcohol solvents, with or without addition of acid catalyst or amine dispersing agent. Attempts were also made to encapsulate the Al_2O_3 spheres using a co-vapor-hydrolysis method as depicted in Figure 4.

The best results were realized by carrying out the hydrolysis reactions in the manner shown in Figure 5.

Chemical modification of the alumina surface by pretreatment with hydrofluoric acid facilitated adhesion of ZrO_2 to the surface. Coupled with this was the acid catalyzed hydrolysis of the metal alkoxide to produce fine metal oxide particles for the coating material.

Alumina microspheres (0.250) were immersed in either 28.9 M HF (48%) or 12.1 M HCl for 30 minutes or 60 minutes, respectively. The excess acid was removed by vacuum siphoning under an inert gas such as argon or nitrogen. The particles were dried at 150°C and allowed to cool to room temperature under an inert gas. The metal alkoxide precursor $Zr(OPr)_4$ was added to the activated, dry alumina particles and agitated. 25 mL of hydrolyzing agent was added and again the system was agitated and allowed to stand for 10 minutes. After this time, the supernatant solution was removed and the particles were calcined for 10 hours at 1,200°C.

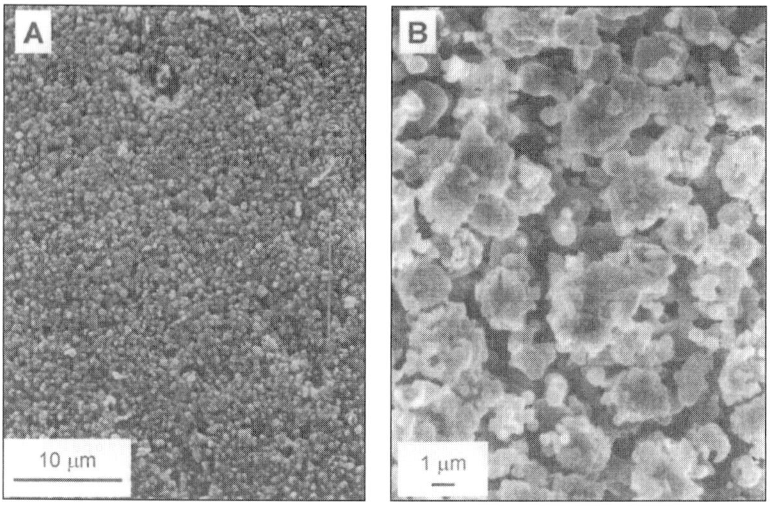

FIGURE 6 SEM images of ZrO_2 coating on a portion of an alumina microsphere (A) and ZrC product after carbiding in CH_4

The hydrolyzing solutions were prepared as follows: 1.00 M stock solutions of acetic, phosphoric, sulfuric, hydrofluoric, hydrochloric, boric and nitric acids. From the stock solutions, dilutions were made to obtain concentrations of 0.50 M, 0.25 M, 0.125 M and 0.063 M. 25 mL aliquots were taken to initiate the hydrolysis of the metal alkoxide-alumina microsphere system.

Depending on the ratio of reagents to the immersed spheres, the thickness of ZrO_2 coating can be controlled in the range of 1–20 μm and the close-packed coating nuclei can have spherical, cubic or needle-like morphologies.

SEM micrographs (Figure 6) of the spheres coated with ZrO_2 and carbided in methane, as well as atomic absorption analysis of the supernatant from carbide-coated spheres immersed in hot nitric acid, revealed that hydrolysis to form ZrO_2 carried out in 0.1 M acetic acid produced the most dense ZrC coating.

CONCLUSIONS

Silicone carbide nanoparticles can be coated with an organic polymer such as a polycarbonate by in situ coating in aqueous solvent with added dispersant.

Aluminum nitride can be passivated to moisture with boron nitride using a combination of polymer coating and treatment with boron oxide. The resulting powder retains very high thermal conductivity when filled in a siloxane rubber.

Aluminum oxide spheres are readily encapsulated with zirconium oxide by a controlled sol-gel method. The oxide coating can be carbided in methane to make the cores stable in strong mineral acid.

ACKNOWLEDGMENTS

The authors gratefully acknowledge discussions with Michael Sennett at US Army Natick Lab, with Tom Mroz and Richard Hexemer at Advanced Refractory Technologies, Inc., and with Wayne Wales at Martin-Marietta Company. Financial support was provided by

US Army Contract Office, NYSERDA and Martin-Marietta Company. Facilities were provided by the Center for Advanced Materials Processing at Clarkson University, sponsored by the New York State Science and Technology Foundation.

REFERENCES

Avella, M., E. Martuscellu, M. Raimo, R. Partch, S. Gangolli, and B. Pascucci, *J. Mater. Sci.* 1997, 32, 2411–16.

Bowen, P., J. Highfield, A. Mocellin, and T. Ring, Degradation of Aluminum Nitride in an Aqueous Environment, *J. Amer. Ceram. Society,* 73 (1990), 724.

Brown, S., and R. Partch, *J. Adhesion,* 1998, 67, 259–65.

Bryan, P., J. Bringley, and R. Partch, Coated Radiographic Phosphors and Radiographic Phosphor Panels, (1997) USP 5646412.

Didenko, Yi, and K. Suslick, *Nature,* 2002, 418, 394.

Dilsiz, N., R.E. Partch, E. Matijevic, and E. Sancaktar, Silver Coating of Spindle- and Filament-Type Magnetic Particles for Conductive Adhesive Applications, *Journal of Adhesion Science and Technology,* 11, No. 8 (1997), 1105.

Goodwin, J., Ed., Colloidal Dispersions, Royal Society of Chemistry Special Publication No. 43, Dorset Press: London, 1981.

Groat, E.A., T.J. Mroz, S.G. Gangolli, and R.E. Partch, Preparation and Characterization of Coated SiC Whiskers; "Advances in Ceramic-Matrix Composites," Ceramic Transactions Vol. 38, N.P. Bansal ed., 1993.

Howard, K., D. Carrol, and S. Biskupsik, Method of Making Moisture Resistant Aluminum Nitride Powder, (1996), EP00644836.

Hsueh, C., P. Becher, and P. Angelini, *J. Am. Ceram. Soc.,* (1996), 1988, 71, 929–33.

Huang, C.-L., R.E. Partch, and E. Matijevic, Coating of Uniform Inorganic Particles with Polymers. II. Polyaniline on Copper Oxide, *J. Colloid Interface Science,* 170 (1995), 275–283.

Jackson, T., A. Virkar, K. Moore, R. Dinwiddie, and R. Cutler, High Thermal Condutivity Aluminum Nitride Ceramics, *J. Amer. Ceramic Society,* 80 (1997), 1421.

Li, Y., T. Oiu, J. Xu, and X. He, Surface Modification of Aluminum Nitride Powder, *J. Mater Sci. Lett.,* 15 (1996), 1758.

Mayville, F., R. Partch, and E. Matijevic, Preparation of Uniform Spherical Titania Particles Coated with Polyurea by the Aerosol Technique, *Journal of Colloid and Interface Science,* 120 (1987), 135.

Morley, N.J., and M.S. El-Genk, Thermal-Hydraulic Analysis of the Pellet Bed Reactor for Nuclear Thermal Propulsion; Nuclear Eng. Design, 149 (1994), 387–400.

Mroz, T., Aluminum Nitride, *Amer. Ceram. Society Bull.,* 73 (1994), 77.

Ogawa, T., K. Fukuda, S. Kashimura, T. Tobita, F. Kobayashi, S. Kado, H. Miyanishi, I. Takahashi, and T. Kikuchi, Performance of ZrC Coated Particle Fuel in Irradiation and Postirradiation Heating Test, *J. Am. Ceram. Soc.,* 75 (1992), 2985–2990.

Ottewill., R., Ed., Colloidal Dispersions, Faraday Discussions of the Chemical Society No. 90, Arrowsmith Ltd.: London, 1990.

Oyama, H.T., R. Sprycha, Y. Xie, R.E. Partch, and E. Matijevic, Coating of Uniform Inorganic Particles with Polymers, I, *Journal of Colloid and Interface Science,* 160 (1993), 298.

Partch, R., and E. Matijevic, Synthesis of Monodispersed Colloids by Chemical Reactions in Aerosols in Fine Particles: Synthesis, Characterization and Mechanisms of Growth, T. Sugimoto, Ed., Marcel Dekker (2000).

Partch, R.E., and S. Brown, Aerosol and Solution Modification of Particle-Polymer Interfaces, *J. Adhesion,* 67 (1998), 259–276.

Partch, R., S. Gangolli, D. Owen, C. Ljungqvist, and E. Matijevic, Conducting Polymer Composites. II. Polypyrrole-Metal Oxide Latexes, ACS Symposium Series, #492 (1997), pp. 368–369.

Partch, R., Fine Particle Synthesis and Coating, in Materials Synthesis and Characterizations, Ch. 1, D. Perry, Ed., Plenum Press (1997), pp. 1–17.

Partch, R., S. Gangolli, E. Matijevic, W. Cai, and S. Arajs, Conducting Polymer Composites. I. Surface Induced Polymerization of Pyrrole on Iron(III) and Cerium(IV) Oxide Particles, *Journal of Colloid and Interface Science,* 114 (1991), 27.

Randive, R., C. Syvinski, and R. Partch, Gradient Functionalization of Metal Oxide Particles with Crown Ether, Fullerene and Electroluminescent Molecules, in preparation.

Sato, T., and R. Ruch, Stabilization of Colloidal Dispersions by Polymer Adsorption, Surfactant Science Series, Marcel Dekker: New York, NY, 1980.

Stephenson, R.C., and R.E. Partch, Metal Oxide and Metal Carbide Thin Film Coating on Large Spherical Particles, Thin Films: Surface and Morphology, Vol. 458, Proceedings Volume of the Materials Research Society, 458 (1996), 435.

Takada, J., New Method for the Preparation of Polyamide Coated AlN Powder, *J. Soc. Powder Technol. Jpn.,* 29 (1992), 831.

Wang, H., S. Bachman, and R. Partch, Covalently Tethered Vinylic and Peroxido Functionalities to Silica Particles and Polymerization Coating Thereof, in preparation.

Wirth, D., in Engineered Materials Handbook (1991), Vol. 4, S. Schneider, Jr., Ed., 1107.

Incorporation of Particles in Metal Matrix Composite Coatings Using Pulsed Electrodeposition

Michael L. Free*, **Lei Chen†**, **and Ravindra Bhide***

The incorporation of particles into metal matrix materials is important to many technologically important areas. Particles that are incorporated into a metal matrix have a significant effect upon the properties of the resulting composite. Proper selection of particles and incorporation densities can lead to optimal composite properties that can be modified as needed for various applications. This paper discusses the parameters that affect the particle incorporation density in electrodeposited metal matrix composite coatings as well as a deterministic model that can be used to predict the effects of such parameters on particle incorporation density.

INTRODUCTION

Electrochemically fabricated metals, alloys, and metal-matrix composites have wide ranging applications. Electrodeposition of composite coatings has attracted continuous interest since the 1970s because of its low manufacture cost as well as the capability of easily tailoring coating properties by modulating the dispersed phase in the matrix (Guglielmi 1972; Cellis and Roos 1977; Franklin and Mathew 1989).

Since electrochemical composite coatings combine the properties of matrix metal or alloy with those of the dispersed particles, which may be difficult or impossible to be integrated into thin films using conventional approaches, many desirable properties such as high wear resistance, corrosion resistance, lubrication, anti-oxidation ability, electro-catalysis, etc., may be achieved by such coatings as needed (Musiani 2000; Bapu and Mohan 1995; Foster, Cameron, and Carew 1985; Kunugi et al. 1990; Kunugi et al. 1991; Iwakura, Furukawa, and Tanaka 1992; Tavares and Trasatti 2000; Kawai et al. 1990). In general, electrochemical codeposition is a good approach for low cost mass production of metal matrix composite coatings.

* Dept. of Metallurgical Engineering, University of Utah, Salt Lake City, Utah.

† United Technologies, East Hartford, Conn.

The properties of an electrodeposited composite depend strongly on the distribution and quantity of dispersed particles in the matrix. Several parameters can affect both particle incorporation and matrix growth and structure. These parameters include particle concentration, type and size, bath constituents, temperature, pH, current density, agitation, and particle surface charge. Elucidation of the general and specific mechanisms governing metal and particle codeposition is indispensable for effective plating parameter selection and process control. Despite the growing applications of electrodeposited composites, the underlying principles of codeposition have eluded systematic analysis because the heterogeneous electrodeposition reaction is complicated by the interaction of alien particles with the electrode interface.

In the early days of composite plating three mechanisms were proposed: Whithers's electrophoresis in 1962 (Whithers 1962), Martin's mechanical entrapment in 1964 (Williams and Martin 1964), and Brandes's particle adsorption on cathode in 1967, which are simplistic and not very useful in interpreting and predicting particle codeposition. One of the first attempts with significant advancement in elucidating the codeposition mechanism is the pioneering work done by Guglielmi (Gugliemi 1972). Gugliemi concluded that codeposition is determined by a two-step adsorption process.

This model has successfully predicted the weight percent incorporation of particles for a wide variety of codeposition systems in which only current density and particle concentration have been varied. These systems include Ni-SiC, Ni-TiO_2, Cu-Al_2O_3, Ag-Al_2O_3, Cu-SiC, and Cu-P. However, this model failed to explain experimentally observed mass transport effects because it excludes the parameters related to mass transport, bath constituents, and particle size.

Celis et al. (Celis, Roos, and Buelens 1987) proposed a five stage mechanism that includes (*i*) adsorption of ionic species on particles, (*ii*) particle movement towards boundary layer by convection, (*iii*) diffusion of particles through diffusion boundary layer, (*iv*) adsorption of particles on cathode surface, and (*v*) reduction of ions associated with particles and incorporation. The most important assumption of this model is that particle incorporation is determined by the probability that a fraction (k/K) of K absorbed species on the particle is reduced upon contact with the growing metal. Again, this model is similar to earlier models in the sense of not being able to predict the particle content directly from the experimental conditions. The most recent model was proposed by Fransaer, Celis, and Roos (1987). The basis of this model is the characterization of particle trajectory during codeposition. According to this model, the particle incorporation will occur if the normal-to-tangential ratio of force acting on the particle at the time the particle reaches the electrode surface is acceptable.

In this paper a deterministic model for predicting the particle incorporation density in electrodeposited composite is proposed based upon well-controlled experiments and applicable theory in an effort to improve upon previous modeling efforts. All experiments were performed by pulsed electroplating in aqueous electrolyte slurries in order to provide the necessary control of mass transport and surface finish without the need for special electrolyte additives.

EXPERIMENTAL PROCEDURES

All electrodeposition experiments were performed using a P3 potentiostat made by Gamry Instruments that was controlled using LabView 5.1 software made by National Instruments. Deposition solutions each contained 0.2 m NiCl and 0.6 m ZnCl. All potentials were made relative to a standard calomel reference electrode. All electrochemical experiments were performed using a traditional three electrode set up with reference,

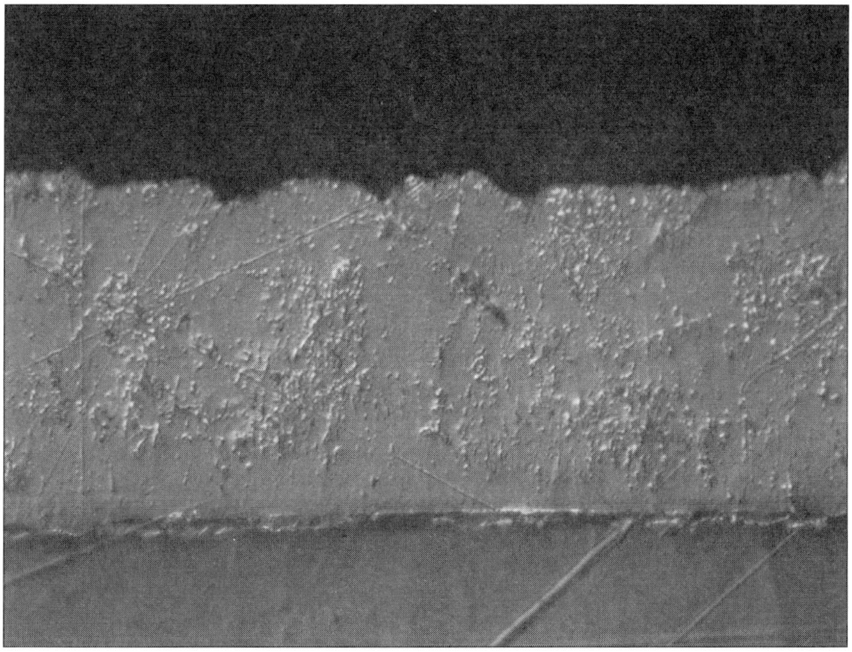

FIGURE 1 Metal matrix composite coating containing alpha alumina particles dispersed in a zinc-nickel alloy matrix produced by pulsed electrodeposition (–1.8 V (SCE) deposition potential, –1.1 V rest potential, 4 ms on, 20 ms off, 22°C, 20 g alumina/liter, 30 min.). The entire image is approximately 150 μm across in the horizontal direction.

working, and counter electrodes. The working electrode consisted of AISI 1018 steel. The counter electrode was made of platinum mesh. Each electrodeposition experiment was performed at 22°C with gentile agitation using a magnetic stirring device (100 RPM to maintain the particle suspension).

Electrodeposited coatings were evaluated for metal content by galvanic stripping. Galvanic stripping was performed by placing the coated electrode in a 170 g/l KCl, 30 g/l H_3BO_3 solution at 50 mA/cm^2 at 22°C until the potential increased to the iron removal potential. The deposited metal was calculated on the assumption of 100% current efficiency and 10% nickel content in the final coating. A cross-sectional image of a sample coating is presented in Figure 1.

Particle content was determined by sedimentation spectroscopy. Electrodeposited composite coatings were exposed to a solution containing 3 vol.% H_2SO_4 and 1×10^{-4} m sodium dodecyl benzene sulfonate until the matrix dissolved. The resulting solution, which contained the electrodeposited particles, was ultrasonically dispersed then spectroscopically analyzed during the sedimentation process. The concentration of particles was determined from the absorbance of the particles contained in each size class, based upon absorbance calibration experiments, an example of which is shown in Figure 2. The size of the particles was determined by the time necessary for the particles of each size class to settle past a narrow spectroscopic window that is approximately 1 mm in height. The absorbance and time values were obtained from a plot of absorbance versus time, an example of which is presented in Figure 3. Note that each section in which the absorbance changes with time is a separate size class. The formula used to calculate the concentration of particles in size class *i* is given as:

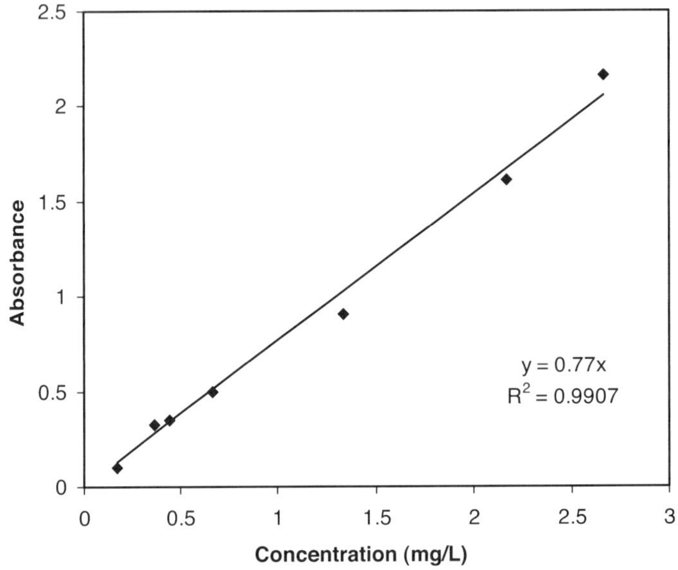

FIGURE 2 Comparison of absorbance versus particle concentration for alpha alumina particles suspended in water in a 1 cm × 1 cm × 4 cm cuvette as measured by an Ocean Optics S2000 spectrophotometer

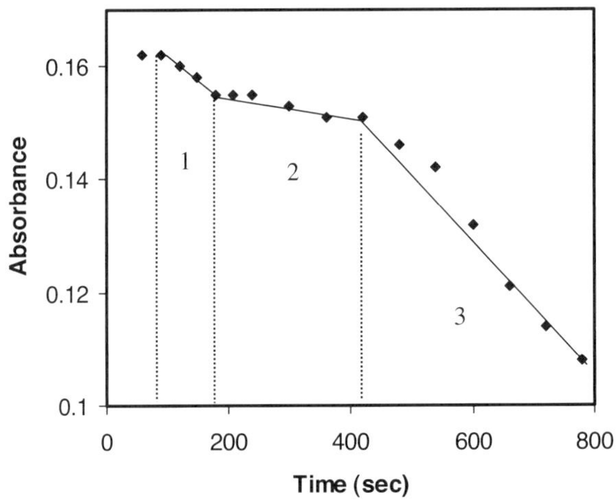

FIGURE 3 Comparison of absorbance and time during particle settling of 1 micrometer alumina particles in water. Notice that there are three size classes of particles—each with different average settling velocities and slopes that correspond to size and concentration information. Note that the total change in absorbance in each section is proportional to the concentration of particles in that section. The time between the beginning and end of each section is the settling time for that size class to settle a distance equivalent to the height of the detector window. The average size for each section or size class can be determined using the initial time for the section, the height of the liquid above the detection window, and Stokes Law.

$$C_{io} = \frac{(A_{i(init)} - A_{i(final)})6\rho_s\eta l}{(t_{i(final)} - t_{i(init)})dk_i r_i(\rho_s - \rho_l)g} \qquad \textbf{(EQ 1)}$$

where $A_{i(init)}$ is the absorbance at which size class i begins to settle past the spectrometer slit, $A_{i(final)}$ is the absorbance at which size class i finishes settling past the slit, ρ_s is the density of the solid, ρ_l is the density of the liquid, η is the viscosity of the liquid, l is the slit height, $t_{i(final)}$ is the time at which size class i finishes settling past the slit, $t_{i(init)}$ is the time at which size class i begins to settle past the spectrometer slit, d is the path length of the light passing through the liquid, g is the gravitational constant, r_i is the radius of the particles of size class i, and k_i is an absorptivity coefficient that is equal to the slope of an absorbance versus mass-based concentration plot (see Figure 2) multiplied by 1.33 times the particle radius times the particle density divided by the path length. The radius of particles of size class i is determined based upon Stoke's Law according to the formula:

$$r_i = \sqrt{\frac{l9\eta}{2(\rho - \rho_l)g(t_{i(final)} - t_{i(init)})}} \qquad \textbf{(EQ 2)}$$

MODEL DERIVATION

Electrodeposition of metal matrix composite coatings or electroformed parts can be accomplished through proper control of deposition parameters. The following model is derived to determine the mass ratio of particles to matrix within the coating, which is referred to herein as the particle incorporation density and is reported as a percentage rather than as a fractional ratio since the volume fraction of particles is generally less than 10%.

The deposition of particles in a metal matrix that is electrodeposited is controlled by several factors. Some of the most important factors that determine the particle incorporation density are:

1. The rate of mass transport of particles through the mass transfer boundary layer to a distance very near the surface that allows interfacial forces to become dominant
2. The rate of mass transport of particles within the interfacial region that is typically on the order of nanometers to the surface
3. The ratio of the rate of particle transport to the rate of metal matrix deposition

Because the incorporation density is dependent upon each of these factors, the overall particle incorporation density can be expressed in terms of probabilities and ratios as:

$$\rho_{p/c} = P_{bdlMT}P_{ifMT}\frac{J_p t_{pt}}{J_m t_{md}} \qquad \textbf{(EQ 3)}$$

in which $\rho_{p/c}$ is the particle incorporation density, P_{bdlMT} is the probability of the particle being transported through the mass transfer boundary layer, P_{ifMT} is the probability of the particle being transported through the interfacial region to the surface, J_p is the maximum net particle flux rate, t_{pt} is the time for particle transportation, which is often different than the matrix deposition time for pulse deposition, J_m is the matrix flux, and t_{md} is the time of matrix deposition.

EFFECT OF MASS TRANSPORT ON PARTICLE INCORPORATION

The probability of a particle being transported through the mass transfer boundary layer is related to its transport to the boundary layer, which for this study involved a rotating disk. In essence, the fluid is pumped toward the center of the rotating disk, then gradually changes its velocity vector to one that is directed outward and parallel to the disk surface. As particles become entrained in the fluid flow lines that move toward the surface, their momentum will tend to continue their motion toward the surface and outside of the liquid flow lines. The general fluid flow pattern will successfully redirect most of the particles radially outward from the rotating disk. However, some of the particles that continue their path toward the surface at various angles may have sufficient momentum to propel them through the mass transfer boundary layer to the interfacial region of the electrode. The particle pathways followed by particles through the boundary layer will be distributed in relation to their velocity profile and momentum. However, the probability of the particles traversing through the boundary layer to the interfacial region will depend upon their initial velocity, which is directly related to the rotational speed of the disk and the time the particle is able to spend traversing through the boundary layer. The velocity of the particle will vary with time due to the dampening effect of viscosity and the pull of gravity. The velocity of the particle at any time, t, in the boundary layer region can be found using a force balance:

$$\text{Net Force} = \text{Drag Force} + \text{Gravity Force} \quad \textbf{(EQ 4)}$$

For laminar flow, this equation can be rewritten as (Kelly and Spotiswood 1982):

$$m\frac{dV}{dt} = -3\pi d_o V\mu - mg \quad \textbf{(EQ 5)}$$

However, since the settling velocity of fine particles is much smaller than the velocity of particles in stirred suspensions, the last term in the expression can usually be dropped without significantly altering the result. Thus, the solution to the force balance equation can be expressed as:

$$V = V_o \exp\left(\frac{-3\pi d_o \mu t}{m}\right) \quad \textbf{(EQ 6)}$$

The probability of particle incorporation due to mass transport through the boundary layer is likely to be proportional to the final velocity near the interface. The time necessary to travel the length of the boundary layer can be calculated by integrating the velocity times the time increment, delta t, from zero to the time needed to reach the interfacial region near the surface. This relationship can be expressed mathematically as:

$$\int_0^{t_\delta} V dt = \delta \quad \textbf{(EQ 7)}$$

The resulting equation can be solved and rearranged to:

$$t_\delta = \frac{-m}{3\pi d_o \mu} \ln\left(1 - \frac{\delta 3\pi d_o \mu}{m V_{init}}\right) \quad \textbf{(EQ 8)}$$

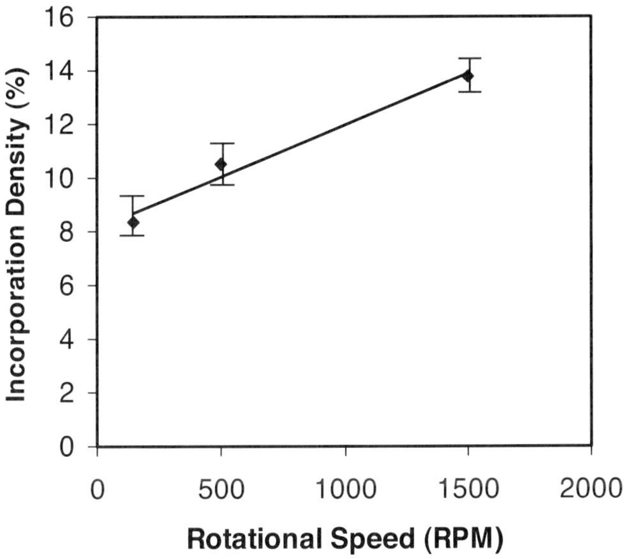

FIGURE 4 Particle incorporation density versus rotational speed for pulsed electrodeposition experiments performed with a 2 ms on time, 10 ms off time, –1.6 V (SCE) on potential, and 20 g alumina powder (1 μm) per liter

This expression for the time needed to traverse the boundary layer is then substituted into the velocity equation to determine the velocity at the interfacial region next to the surface, which is approximately a distance of δ from the outer boundary layer:

$$V = V_{init} - \frac{\delta 3\pi d_o \mu}{m} \quad \text{(EQ 9)}$$

Because V_{init} is directly proportional to the rotational speed, and because the boundary layer thickness is equal to $1.61\nu^{1/6}D^{1/3}\omega^{1/2}$ (Wadsworth 1979), the velocity can be expressed as:

$$V = k\omega - \frac{k'}{\sqrt{\omega}} \quad \text{(EQ 10)}$$

where k and k' represent constants. Note that at low rotational velocities, the term on the far right dominates, whereas the $k\omega$ term dominates at high rotational velocities. Converting the expression into a probability of particle incorporation into the metal matrix at high rotational speeds assuming the incorporation probability is directly proportional to the velocity at the interfacial boundary next to the surface leads to:

$$P_{bdlMT} = k''V \approx k_2\omega \quad \text{(EQ 11)}$$

Moreover, the incorporation density should be proportional to the probability such that a plot of incorporation density versus rotational speed should reveal a linear relationship. The data presented in Figure 4 show such a relationship, thus corroborating the previous assertions with regard to mass transfer effects. Note that the intercept in Figure 4 is not zero because the deposition of particles will occur due to the stirring that is needed to maintain the particles in suspension even when the electrodeposition surface is not rotated.

EFFECT OF ELECTROSTATIC ATTRACTIVE FORCES ON PARTICLE INCORPORATION

The effect of electrostatic attraction between the particles and the electrodeposition surface is related to the charge and electric field. It is assumed that the charge on the particle and the charge on the surface are opposite in sign, since similar charges would result in repulsion and reduced incorporation. The electrostatic force between a charged particle and another entity is proportional to the electric field between the entities. Thus, if the probability of particle incorporation due to electrostatic attractive forces between the particle and the surface is proportional to the associated force (Hudson and Nelson 1982):

$$P_{ifMT} \propto F = q\frac{E}{s} \qquad \textbf{(EQ 12)}$$

where q is the charge on the particle, E is the voltage, and s is the separation distance. Assuming the particles in the electrodeposition slurry have the same charge in each of the experiments, which are each performed in the same aqueous environment, this leads to a constant value of q for all particles used in this study. If the distance between the particle and surface is arbitrarily assigned a constant value for all tests, the probability of particle incorporation due to electrostatic attraction can be expressed as:

$$PifMT = kifMT(\psi_p - \psi_s) \approx k_{ifMT}(E_c) \qquad \textbf{(EQ 13)}$$

(assuming the particle and surface have opposite charges) where k_{ifMT} represents the associated electrostatic attraction constant associated with mass transfer through the interfacial region, ψ_p represents the surface potential of the particle, and ψ_s represents the surface potential of the electrodeposition surface. E_c represents the cathodic electrodepostion overpotential, which for practical electroplating is much larger than the particle surface potential, hence the approximate equality between the probability of incorporation and the overpotential.

The relationship between the electrostatic force and particle incorporation density can be evaluated only after consideration of the effect of charge on the rate of matrix deposition, since increasing the electrostatic force by changing the deposition potential also changes the matrix deposition rate. Because the incorporation density is proportional to the ratio of the flux of particles to the flux of metal matrix, the effect of parameters on the particle flux, such as the electrostatic force, can be evaluated by multiplying the incorporation density by the matrix flux. The matrix flux is directly proportional to the anodic stripping time. Consequently, a comparison of the incorporation density times the matrix stripping time should be linearly related to the potential difference between the particles and the electrodeposition surface, which can be approximated by the cathodic overpotential. Figure 5 shows such a comparison and a relatively linear relationship, suggesting that the electrostatic effect on incorporation is related to the electrodeposition overpotential or the difference between the equilibrium surface potential and the deposition potential.

EFFECT OF THE PARTICLE AND MATRIX FLUX RATIO ON PARTICLE INCORPORATION

The final term in the expression for the particle incorporation density is the relative flux ratio of particle deposition to that of the matrix deposition. The relative flux and deposition time for particles is expressed as:

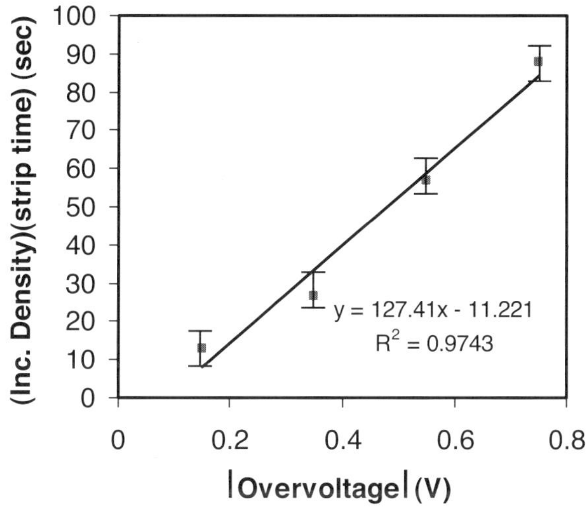

FIGURE 5 Particle incorporation density times stripping time versus cathodic overpotential for pulsed electrodeposition experiments performed with a 2 ms on time, 10 ms off time, 150 RPM, and 20 g alumina powder (1 μm) per liter

$$J_p t_{pt} = k_p(t_{on} + t_{off}) \quad \text{(EQ 14)}$$

where k_p is the flux constant that includes the effect of particle concentration and particle transport rate, t_{on} is the time of the primary deposition pulse, and t_{off} is the time at the off potential between the primary deposition potentials.

The matrix deposition rate is proportional to the current flow, which is equal to the voltage divided by the sum of the resistances (solution, charge transfer, mass transfer, and particle resistance to matrix flux). However, since the charge transfer resistance is a two-part exponential function of the overvoltage, it is most appropriate to leave the matrix flux term as a variable that can be accounted for using the matrix deposition or stripping data. In other words, no change will be made to the matrix flux terms.

The ratio of particle and matrix deposition fluxes can thus be expressed as:

$$\frac{J_p t_{pt}}{J_m t_{md}} = \frac{k_p(t_{on} + t_{off})}{J_m t_{on}} \quad \text{(EQ 15)}$$

Thus, the overall equation for particle incorporation, which includes the effects of each of the parameters discussed herein, is given as:

$$\rho_{p/c} \cong k_2 \omega k_{ifMT} E_c \frac{k_p}{J_m} \frac{(t_{on} + t_{off})}{t_{on}} \quad \text{(EQ 16)}$$

Consequently, if all variables except for on and off time are held constant, the effect of the on/off time changes on the incorporation density can be evaluated by plotting the incorporation density versus the ratio of the on plus off times divided by the on time as shown in Figure 6. The data in Figure 6 indicate that the relationship between the on + off/on times is linearly related to the incorporation density as predicted.

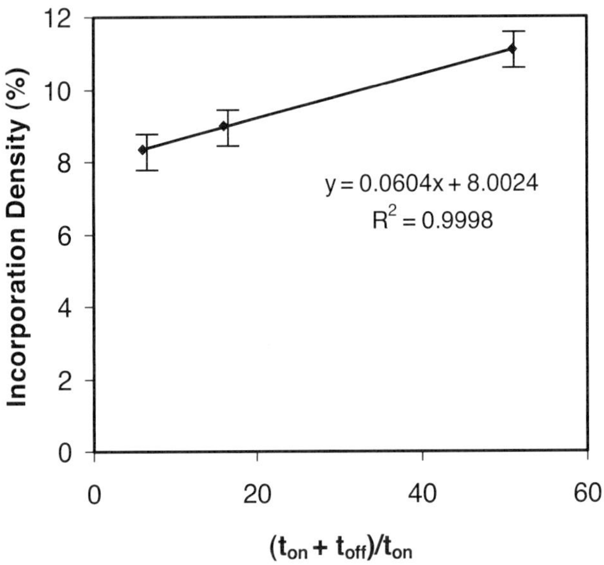

FIGURE 6 Particle incorporation density versus (on + off)/on times for pulsed electrodeposition experiments performed with a –1.6 V (SCE) on potential, 150 RPM, and 20 g alumina powder (1 μm) per liter

Because the overall relationship between the incorporation density and the various parameters contains several constants, it is expected that the constants may be combined to form a simplified overall expression such as:

$$\rho_{p/c} \cong K\omega E_c \frac{(t_{on} + t_{off})}{t_{on} J_m} \quad \textbf{(EQ 17)}$$

Accordingly, a plot of the incorporation density multiplied by the matrix flux should reveal a linear relationship with the remaining terms on the right-hand side of the equation. The data plotted in Figure 7 show that such a relationship is reasonably linear, suggesting that the final simplified expression can provide a reasonable prediction of the particle incorporation into the metal matrix during pulsed electrodeposition. The ability of the final expression to predict the particle incorporation is particularly useful in such systems, because there are many relevant variables that affect the deposition process.

CONCLUSIONS

Pulsed electrodeposition of metal matrix composite coatings with targeted particle incorporation densities can be accomplished effectively and predicted reasonably by appropriate control of processing parameters. The processing parameters that have a significant influence on the incorporation density that were evaluated in this study include the mass transfer of particles through the boundary layer, electrostatic attraction in the interfacial region that is controlled principally by the electrodeposition potential, and the on/off pulse times. The prediction of the particle incorporation density was demonstrated by

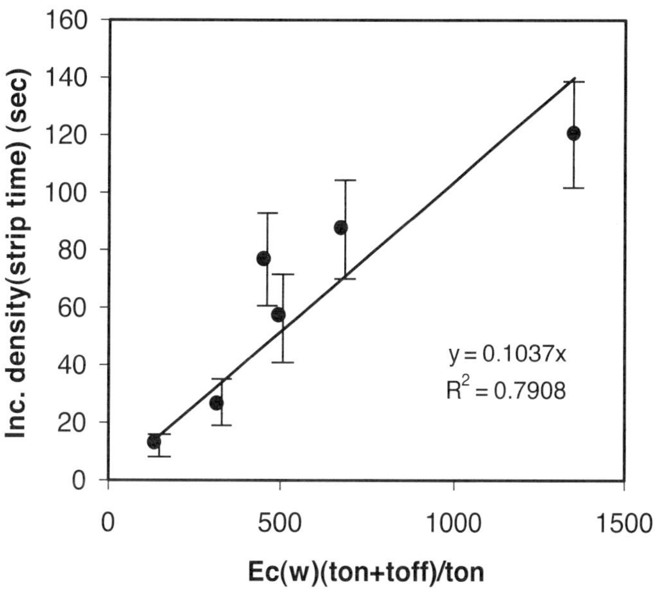

FIGURE 7 Particle incorporation density times stripping time versus the cathodic overpotential times the rotational velocity times the sum of the on and off times divided by the on time for pulsed electrodeposition experiments performed for all ranges of conditions with 2 ms on time, 10 ms off time, and 20 g alumina powder (1 μm) per liter

assigning probability weighting factors to each of the relevant parameters based upon applicable theories. The probability weighting factors were combined into an overall equation. The utility of the applicable theories was demonstrated for each of the relevant parameters, and the overall equation was compared to measured data. The agreement between the individual probability weighting factors as well as the overall equation and the measured data was reasonable. It is anticipated that the overall equation will become an important tool in predicting and modeling particle incorporation in metal matrix composite coatings produced by pulsed electrodeposition.

REFERENCES

Bapu, G.N.K.R., and S. Mohan. 1995. *Plat. Surf. Finish.*, 4, 86.

Celis, J.P., J.R. Roos, and C. Buelens. 1987. *Electrochemical Soc.*, 134, 1402.

Celis, J.P., and J.R. Roos. 1977. *J. Electrochem. Soc.*, 124, 1508.

Foster, J., B.P. Cameron, and J.A. Carew. 1985. *Tans. Inst. Met. Finishing*, 63, 115–119.

Franklin, T.C., and S.A. Mathew. 1989. *J. Electrochem. Soc.*, 136 (12) 3627.

Fransaer, J., J.P. Celis, and J.R. Roos. 1987. *Electrochemical Soc.*, 139, 413.

Guglielmi, N. 1972. *J. Electrochem. Soc.*, 119 (8) 1009.

Hudson, A., and R. Nelson. 1982. University Physics, Harcourt Brace Jovanovich, Publishers, New York, pp. 517–614.

Iwakura, C., N. Furukawa, and M. Tanaka. 1992. *Electrochim. Acta.*, 37, 757.

Kawai, K., N. Mihara, S. Kuwabata, and H. Yoneyama. 1990. *J. Electrochem. Soc.*, 137, 1793.

Kelly, E.G., and D.J. Spottiswood. 1982. *Introduction to Mineral Processing*, John Wiley and Sons, New York, pp. 62–94.
Kunugi, Y., T. Fuchigami, T. Nonaka, and S. Matsumra. 1990. *J. Electroanal. Chem.*, 287, 385.
Kunugi, Y., T.R. Kumada, T. Nonaka, Y.B. Chong, and N. Watanabe. 1991. *J. Electroanal. Chem.*, 313, 215.
Musiani, M. 2000. *Electrochimica. Acta.*, 45, 3397–3402.
Tavares, A.C., and S. Trasatti. 2000. *Electrochim. Acta*, 45, 4195–4202.
Wadsworth, M.E. 1979. Principles of Leaching in *Rate Processes of Extractive Metallurgy*, ed. H.Y. Sohn and M.E. Wadsworth, Plenum Press, New York, pp. 133–197.
Whithers, J.C. 1962. *Pord. Fin.*, 26, 62.
Williams, R.V., and P.W. Martin. 1964. *Trans. Inst. Met. Finish.*, 42, 182.

A Review of the Various Applications of Gypsum Fillers

Qingxia Liu*, **S. Veeramasuneni***, **and Susan Dichter***

Gypsum ($CaSO_4$-$2H_2O$) is an abundant and naturally occurring mineral. Under elevated temperatures (250 to 300°F) gypsum loses its crystalline water and is converted to hemihydrate ($CaSO_4$-$0.5H_2O$). The hemihydrate form is also called plaster that is widely used in wallboard, joint compound, and many other industrial applications. At temperatures higher than 450°F (up to 1,200°F), gypsum loses all crystalline water and is converted to anhydrite form ($CaSO_4$). The loss of crystalline water at elevated temperatures is an endothermic reaction. This property makes gypsum an attractive filler and fire retardant in certain thermosetting plastics. The applications of gypsum in plastics, food, pharmaceutical, and paper coatings will be reviewed in this paper. The future of synthetic gypsum, a byproduct of various industrial processes, will also be discussed.

INTRODUCTION

The use of gypsum fillers in plastics, food, pharmaceutical, and paper coating applications is reviewed in this paper. Brief description of synthetic gypsum, a byproduct of various industrial processes, is also included.

Fillers, as the name implies, have commonly been employed to reduce costs or to extend a product with an evident change or modification of the properties of the unfilled materials such as hardness, rigidity, viscosity, or color. Gradually, the realization grew that by the selective use of fillers, certain properties of the unfilled material could be enhanced or even exceeded and a reinforcement of properties was possible. Hence, fillers are today employed to obtain any, or a combination, of the effects such as rigidity, stiffness or hardness, thermal expansion and shrinkage, heat resistance, electrical characteristics, strength and creep, rheological properties (flow, thixotropy, body), processability (lubrication, flow, mixing, dispersion, etc.), appearance, density, cost, etc. (1,2). Calcium sulphate hydrates belong to an important family of inorganic mineral fillers and

* USG Corporation, Libertyville, Ill.

are used in various industrial applications. A brief review of calcium sulphate hydrates is presented in the next few paragraphs.

There are five different phases to be found in the calcium sulphate hydrate system, namely, calcium sulphate dihydrate, calcium sulphate hemihydrate, and three anhydrous phases: soluble calcium sulphate anhydrite, insoluble calcium sulphate anhydrite, and a high temperature calcium sulphate anhydrite phase. The α- and β-hemihydrates, and the corresponding soluble anhydrites, are not true polymorphs as they differ only in crystal size (3).

Calcium sulphate dihydrate occurs naturally as the mineral gypsum in a variety of forms including selenite, alabaster, and satin-spar. It is the most common sulphate mineral (4) and thus, traditionally, it has been extracted from the ground. In recent times, gypsum has also been generated as a byproduct of industrial processes. Two other mineral forms of calcium sulphate are commonly found: bassanite (calcium sulphate hemihydrate) and anhydrite (insoluble calcium sulphate anhydrite). In the German and Japanese literature, the anhydrite phases are referred to as anhydrite III, anhydrite II, and anhydrite I, respectively (5–7). Dihydrate and anhydrite II occur in a natural form or produced in chemical processes. The metastable phases hemihydrate and anhydrite III and the anhydrite II are obtained from the dihydrate by dehydration. These phases are of industrial importance. Anhydrite I remains stable only at temperatures above 1,180°C and is of no significance industrially.

As discussed previously, there are many forms of calcium sulphate hydrates that occur either in a natural form or are produced in chemical processes. However, the two commonly used gypsum fillers are the dihydrate and anhydrite forms. Gypsum (calcium sulfate dihydrate) and its anhydrite (calcium sulfate) have found a wide range of applications in agriculture, pharmaceuticals, food, and plastic industry. In food and agriculture applications, gypsum is used as a cost effective source of calcium or sulfur. Calcium sulfate dihydrate contains approximately 23% calcium while the anhydrous form contains 29% calcium. In the plastics industry, gypsum finds application as a flame retarding filler in room temperature cured polyesters typically used by acrylic or gel coat tub/shower manufacturers. Anhydrous calcium sulfate has extremely high whiteness and can be successfully used as a TiO_2 extender in color concentrates (8).

The dihydrate forms of gypsum have a fairly neutral pH and about 20% crystalline water. Although exact temperatures will vary depending on vapor pressure, relative humidity, etc., gypsum starts to lose its crystalline water as it is heated above 150°F. The dehydration of gypsum to hemihydrate occurs at temperatures between 230~248°F with an endothermic energy of about 414 J/g. Further dehydration of hemihydrate to soluble anhydrite occurs at a temperature of about 390°F with an endothermic energy of 135 J/g. The high capacity of heat adsorption during gypsum dehydration allows for the many applications of gypsum in fireproofing. Gypsum, having a specific heat of 0.27 Cal/°C g^{-1} and thermal conductivity of 0.42 w $m^{-1}K^{-1}$ (at temperature between 282–373°F), is a flame and fire retardant.

Gypsum fillers are characterized by high purity, whiteness, softness, low abrasion, resistance to mild acids, suitability for food contact, good electrical properties (anhydrous), no-health hazard, and flame retardant. Table 1 lists typical properties of calcium sulfate fillers made by USG (9).

APPLICATIONS IN PLASTICS

The dihydrate form of calcium sulfate is used in high volumes as a flame retarding filler in polyester backed tub/showers and spas. Gypsum functions as a cost effective

TABLE 1 Typical properties of calcium sulfate fillers

	Terra Alba	LP Filler #2 (East)	SNOW WHITE® Filler	CA-5™ Filler	CAS-20-4™ Filler
Chemical formula	$CaSO_42H_2O$	$CaSO_42H_2O$	$CaSO_4$	$CaSO_4$	$CaSO_4$
Specific gravity (g/cm^3)	2.32	2.43	2.96	2.96	2.96
pH	7.3	7.6	10.4	10.4	10.8
Median particle size (microns)[1]	12	35	7	2	4
Top particle size (mesh)[2]	–100	-50	–100	–100	-625
Combined water(%)[3]	20.30	16.70	0.35	0.44	0.20

[1] Micromeritics Sedigraph Particle Size Analyzer

[2] Alpine Jet Air Sieve Model 200

[3] Computrac Max 50 Moisture Analyzer

replacement for alumina trihydrate. Particle size is controlled carefully to allow high loading levels in the polyester resin formulation (10).

Anhydrite can be processed into an extremely white, finely divided particulate filler. It is manufactured from high purity gypsum rock using fine grinding and controlled calcination, which greatly reduces its moisture affinity. The narrow particle size distribution of anhydrite gives optimum impact properties for thermoplastic applications. Anhydrite can be used at high concentrations and consistently provides an excellent balance of properties and good surface appearance. The low Mohs hardness of anhydrite minimizes abrasion to processing equipment. Because of its high brightness, anhydrite is also used as a partial replacement for titanium dioxide in thermoplastic color concentrates. Color and opacity readings are maintained at replacement levels of 15–20% (8).

Anhydrite filler is ideal for many rigid poly vinyl chloride (PVC) applications. These include vinyl siding, window frames, moldings, conduit, and pipe. Anhydrite provides PVC processors various benefits including whiteness, low abrasion, excellent impact properties, good electrical properties, good processability, and cost savings potential.

In Table 2 the properties of anhydrite (CAS-20™-4) filler as compared with calcium carbonate ($CaCO_3$) are shown. This independent study was conducted on two typical vinyl siding substrate formulations, one containing 8 parts per hundred resin (phr) of calcium carbonate and one containing 8 phr of calcium sulfate (CAS-20™-4). Dry blends were prepared on site by a vinyl siding manufacturer. All other additives were held constant. Profiles, measuring 7" × 0.04", were processed on a 1.57" American Maplan, twin screw extruder (11).

These tests demonstrate that, at comparable loading levels, anhydrite (CAS-20™-4) can provide significantly higher modulus and equivalent impact strength versus ultrafine treated carbonate.

FOOD AND PHARMACEUTICAL APPLICATIONS

Calcium sulfate serves the food, beverage, and pharmaceutical industries as an economical source of supplemental calcium. It is used in enriched flour and breads, cereals, baking powder, yeast food, bread conditioners, canned vegetables, and artificially sweetened jellies and preserves. In beer manufacturing, the calcium ions, together with the needed buffering action as provided by proper water correction, promotes proper gelatinization of the starch, as well protein degradation and starch conversion. Thus, yield in the main mash is increased. In addition, the color is improved, and better precipitation and flocculation of

TABLE 2 Results of CAS-20-4 filler vs. calcium carbonate ($CaCo_3$) in PVC

ASTM D638	8 phr $CaCO_3$	8 phr $CaSO_4$
Tensile strength at yield, psi	6,750	6,810
Tensile strength at yield, kPa	46,540	46,950
Tensile modulus x 10^5, psi	3.45	4.46
Tensile modulus x 10^6, kPa	2.38	3.07
Elongation at yield %	3.9	3.8
ASTM D3029		
Falling weight impact, in. -lbs.	102.7	100.0
Falling weight impact, J	11.6	11.3

TABLE 3 Typical analysis of USG F&P fillers

	Terra Alba	SNOW WHITE®
Total calcium	23.1%	29.2%
CaO	32.31%	40.92%
SO_3	45.22%	57.46%
$CaSO_4$	0.39%	97.68%
$CaSO_42H_2O$	97.1%	—
$CaCO_3$ $MgCO_3$	1.52%	0.77%
SiO_2 and insoluble	0.24%	0.13%
Water loss at 250°C	20.31%	0.33%
Brightness index*	84.4	97.1
Oil absorption†	23.5	26.5
Specific gravity	2.32	2.96
Bulk density (pcf)		
loose	42	44
compacted	70	80
Solubility (70°F) per 100 cc water	0.26 grams	0.26 grams
pH 10% slurry	7.3	10.4
Refractive index	1.52	1.56
Through 100 mesh	100%	100%
Through 325 mesh	93%	97%
Avg. particle size (microns)	12–15	7–9

* Brightness Index was determined on a Beckman DU Spectrophotometer using magnesium oxide as the standard.
† Oil adsorption is the amount of linseed oil in cubic centimeters required to wet 100 grams of filler.

undesirable protein complexes are achieved. The result is a paler, smoother-tasting beer with improved stability and shelf life. For pharmaceutical applications, calcium sulfate is extensively used as a diluent serving as an excellent inert extender while it supplies dietary calcium. USG makes two fillers for food grade applications: Terra Alba and SNOW WHITE®. Table 3 lists the typical analysis of these two products (12).

The use of gypsum in combination with calcium phosphate provided additional calcium in cheese, and a cheese containing gypsum and calcium phosphate tastes much better than the one using a single calcium sulfate or calcium phosphate (13).

TABLE 4 Annual pigment consumption for paper coatings in North America

	North America	
	1000 × Metric Tons (Mt)	**Percentage (%)**
Clay	2,280–2,660	60–66
GCC	760–950	20–25
PCC	80–110	2–3
TiO2	80–110	2–3
Talc and others	80–110	2–3
Gypsum	0*	0
Total	3,800–4,000	100

* In contrast, the annual consumption of gypsum for paper coatings in Europe is ~100,000 Mt.

SNOW WHITE® filler, because of its whiteness, is used successfully to extend TiO_2 in white cake icings at a much lower cost. This filler also functions as a stabilizer and supplies enrichment calcium. It is acceptable for mixture with pigments and colorants used in food containers.

GYPSUM AS FILLER FOR PAPER COATINGS

Gypsum has been used as a paper filler and as a paper coating pigment in Europe (14). Kemira Chemicals Inc. has a gypsum filler plant (Cargas Blancas) in Spain with a capacity of 75,000 metric tons per year, and a gypsum pigment plant for paper coating in Finland with a capacity of 100,000 metric tons per year. In paper coating applications, synthetic gypsum (calcium sulphate dihydrate) was wet-ground to about 2 microns with controlled particle size distribution (either normal or narrow size distribution, depending on the applications). After post-treatment for stabilization, a gypsum slurry was blended with kaolin clay in the paper coating formulation. M-real Company, the third largest paper company in Europe, has used the gypsum slurry at the Kangas mill for paper coatings to produce the Galerie One printing paper. For paper filler applications, precipitated calcium sulfate (PCS) was made through the hydration of hemihydrate. Crystal modifier and stabilizer may be used to control the crystal shape and stability. PCS was used in either a slurry form or a dried powder form as paper filler. The advantages of using gypsum in paper coatings are lightweight, high printing gloss, better whiteness, and cost effectiveness. Table 4 lists the pigments used in North America for paper coatings.

Technology of Manufacturing Coating Grade Gypsum

Two methods can be used to manufacture a gypsum coating pigment. In the first method, crystallization, the raw material is hemihydrate that is crystallized into the dihydrate form. By controlling the condition of crystallization, both particle size distribution and pigment shape can be influenced. After crystallization, water is removed from the slurry by filtering. Finally, the gypsum pigment is either dispersed into pumpable slurry or dried into a powder for transportation. In Sweden, gypsum pigment was manufactured at the turn of the 1990s with a fairly large pilot device, and the pigment was used successfully in the coating of fine paper. The second method is grinding. Gypsum can be ground to produce a pigment which is either a dry powder or a slurry. In a wet-grinding process, gypsum is first converted into a slurry form with water, and then a magnetic separator is used to remove potential ferrous mineral contaminants. After this, the rheology of the gypsum slurry is adjusted with dispersing chemicals, and the slurry is ground with ball mills. After post treatment, the finished product is then pumped into containers equipped with a mixer. The slurry product is transported to customers in tanker trucks, and the powder product is shipped by railcar.

The properties of coatings and coated papers containing gypsum:

1. Gypsum has similar ISO-brightness to calcium carbonate, but better than kaolin clay. The specific density of the dihydrate form of gypsum is less than calcium carbonate and clay. This means that by using gypsum as coating pigment, it is possible to keep the thickness of the coating layer unchanged while reducing the base weight of paper.
2. The solubility of gypsum in water changes slightly with pH. This suggests that gypsum can be used for either acidic or alkaline pulp. In comparison, calcium carbonate can only be used in neutral and alkaline pulp.
3. Gypsum coatings increase the effectiveness of optical brightening agents.
4. Gypsum coatings increase opacity, bulk, and offset printability.
5. Gypsum used for paper coatings by M-real is a by-product from the production of phosphoric acid. It is cost-effective and environmentally benign.

SYNTHETIC GYPSUM (15–17)

Synthetic gypsum is a material that is a byproduct of an industrial process. It falls into four main categories: desulphogypsum, phosphogypsum, titanogypsum, and gypsum byproducts from the production of synthetic organic acids. Gypsum manufacturers are increasingly using desulphogypsum as a substitute for mined gypsum. In most power plants, the burning of coal produces undesirable emissions of sulfur, a leading cause of acid rain. A wet lime-limestone scrubber is a common method for removing this pollution from the air. It works via a device installed on the smoke stack of the coal-burning furnace. The calcium and water in the wet limestone combine with sulfate in the exhaust to create calcium sulfate (gypsum) and water. This material is called either "synthetic" or "chemical" gypsum and can be readily used to manufacture gypsum products. The U.S. Bureau of Mines estimates that roughly 20 million tons of synthetic gypsum was generated by electric utilities equipped with wet lime-limestone scrubbers. USG Corporation uses more than 2.75 million tons annually in the production of acoustical ceiling panels, wallboard, and other gypsum products (18). Since, synthetic gypsum is environmentally benign and has unique performance characteristics coupled with cost benefits and availability, it could be a material of choice for the applications mentioned in this paper.

CONCLUSIONS

Gypsum is used as a filler in diverse markets ranging from food to engineering plastics. The general features of gypsum such as high purity, whiteness, softness, resistance to mild acids, suitability for food contact, good electrical properties, no-health hazard, flame retardancy, etc., makes it a unique and versatile filler material. However, the full potential of gypsum as a functional filler has not been realized and there is an ample opportunity to expand the gypsum fillers market into many new areas. The growth rate of gypsum fillers will be dependent on developing new products to meet specific market needs. Continued research and innovation are the keys to maintaining and growing the gypsum fillers market share.

REFERENCES

1. George, W. 1993. *Fillers*. Canada: ChemTec Publishing.
2. William, C.W. 1971. *Fillers for Plastics*. London: Iliffe Books, an imprint of the Butterworth Group.
3. Hand, R.J. 1997. Calcium sulphate hydrates: a review. *British Ceramic Transactions*. 96 (3): 116.
4. Deer, W.A., R.A. Howie, and J. Zussman. 1992. *An introduction to the rock-forming minerals*. London: Longman.
5. Lehmann, H. and H. Holland. 1966. *Tonind.-Ztg*. 90: 20.
6. Kuzel, H.-J. and M. Hauner. 1987. *Zem-Kalk-Gips*. 40: 628.
7. Yasue, T., Y. Kojima, and Y. Arai. 1989. *J. Ceram. Soc. Jpn*. 97: 728.
8. *USG CAS-20™-4 Anhydrous Calcium Sulfate for Titanium Dioxide Extension*, Technical Data Sheet for USG's products. 2001, IG-1499/10-01.
9. *Calcium Sulfate Fillers*, Technical Data Sheet of USG's Products, IG-596/8-92.
10. Owens Corning Fiberglas "*Characterization of Calcium Sulfate Extended Polyester Resin*," 42nd Annual Conference, Composites Institute, The Society of the Plastics Industry.
11. *CAS-20™-4 Calcium Sulfate Filler in Rigid PVC*, Technical Data Sheet of USG's Products, IG-661-rev. 6-95.
12. *USG Terra Alba–F&P Grade SNOW WHITE®Filler–F&P Grade, High-Purity Additives for High-Quality Products*, Technical Data Sheet of USG's Products, IG-172/rev. 3-94.
13. Brafford, et al. U.S. Patent No. 6,326,038 (4 December 2001).
14. Kari Lehtinen, 2000. Gypsum. In *Pigment Coating and Surface Sizing of Paper*. Book 1. Edited by Esa Lehtinen. Helsinki, Finland: Fapet Oy. 153–166.
15. Wirsching, F. 1991. FGD gypsum in research, development, production, and application. *Proceedings 2nd International Conference on FGD and Chemical Gypsum*.
16. Weist, J., L.M. Luckevich, R. Collins, R. Kuntze, and F. Theodore. 1994. *The Gypsum Industry and Flue Gas Desulfurization (FGD) Gypsum Utilization: A Utility Guide*. EPRI Report TR-103652.
17. Barber, J.W., B.A. Hudgens, C.D. Byers, and V.R. Nathan. 1991. Characterization of synthetic gypsum. *Proceedings 2nd International Conference on FGD and Chemical Gypsum*.
18. The Gypsum Construction Handbook. 2000. USG Corporation, Chicago, IL 60606.

Surface Modification of Filler Having a Negative Coefficient of Thermal Expansion

Eric Swanson*, William Cross*, John Starkovich†, and Jon Kellar*

Zirconium tungstate (ZrW_2O_8) is a unique material with a negative coefficient of thermal expansion. In order to fully utilize this material as a filler, a surface treatment needs to be developed to better improve the wetting and dispersion capabilities of the filler. This paper describes an initial study to characterize the filler as well as investigate fatty acid and silane surface treatments for eventual compounding with thermosetting matrices. It was found that ZrW_2O_8 has a point-of-zero charge between pH 2.7–3.0, and that both fatty acid and organofunctional silane treatments appear to be viable surface treatments for this unique filler.

INTRODUCTION

Zirconium tungstate (ZrW_2O_8) is a material of interest because of its rare characteristic of having a negative coefficient of thermal expansion (CTE). The CTE of ZrW_2O_8 is -8.7×10^{-6} $\Delta L/L$ per °C (Sleight, 1998) and is linear over a temperature range of 2 K to 1,050 K (David, Evans, and Sleight, 1999). In addition to having a negative CTE, the thermal expansion with change in temperature is isotropic (Sleight, 1996). Thus, given that most polymers have positive CTEs, one could envision a scenario whereby a ZrW_2O_8 filled polymer would be dimensionally stable over a wide temperature range. More specifically, when added to a matrix in sufficient volume percent, a composite could be produced that neither expanded nor contracted with changes in temperature. Such a composite could be very useful in a number of industrial applications.

ZrW_2O_8 surface treatments are very important for improving the dispersion and wetting of the particles in the polymer matrix. Organo-functional silane and fatty acid surface

* Dept. of Materials and Metallurgical Engineering, South Dakota School of Mines and Technology, Rapid City, S.D.

† TRW Space & Technology.

treatments are being considered for this application. Silane coupling agents are commonly used for a variety of surface treatments in industry, particularly to improve the processing and performace of fiber and filler surfaces for use in polymer matrix composites (PMCs) (Plueddemaun, 1991). Often, the silane modifies the surface energy of the substrate, allowing for better dispersion and wetting, and also serves as a strong coupling agent to the polymer matrix of interest. Fatty acids are characterized by polar carboxyate head groups and, with variations such as oleic and stearic acid, a long alkyl chain. Fatty acids have been shown to be good surface modifiers for a variety of filler surfaces, most notably $Mg(OH)_2$ (Wensel, 1994; Johnson et al., 1999).

EXPERIMENTAL PROCEDURE

Zirconium Tungstate Powder

The material used in these experiments was synthesized for and supplied by TRW Space and Electronics. The average particle size was specified to be approximately one micron and a BET surface area of 0.75 m^2/g.

Zeta Potential

While there has been some general characterization of ZrW_2O_8 completed (Sleight, 1996; Sleight, 1998), the surface charge behavior in aqueous solutions has not been reported. Given the importance of surface charge to any solution-based processing, the zeta potential of ZrW_2O_8 was measured as a function of solution pH. The analysis was performed using Zeta-meter 3.0 using the standard technique (Zeta-meter Inc., 1968). Aqueous ZrW_2O_8 slurries at set pH values were used. The slurry of ZrW_2O_8 powder was prepared by mixing the powder with deionized water. The slurry pH was then raised to 10.0 ± 0.15 by adding 0.1 M NaOH solution. Electrophoretic mobility measurement of the ZrW_2O_8 powder was performed. For the remaining slurry, the pH was then lowered in increments of 0.5 ± 0.2, through the addition of 0.1 M HCl, until a pH of 7.0 ± 0.15 was reached. After each 0.5 ± 0.2 unit step in pH, the electrophoretic mobility of the ZrW_2O_8 was measured. Next, a second slurry of ZrW_2O_8 was prepared with the initial pH lowered to 1.5 ± 0.15 with 0.1 M HCl. Similar to the pH 10.0 slurry, the electrophoretic mobility of the powder at pH 1.5 was measured. The pH was then raised in increments of approximately 0.5 ± 0.2, through the addition of 0.1 M NaOH solution, until a pH of 7.0 ± 0.15 was reached. Once again, the electrophoretic mobility of the powder was determined at each 0.5 ± 0.2 unit pH increment.

Surface Modification with Stearic Acid

Dry blending is a simple, inexpensive method for applying some surface modifications; this method was used with neat stearic acid. At room temperature, 2.8 mg of stearic acid were mixed with 0.5 g of ZrW_2O_8 for 30 minutes using a magnetic stirrer. At a temperature of 82°C, 2.9 mg of stearic acid were mixed with 0.5 g of ZrW_2O_8 for 60 minutes. The treated samples were then examined using the Fourier transform infrared (FTIR) spectroscopy.

Adsorption of Sodium Oleate

Oleate modified ZrW_2O_8 samples were prepared using three aqueous concentrations: 1×10^{-4}, 1×10^{-3}, and 5×10^{-3} M. Sodium oleate powder used in the these experiments was from J.T. Baker Chemical Company. Sodium oleate solutions were prepared by adding the required amount of sodium oleate to 100 mL of deionized water and were stirred for 30 minutes to achieve complete dissolution. The pH of each solution was then adjusted to a pH of 10.0 through the addition of 0.1 M NaOH. During the mixing of the oleate solution, a slurry of ZrW_2O_8 was prepared by adding 0.5 g of ZrW_2O_8 to 100 mL of deionized water. Then, 0.1 M NaOH was added to the slurry until a pH of 10 was reached. The two solutions were then mixed and stirred for 30 minutes to allow the oleate to adsorb onto the ZrW_2O_8 surface. The particles were then filtered and dried at 100°C for 16 hours. After examination by FTIR spectroscopy, the treated particle were rinsed in an aqueous NaOH solution with pH adjusted to 10.0, and examined again by FTIR spectroscopy in order to determine if the oleate had adsorbed to the surface of the ZrW_2O_8 particles or if it was physically trapped during the filtration process.

Surface Modification with 3-Aminopropyltrimethoxysilane

The silane used in this portion of the study was 3-aminopropyltrimethoxysilane (APTS). Solutions of APTS, 0.5% and 1.0 % by weight, were prepared by adding 1.0 g and 2.0 g, respectively, to 100 mL of an aqueous solution of acetic acid with a pH of 3.0. After the addition of APTS, the solution pH was readjusted to 4.5 ± 0.15 through the addition of glacial acetic acid. The solution was then mixed with a 100 mL slurry containing 0.5 g of ZrW_2O_8 and deionized water, with the pH adjusted to 4.5 ± 0.15 by addition of acetic acid. A pH of 4.5 was chosen as it has been shown to be a good pH for the adsorption of silanes to many inorganic substrates (Arkles, 1993). This slurry was then stirred for 30 minutes to allow the APTS to adsorb onto the surface of the ZrW_2O_8. The ZrW_2O_8 particles were then filtered and dried at 100°C for 16 hours. After examination by FTIR spectroscopy, the particles were rinsed in an aqueous acetic acid solution, with a pH of 4.5 ± 0.15 and examined again by FTIR spectroscopy in a method similar to the oleate samples.

Infrared Spectroscopy

Examination of the treated particles was performed using diffuse reflectance infrared spectroscopy on a Digilab FTS 6000 FTIR spectrometer. A globar mid-IR source was used with a germanium-coated KBr beamsplitter and a Peltier-cooled deuterated tri-glycine sulfate (DTGS) detector. The spectrum for each sample was obtained by the co-addition of 64 scans in the mid-IR region of the electromagnetic spectrum. The diffuse reflectance IR spectra can be used to determine the nature and extent of surface modifications. Non-absorbing potassium bromide was used as the background spectrum to ratio the treated ZrW_2O_8 spectra against. Once ratioed against potassium bromide, a spectral subtraction of untreated ZrW_2O_8 spectrum from the treated particle spectrum was performed to determine any surface modifications made to the ZrW_2O_8 particles.

RESULTS AND DISCUSSION

The results attained for the characterization and modification of ZrW_2O_8 particles are discussed in the following sections.

Zeta Potential and Point-of-Zero-Charge

The surface charge behavior of ZrW_2O_8 is of particular interest since it allows understanding of the role of surface charge with respect to surfactant adsorption. Figure 1

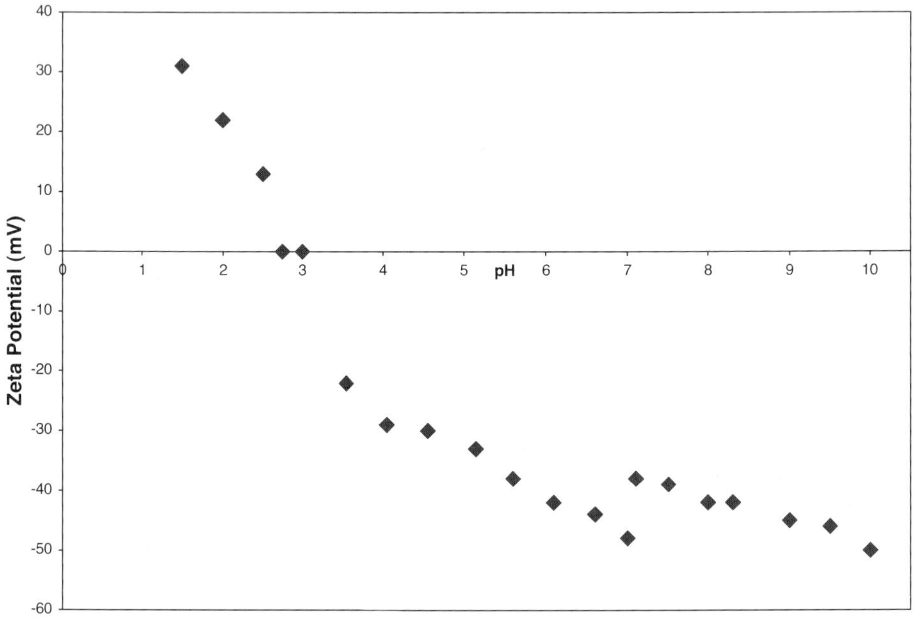

FIGURE 1 Zeta potential as a function of pH for ZrW_2O_8 powder in aqueous solution

shows the variation in ZrW_2O_8 zeta potential as a function of pH. The point-of-zero-charge (PZC) of ZrW_2O_8 was found to be between pH 2.7 and 3.0. This result is of particular interest as the ZrW_2O_8 PZC is approximately the same as that for SiO_2 and most silicate-based glasses (Kellar, Winter, and Zheng, 1993). Furthermore, one can infer that from a surface charge perspective, many surface modifiers for glasses (e.g., silanes) should be amenable for ZrW_2O_8.

Spectroscopy

Diffuse reflectance IR spectroscopy is a robust method for analysis of particulates and was therefore used in this study of ZrW_2O_8.

Stearic Acid Treated Particles. Figure 2 shows the IR spectra of neat ZrW_2O_8. The two areas of particular interest for determining if the occurrence of stearic acid adsorption are in the ranges of 3,100–2,800 cm^{-1} and 1,650–1,550 cm^{-1}. The carboxylate function group stretching often appears in the range of 1,650–1,550 cm^{-1}. To help determine anticipated carboxylate band location for any chemisorbed zirconium oleate products, zirconium Versalate™ (zirconium 2-ethylhexanoate) was used as a standard. The zirconium Versalate™ were obtained from Shepherd Chemical Co., Cincinnati, Ohio. Figure 3 shows the region of 1,650–1,550 cm^{-1} for stearic acid treated ZrW_2O_8 spectra versus zirconium Versalate™. Around 1,570 cm^{-1} there is no definitive peak to indicate any zirconium carboxylate groups on the stearic acid treated ZrW_2O_8. Therefore, the stearic acid did not react with the zirconium tungstate under the conditions used, and dry blending does not appear to be a promising route for modifying the surface of the ZrW_2O_8.

Oleate Treated Particles. The two areas of particular interest for determining if any sodium oleate adsorption occurred are also in the ranges of 3,100–2,800 cm^{-1} and

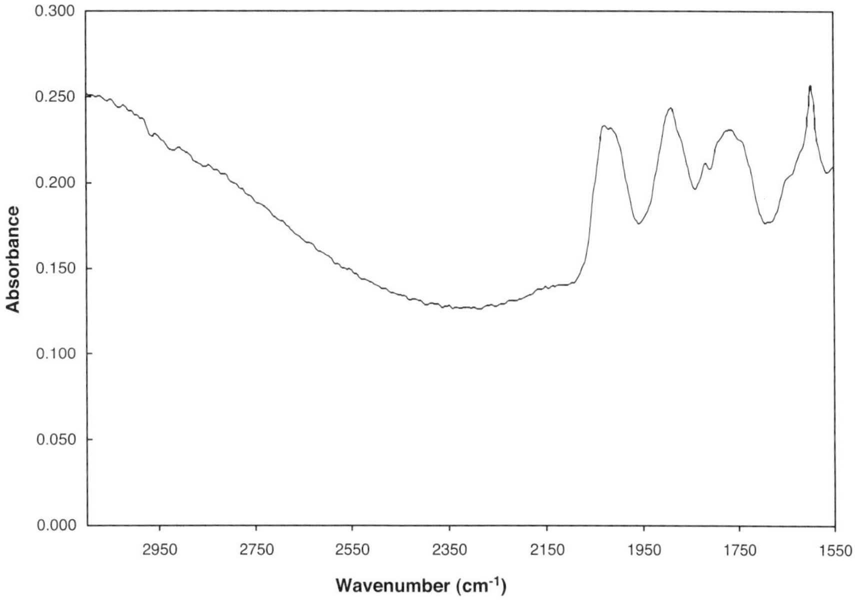

FIGURE 2 Neat ZrW_2O_8 spectra

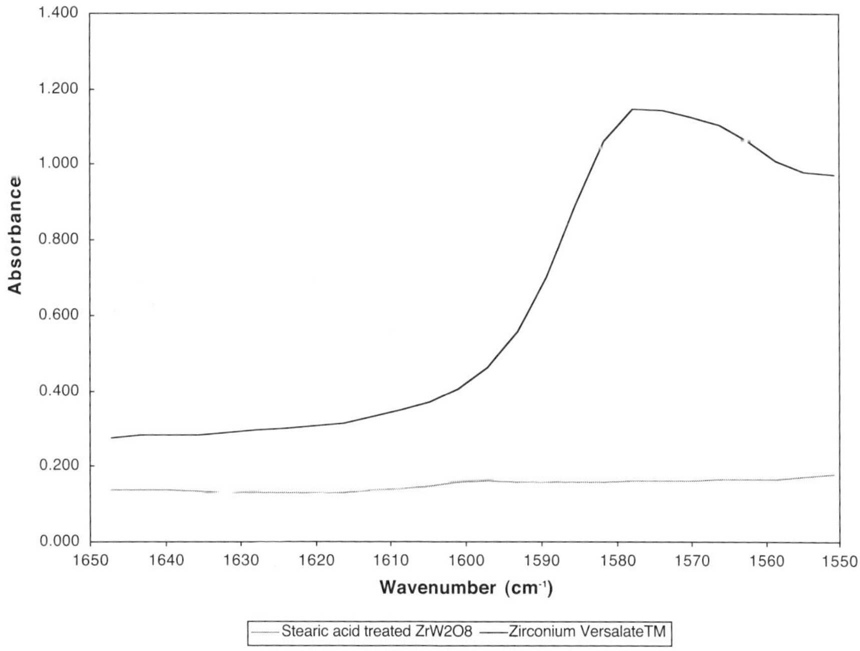

FIGURE 3 Carboxylate region of zirconium Versalate™ (top) and stearic acid (bottom)

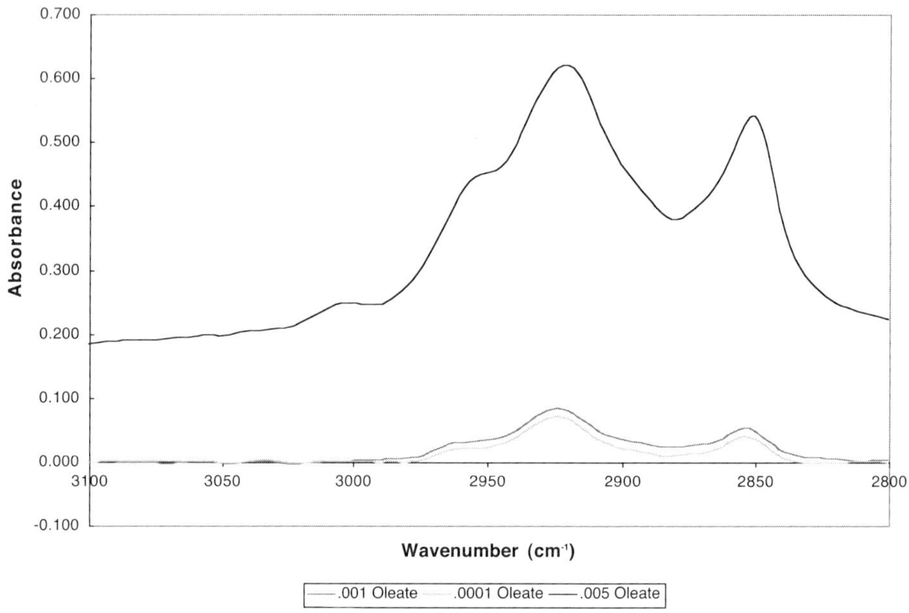

FIGURE 4 Assymmetric CH_2 stretching band region for different initial concentrations of sodium oleate treated ZrW_2O_8; 0.005 (top), 0.0001 (bottom)

1,650–1,550 cm^{-1}. Figure 4 shows a comparison of the spectra for the different initial concentrations of sodium oleate treated ZrW_2O_8 samples in the range of 3,100–2,800 cm^{-1}. Peaks for the relatively strong asymmetric CH_2 stretching band appear around 2,924 cm^{-1} and have 0.62 absorbance units for ZrW_2O_8 powders subjected to 0.005 M sodium oleate, 0.084 absorbance units for 0.001 M sodium oleate, and 0.072 absorbance units for 0.0001 M sodium oleate. These results all indicate that oleate is present on the ZrW_2O_8 surfaces at each concentration of oleate. However, the oleate in the IR spectra is only indicative of presence of oleate with the sample, not necessarily adsorbed oleate. In other words, the oleate observed may simply be trapped with the ZrW_2O_8 during filtering and neither chemically or physically adsorbed. To begin to ascertain whether or not adsorption had occurred, the rinsing procedure described in the experimental procedure section was conducted, and these samples analyzed by infrared spectroscopy. Figure 5 compares the rinsed sample to the original 0.0001 M sample. In Figure 5, the absorbance decreases to approximately 0.007 absorbance units after washing, which is about 10% of the original absorbance. This implies that a large amount of the oleate present was trapped during the filtration process. The investigation next focused upon possible chemical reactions between the adsorbed oleate and the ZrW_2O_8. Typically the region 1,650–1,550 cm^{-1} is examined for possible chemisorbed carboxylate products. In addition to the zirconium Versalate™, IR spectra were obtained for sodium oleate and neat ZrW_2O_8 powder. These spectra, in the region 1,650–1,550 cm^{-1}, are all shown in Figure 6. Due to intrinsic absorbance bands associated with ZrW_2O_8, as well as similarities and overlap in the zirconium and sodium carboxylate band locations, nothing definitive could be said concerning the chemical nature of the adsorbed oleate species. Current efforts are directed toward utilizing x-ray photoelectron spectroscopy for elucidation of the chemical nature of the adsorbed oleate species.

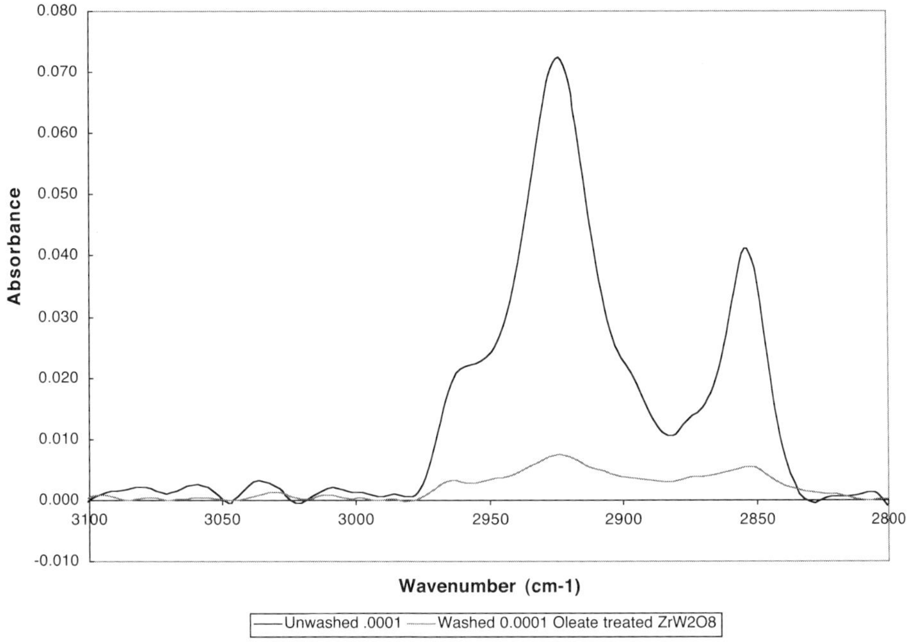

FIGURE 5 Assymmetric CH_2 stretching band region for washed (bottom) and unwashed (top) sodium oleate treated ZrW_2O_8

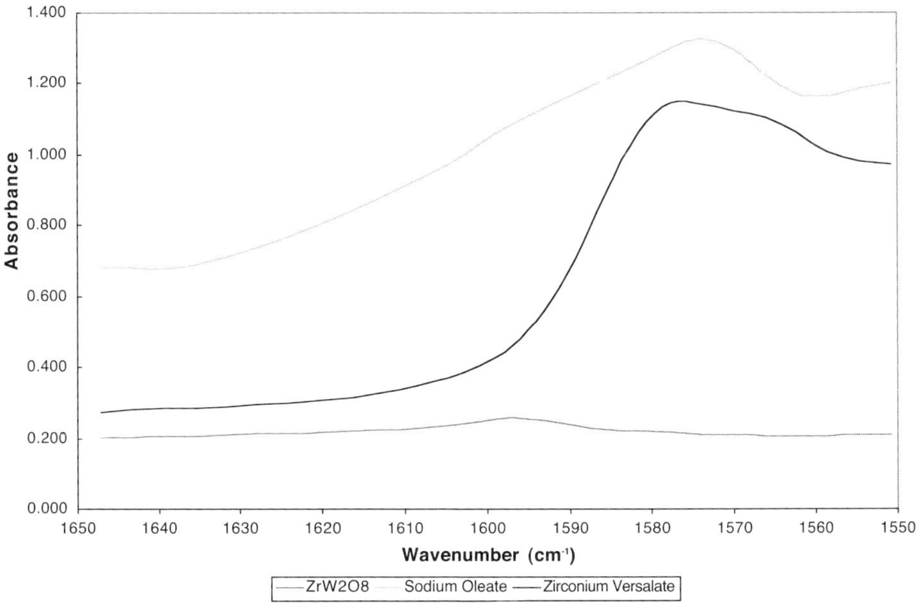

FIGURE 6 Carboxylate group absorbance region, ZV (mddle), ArW2O8 (bottom), oleate (top)

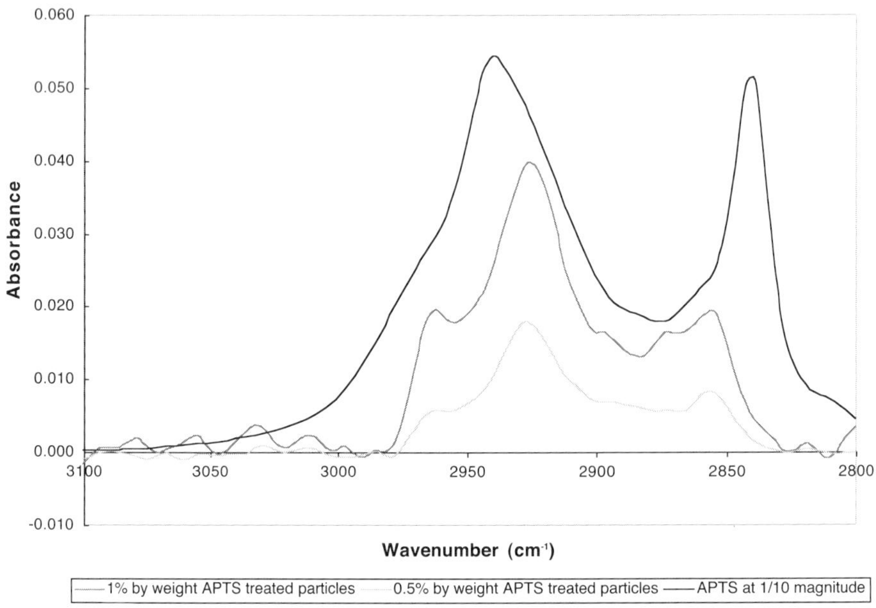

FIGURE 7 Assymmetric CH_2 stretching band region for APTS (top) and APTS treated ZrW_2O_8; (0.5% bottom), (0.1% middle)

APTS Treated Particles. In order to determine possible silane reaction with the ZrW_2O_8 particles, the wavenumber region of 3,100–2,800 cm^{-1} was observed. Figure 7 shows a comparison of the 0.5 and 1.0% by weight APTS treatments. The asymmetric-CH_2 stretching band appears around 2,928 cm^{-1}, with 0.039 absorbance units for 1.0% by weight silane and 0.018 absorbance units for 0.5% by weight silane. If one assumes a Beers Law type relationship, one can estimate that the amount deposited is roughly twice as much at the higher APTS concentration. Figure 8 shows a comparison of the washed 1.0% by weight silane sample with the unwashed sample of the same concentration. Very little is changed in the absorbance values for the rinsed silane treatment. While this does not confirm the complete adsorption of silane onto the ZrW_2O_8, it does suggest that the APTS is more tenaciously bound to the ZrW_2O_8 than the oleate. The fact that little APTS was removed by washing is not surprising given the tendency for adsorbed silanes to form oligomerized networks after adsorption (Plueddemann, 1991).

CONCLUSION

The use of ZrW_2O_8 in future applications will only come about through understanding the material's characteristics and chemical interactions. By knowing the surface charge behavior of ZrW_2O_8, correlations can be made between the surface charge and surfactant adsorption behavior. Through FTIR spectroscopy, several observations have been made. First, ambient and elevated temperature treatment of ZrW_2O_8 with dry stearic acid does not result in appreciable adsorption of stearic acid. APTS deposited onto ZrW_2O_8 from aqueous solutions do exhibit evidence of chemisorption when analyzed by infrared spectroscopy and washing. Examination of the carboxylate stretching region for

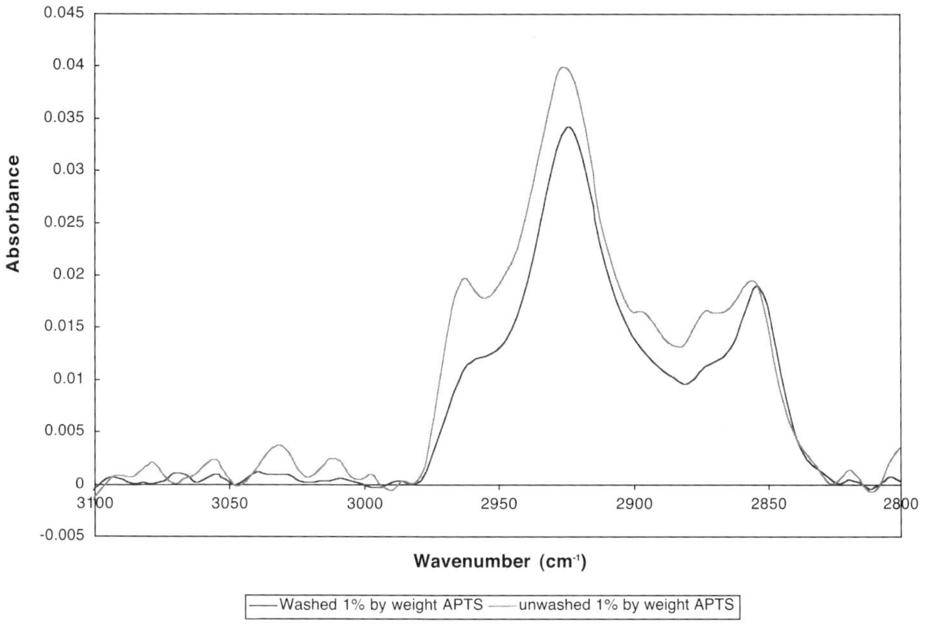

FIGURE 8 Assymmetric CH_2 stretching band region for washed (bottom) and unwashed (top) 1% by weight APTS treated ZrW_2O_8

oleate treated ZrW_2O_8 proved to be inconclusive in terms of showing a chemisorbed oleate complex. Further, characterization of the chemistry of both the oleate and APTS adsorption will be pursued through the use of x-ray photoelectron spectroscopy. Future work will also involve compounding and property testing of ZrW_2O_8-filled polymer matrix composites.

REFERENCES

Arkles, B. 1993. Silane Coupling Agent Chemistry. In *Silicon Compounds Register and Review.* 5th ed. Bristol, PA: United Chemical Technologies, Inc. 59–64.

David, W.I., Evans, J.S., and Sleight A.W. 1999. Structural investigation of the negative thermal expansion mater ZrW_2O_8. *Acta Crystallographica.* B55: 333–340.

Johnson, F.J., Pommer, A.G., Vernon, M.S., Winter, R.M., Cross, W.M., and Kellar, J.J. 1999. Preparation of magnesium hydroxide surfaces as a model for flame retardant fillers. *Minerals & Metallurgical Processing.* 16: 65–68.

Kellar, J.J., Winter, R.M., and Zheng, J. 1993. A comparison of the electrokinetic behavior of silica and glasses. *Composite Interfaces.* 1(5): 403–409.

Plueddemann, E.P. 1991. *Silane Coupling Agents.* 2nd ed. New York: Plenum Press.

Sleight, A.W. 1998. Compounds that contract on heating. *Inorganic Chemistry.* 37: 2854–2860.

Sleight, A.W., et al., U.S. Patent No. 5,514,360 (7 May, 1996).

Wensel, R. 1994. *Influence of surface treatment of magnesium hydroxide in filled polymer systems.* M.S. thesis. South Dakota School of Mines and Technology, South Dakota.

Zeta-meter Inc. 1968. *Zeta-meter Manual: Second Edition.* New York, NY.

Effect of High Reactivity Metakaolin (HRM) on Mechanical, Chemical, and Optical Properties of Cement-based Systems

Sharad Mathur*

The purpose of this paper is to present the benefits of high reactivity metakaolin (HRM) in high performance cement-based systems. HRM, unlike other pozzolans, is manufactured from selected naturally occurring kaolin ore and carefully processed under stringent conditions to result in a consistent product with highly pozzolanic properties. HRM is an ultrafine pozzolan that enhances the mechanical properties (workability and strength) and chemical stability (durability) of Portland cement concrete and cement-based products. HRM is unique with respect to its white color which makes it an ideal choice for architectural applications where visual and aesthetic concerns are critical.

BACKGROUND

Pozzolans are siliceous or siliceous and aluminous materials, which in themselves possess little or no cementitious value, but will, in finely divided form and in the presence of moisture, chemically react with calcium hydroxide at ordinary temperatures to form compounds possessing cementitious properties.

The above statement can be simply expressed in the following chemical reactions

$$\text{Cement} + H_2O = \text{Paste} + Ca(OH)_2$$

$$\text{Pozzolan} + H_2O = \text{slurry}$$

$$\text{Pozzolan} + Ca(OH)_2 + H_2O = \text{cementitious products}$$

Calcium hydroxide, which is a byproduct of cement hydration, is a "weak link" of any cement-based system for the following reasons.

* Engelhard Corp., Gordon, Calif.

1. Calcium hydroxide is soluble in water and, therefore, can be leached from the cement paste, making the system more porous. By reacting with the pozzolan, the calcium hydroxide is stabilized as a cementitious material, increasing the strength and reducing the permeability of the system.
2. Calcium hydroxide after dissolution travels to the surface of the cement-based system and reacts with atmospheric carbon dioxide to form calcium carbonate deposit better known as efflorescence. This leads to an unsightly appearance.
3. Calcium hydroxide can participate in reactions with certain aggregates resulting in alkali-silica reactivity (ASR) or various deleterious chemicals such as sulfates weakening the cement-based system.

By undergoing the reactions previously illustrated, pozzolanic materials are able to convert the calcium hydroxide into cementitious material, increasing the strength and durability of cement-based systems.

Natural pozzolans are typically clay-based materials that originate naturally or are modified by thermal activation such as drying or calcination. They vary in chemical composition and physical characteristics such as particle size and size distribution, surface area, crystallinity, and color, and, therefore, possess varying levels of pozzolanic activity, color, and performance.

METAKAOLINS AND HIGH REACTIVITY METAKAOLIN (HRM)

Metakaolin is a chemical phase that forms upon thermal treatment of kaolinite. Kaolinite's chemical composition is Al_2O_3:$2SiO_2$. $2H_2O$, and as a result of thermal treatment in the range of 800–1,000°C, the water is driven away to form an amorphous aluminosilicate called metakaolin. The temperature range depends on the kaolin (kaolinite with minor impurities) characteristics such as degree of crystallinity and particle size.

Metakaolin is white in color and acts as a pozzolanic material (Caladrone et al., 1994; Caladrone and Gruber, 1995; Zhang and Malhotra, 1995). The reactivity of the metakaolin may also be affected by grinding to a finer particle size (Mathur et al., 2001). The purity of the kaolin affects the overall color and reactivity. For example, according to ASTM 618 the minimum amount of SiO_2, Al_2O_3, and Fe_2O_3 that needs to be present in a Class N pozzolan is 70%. Therefore an impure source of kaolin may be used to result in a pozzolanic material that meets the ASTM C618 requirements, but will not have the same reactivity as high reactivity metakaolin (HRM). The lower amount of siliceous and aluminous material will result in a lesser reactivity, which may be further diminished by a coarse particle size. Also, generally, the color will not be white and the performance may vary depending on the impurity type and level.

High reactivity metakaolin has the highest content of siliceous and aluminous material among the purified kaolins (Caladrone et al., 1994; Caladrone and Gruber, 1995; Zhang and Malhotra, 1995). This is because at least 90% of the product consists of silica and alumina. The HRM is manufactured under controlled conditions to provide consistent product in terms of particle size distribution, surface area, and color and chemical composition. The HRM by virtue of its highest purity is white in color, which is critical to architectural applications. Most importantly, the HRM is thermally activated in a controlled process so the heat treatment results in maximum pozzolanic activity for the product.

Manufacturing Process of High Reactivity Metakaolin

The manufacturing of HRM is optimized for cement-based systems to provide maximum benefits in terms of pozzolanic reactivity, white color, and working properties.

The kaolin is selectively mined for particle size distribution, morphology, and chemical composition. The crude kaolin is dispersed into a slurry form and pumped to the manufacturing site and processed to:

- remove impurities to provide consistent color
- produce desired particle size distribution for reactivity and application needs, and
- maximize pozzolanic activity through optimum thermal treatment.

The particle size distribution of pozzolan is important both for the fresh properties of cement-based systems as well as the speed of pozzolanic reaction. A coarser particle size distribution will result in lower water demand because of its lower surface area. Consequently, at the same water-to-cementitious materials ratio, the coarser particle size will be more flowable. However, the low surface area also results in a slower speed of reaction and, therefore, the pozzolanic reaction proceeds at a slower rate. Thus, a lower early strength and most likely lower hardened structure strength will occur with a coarse particle size distribution of pozzolan. Additionally, because of the reduced reaction rate, the system with the coarse pozzolan may have less chemical resistance and higher permeability than the system with a finer particle size distribution pozzolan.

The crude used in the production of HRM is centrifuged to appropriate particle size. Since the crude is selectively mined for particle size distribution, the particle size distribution of HRM is consistent from lot to lot. The centrifuged product goes through impurity removal steps such as magnetic separation and chemical bleaching to facilitate a consistently white color HRM product. The HRM undergoes optimal thermal treatment in a calciner such that the product exhibits the same maximum pozzolanic activity.

Typical Properties of High Reactivity Metakaolin (HRM)

- $SiO_2 + Al_2O_3 + Fe_2O_3 > 90\%$
- WHITE in color
- Specific gravity: 2.5
- Particle size less than fly ash (FA) and greater than silica fume (SF)

From a quality standpoint, HRM is a consistent product because of controlled manufacturing with respect to

- Chemical composition
- Particle size distribution
- Pozzolanic reactivity
- White color

BENEFITS OF HRM IN CEMENT-BASED SYSTEMS

HRM can provide a variety of benefits in a number of cement-based systems such as the following (Caladrone et al., 1994; Caladrone and Gruber, 1995; Zhang and Malhotra, 1995; Thomas et al., 1997):

- Ready-mixed concrete
- Glass fiber reinforced concrete (GFRC)
- Oil well cements
- Precast concrete
- High performance grouts, mortars, stuccos, and plaster

Typical benefits provided by usage of HRM are:

- Workability enhancement
- Increased strength
- Reduced permeability
- Increased chemical resistance
- Ability to color match and produce high performance architectural products
- Additional enhancement of other pozzolans

We will present examples to illustrate the benefits listed above by HRM.

HRM Workability Enhancement

The workability enhancement due to HRM means the following:

- Increased "creamy" texture for flatwork, masonry, etc. HRM can produce a cohesive mixture that is described by finishers as "buttery" or "creamy." Although the mixture seems "tighter" when formulated correctly, the mixture flows well with minimal effort when moved by trowel.
- Reduced rebound in shotcrete with good finishability using HRM. The cohesiveness of the mixture minimizes rebound but the shear-thinning nature of the mixture allows pumpability and ease of troweling.
- Improved consolidation in precast and ready-mixed concrete. Mixtures containing HRM are cohesive and yet move easily under vibration without segregation. The HRM concrete is less "sticky" than other finely divided pozzolans, allowing the product to consolidate well in forms.

When comparing HRM to silica fume, lower dosages of high range water reducer (HRWR) are generally required. This is illustrated by way of the concrete mixture design (Table 1) where the HRM was substituted for SF in ready mix application as replacement or addition to cement. The design shows a lower amount of HRWR required with HRM when compared to the mixture with silica fume. Both concrete mixtures had similar slump and air content, but the fresh HRM concrete was creamier and easier to work with.

Increased Strength with HRM

As described above, this is a typical benefit of the pozzolanic reaction and is a result of the conversion of calcium hydroxide, which does not contribute to the strength of the paste, to cementitious products. For the mixture design previously mentioned, Table 2 compares the strength achieved due to presence of HRM and SF. HRM resulted in significantly higher strength than the non-pozzolanic control formulation and a higher strength than the concrete containing silica fume.

Reduced Permeability with HRM

Calcium hydroxide in the paste may solubilize and migrate to the surface resulting in increased porosity and permeability of the paste. However, the rapid pozzolanic reaction of HRM stabilizes the calcium hydroxide as cementitious products, therefore, reducing porosity

TABLE 1 Concrete mixture proportions (per cubic meter)

	Control Formulation	Addition of 8% HRM	Addition of 8% Silica Fume	Replacement of 8% of Cement with HRM	Replacement of 8% of Cement with Silica Fume
Type I cement (kg)	386	386	386	353	353
Fine aggregate, (kg)	872	837	837	872	872
19-mm. coarse aggregate, (kg)	1,032	1,032	1,032	1,032	1,032
Densified silica fume (kg)	0	0	33	0	33
High reactivity metakaolin (kg)	0	33	0	33	0
Water (kg)	154	154	154	154	154
Type A water reducer (ml.)	755	755	755	697	697
Type F high range water reducer (l)	3.5	6.4	8.2	4.7	7.1
Slump (mm)	216	267	254	248	248
Temperature (°F)	20	22	22	21	22
Air (%)	2.3	3.0	3.1	2.8	3.5
Unit weight (kg/m^3)	2,420	2,440	2,410	2,410	2,460

TABLE 2 Comparison of compressive strength with time for HRM and SF with the mixture design in Table 1

Testing Age (Days)	Compressive Strength (MPa)				
	Control Formulation	Addition of 8% HRM	Addition of 8% Silica Fume	Replacement of 8% of Cement with HRM	Replacement of 8% of Cement with Silica Fume
7	59.0	79.8 **(+35%)**	70.3 **(+19%)**	72.8 **(+23%)**	66.0 **(+12%)**
28	69.0	94.6 **(+37%)**	88.6 **(+28%)**	85.8 **(+24%)**	84.2 **(+22%)**
56	78.0	104.9 **(+34%)**	96.2 **(+23%)**	94.2 **(+21%)**	93.5 **(+20%)**
90	83.8	113.8 **(+36%)**	108.9 **(+30%)**	112.7 **(+35%)**	105.7 **(+26%)**
180	89.6	119.6 **(+33%)**	114.8 **(+28%)**	119.1 **(+33%)**	115.8 **(+29%)**
365	97.5	127.5 **(+31%)**	124.0 **(+27%)**	129.1 **(+32%)**	121.0 **(+24%)**

( ) numbers in parenthesis represent % strength increase over control formulation.

and permeability. This reduced permeability is effective against penetration of ionic species such as chloride ions, which negatively impact the steel reinforcement in concrete.

Table 3 shows the mixture design that was used to test the permeability of chloride ions in concrete with two different levels of HRM. In this case HRM substituted for part of the cement.

TABLE 3 Summary of mix design and fresh properties (Hooton et al. 1997)

Mixture	1	2	3
w/cm	0.40	0.40	0.40
OPC	100%	92%	88%
HRM	0	8%	12%
OPC (kg/m^3)	380	350	334
HRM (kg/m3)	—	30	46
Coarse agg. (kg/m^3)	1,100	1,100	1,100
Fine agg. (kg/m^3)	655	647	641
Water (kg/m^3)	152	152	152
25 XL water reducer (mL/100kg)	325	325	325
SPN superplasticizer (mL/100kg)	400	600	800
MicroAir entrainer (mL/100kg)	40	40	40
Slump (mm)	135	125	200
Air content (%)	9.5	7	6
Fresh-concrete density (kg/m^3)	2,275	2,353	2,424

Increased Chemical Resistance with HRM

Silica, present in various types of reactive aggregates, can dissolve in the highly alkaline environment of concrete. This alkali-silica reaction (ASR) produces a gel at the paste/aggregate interface. The gel expands in the presence of moisture and produces internal stresses. These stresses can crack the aggregates or cause debonding between the cement paste and the aggregate and can ultimately decrease the performance of the concrete. MetaMax, however, lowers the alkalinity of the pore solution by consuming much of the free lime. This decreases the silica solubility and the propensity for alkali-silica reaction. (See Figure 2.) (Note that, while the reduction in pH due to MetaMax is sufficient to mitigate against ASR, it is not drastic enough to impinge on the oxide coating that helps protect steel reinforcement against corrosion.)

Figure 2 illustrates the impact of HRM on reducing the deleterious expansion from ASR due to reactive aggregates as measured by the CAN/CSA A23.2-14A (similar to ASTM C1293). Clearly, with 15% or higher levels of HRM, the 2 year expansion was reduced by over 80% and less than the CSA limit of 0.04% (Gruber et al. 2001).

Benefits on Optical Properties Due to White Color of HRM

The white color of HRM provides a distinct advantage over other pozzolans, including SF and FA, as it does not darken or change the color of white cement and/or pigmented systems. The white color allows an architect to design high-performance structures with light or vibrant colors. Figure 3 illustrates the preservation of white color in a mortar made with white cement and 15% dosages of various pozzolans. The figure shows that HRM is clearly preferential for architectural products because of its ability to maintain the white color, which will translate into improved pigmentation.

In architectural products HRM also provides the following benefits:

- Increases the workability and durability of the fresh mixture
- Reduces efflorescence because of its reaction with calcium hydroxide
- Increases reflectivity versus other alternatives because of the consistent white color of HRM

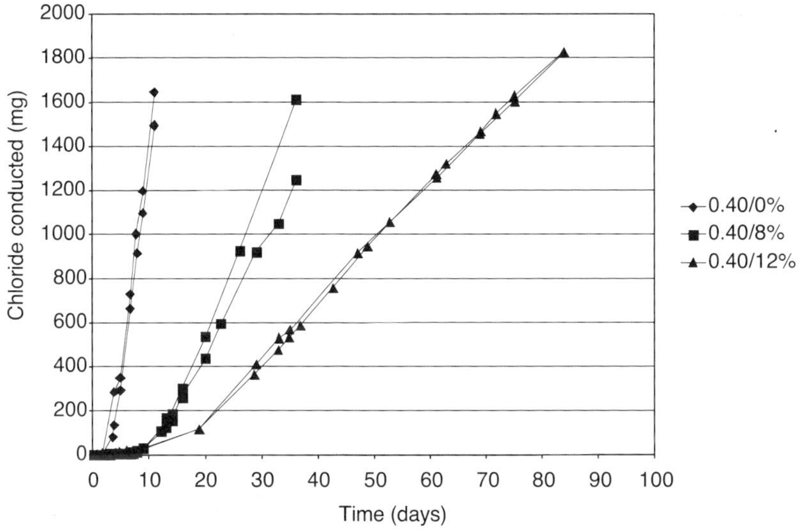

FIGURE 1 Effect of HRM addition on chloride migration (Hooten et al., 1997)

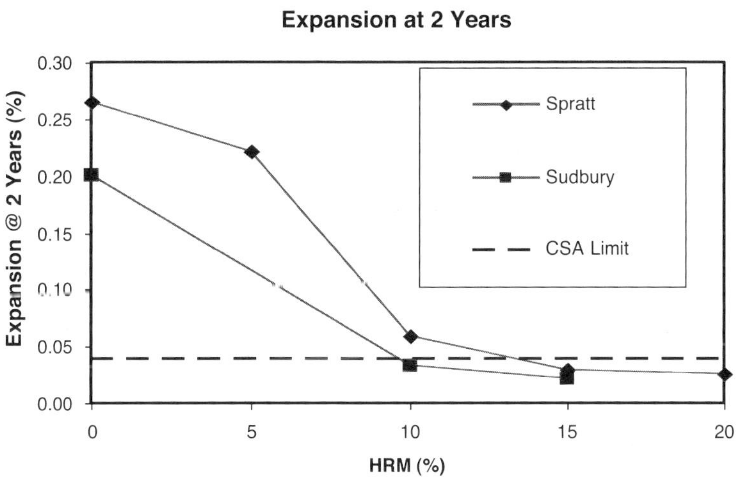

FIGURE 2 Effect of HRM on ASR (Gruber et al., 2001)

Examples of architectural applications are:

- Swimming pool plaster
- High durability structures–parking garages, bridge decks/overlays, industrial floors, specialty precast pieces, and spun piles
- GFRC–thin casting to bring out intricate details. HRM was used in over 40,000 pieces of pre-cast GFRC required for the restoration of Shepard Hall at the City College of New York (see Figure 4). Before selecting the cladding material, the

TABLE 4 Concrete mixture proportions

	FA	HRM Addition	Additional FA
Type I cement (lb.)	500	500	500
Class F FA (lb.)	100	100	145
HRM (lb.)	0	45	0
Fine aggregate (lb.)	1,340	1,340	1,340
Coarse aggregate (lb.)	1,740	1,740	1,740
Type A water reducer (oz.)	15	15	15
Air-entraining agent (oz.)	5	9	9
Type F HRWR (oz.)	33	97	61

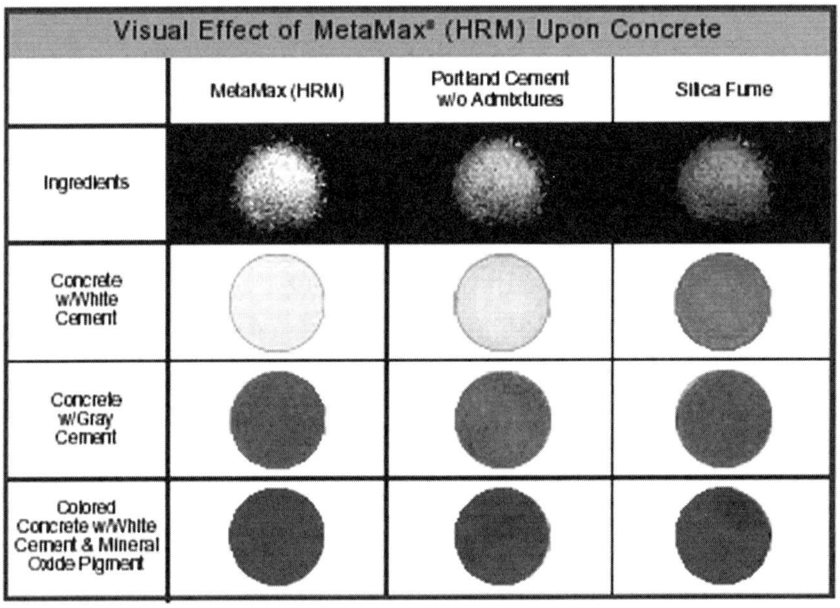

FIGURE 3 Mortars showing the effect of HRM, SF and FA on color (www.engelhard.com)

architects subjected about a dozen concrete products to six months of rigorous accelerated aging tests. Concrete containing HRM met the durability testing and aesthetic requirements and was the most cost-effective at the same time.

In case of swimming pool plasters, HRM provides the following specific benefits:

- Reduced permeability
- Improved chemical resistance
- Improved trowelability
- Reduced mottling
- Control of spot-etching

FIGURE 4 Pre-cast GFRC pieces meeting high performance and aesthetic criterion with HRM (www.engelhard.com)

TABLE 5 Fresh concrete properties with FA, HRM and additional FA

	FA	HRM Addition	Additional FA
Slump (in.)	6½	6¼	7
Air (%)	7.4	7	7.4
Unit weight (lb/ft^3)	141.1	142.3	142.7

In case of GFRC, the following specific benefits are obtained with HRM usage:

- Increased production
- Durable (retains toughness)
- Improved surface appearance
- Less chipping of delicate pieces
- Improves bonding of fibers

Improvement in Performance of Other Pozzolans and Specialty Cements with HRM

Pozzolanic reaction of HRM can supplement other slower reacting products such as FA and ground granulated blast furnace slag (GGBFS). This approach allows the optimization of the mixture design. Also HRM can be successfully used in specialty cements such as oil well and shrinkage compensating.

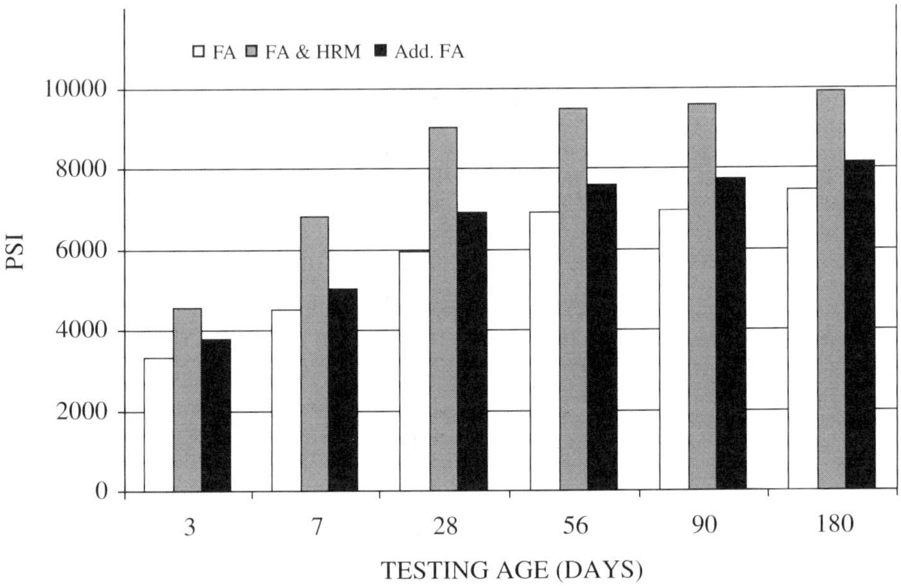

FIGURE 5 Increase in compressive strength with HRM compared to FA as well as additional FA

TABLE 6 Rapid chloride penetration test (RCPT) with FA, HRM and additional FA

Mixture Design	Coulombs Passed
Fly ash	2,500
FA + HRM	800
FA + FA	1,500

To illustrate the benefit of using FA and HRM together rather than just FA, the following concrete design mixture was formulated.

The compressive strength data in Figure 5 shows that addition of HRM leads to a significant strength improvement versus just FA alone or even when additional FA is added to have the same cementitious content as with HRM addition. Compressive strength development of fly ash concrete can be improved by the use of HRM and other properties can benefit from this combination.

The reduction in permeability with HRM is also higher than with FA alone or additional level of FA added to match the amount of cementitious material in the mixture (see Table 6).

USAGE OF HRM

The recommended amount of HRM varies based upon application and need; in general, 10% by mass of cement for increased strength and reduced permeability. The amount of HRM needs to be increased to 15–20% for improved chemical resistance and alkali-silica reactivity (ASR) control. Further, adjustments may be required in formulation such as change in water-reducer and air-entraining agent levels as shown in examples previously.

Unlike cement, HRM has an unlimited shelf life as long as it is kept dry. Direct moisture will not cause HRM to react, but may produce lumps of material to form that may decrease the dispersion in a mixture.

The finer particle size distribution of HRM than cement results in a higher water demand and the best way to offset the higher water demand is to increase the amount of water reducer. In case no other additives are being used, then the minimal amount of water should be added to maintain workability. As pointed out earlier, workability should not be assessed by the appearance of the mixture in the mixing equipment where it would look like a cohesive mass and need of more water. Workability should actually be assessed by troweling some of the mixture.

Finally, proper application and finishing techniques are still required with the usage of HRM just as without HRM.

SUMMARY

HRM is recommended because of the following benefits in a number of cement-based systems.

- Excellent workability
- Architecturally appealing–no darkening or changing of color
- Increased strength
- Reduced permeability
- Control of alkali-silica reactivity
- Enhancement of lower/slower reactivity materials–fly ash, GGBS

All the above listed benefits translate to the following **value** from HRM to the end-user.

- Increased service life
- Improved performance
- Improved appearance
- Increased productivity

REFERENCES

Caladrone, M.A., Gruber, K.A., and Burg, R.G., 1994, "High Reactivity Metakaolin: A New Generation Mineral Admixture," Concrete International, Vol. 16, No. 11, pp. 37–40.

Caladrone, M.A., and Gruber, K.A., 1995, "High Reactivity Metakaolin–A Mineral Admixture for High Performance Concrete," Concrete Under Severe Conditions, CONSEC 95, Sapporo, Japan, Sakai, K., Banthia, N., and Gjorv, O.E., eds., Chapman and Hall, New York, pp. 1015–1024.

Gruber, K.A., Ramlochan, T., Boddy, A., Hooton, R.D., and Thomas, M.D.A., 2001, "Increasing Concrete Durability with High Reactivity Metakaolin," Cement and Concrete Composites, 23, pp. 479–484.

Hooton, R.D., Gruber, K.A., and Boddy, A., 1997, "The Chloride Penetration Resistance of Concrete Containing High Reactivity Metakaolin," Proceedings of the PCI/FHWA International Symposium on High Performance Concrete, New Orleans, Louisiana, pp. 172–183.

Mathur, S., Gruber, K.A., and Reid, H.J., 2001, "Manufacture of Improved Metakaolin by Grinding and Use in Cement-based Composites," US Patent 6,221,148, April 24.

Thomas, M.D.A., Gruber, K.A., and Hooton, R.D., 1997, "The Use of High Reactivity Metakaolin in High Performance Concrete," Proceedings of First International Conference, Azizinaminai, A., Darwin, D., and French, C. eds., ASCE, 1997, Kona, Hawaii, pp. 517–530.

Zhang, M.H., and Malhotra, V.M., 1995, "Characteristics of a Thermally Activated Alumino-Silicate Pozzolanic Material and Its Use in Concrete," Cement and Concrete Research, Vol. 25, No. 8, pp. 1713–1725.

Index